T0134448

Intelligent Systems Reference Library

Volume 149

The aim of this series is to publish a Reference Library, including novel advances and developments in all aspects of Intelligent Systems in an easily accessible and well structured form. The series includes reference works, handbooks, compendia, textbooks, well-structured monographs, dictionaries, and encyclopedias. It contains well integrated knowledge and current information in the field of Intelligent Systems. The series covers the theory, applications, and design methods of Intelligent Systems. Virtually all disciplines such as engineering, computer science, avionics, business, e-commerce, environment, healthcare, physics and life science are included. The list of topics spans all the areas of modern intelligent systems such as: Ambient intelligence, Computational intelligence, Social intelligence, Computational neuroscience, Artificial life, Virtual society, Cognitive systems, DNA and immunity-based systems, e-Learning and teaching, Human-centred computing and Machine ethics, Intelligent control, Intelligent data analysis, Knowledge-based paradigms, Knowledge management, Intelligent agents, Intelligent decision making, Intelligent network security, Interactive entertainment, Learning paradigms, Recommender systems, Robotics and Mechatronics including human-machine teaming, Self-organizing and adaptive systems, Soft computing including Neural systems, Fuzzy systems, Evolutionary computing and the Fusion of these paradigms, Perception and Vision, Web intelligence and Multimedia.

More information about this series at http://www.springer.com/series/8578

George A. Tsihrintzis · Dionisios N. Sotiropoulos
Lakhmi C. Jain
Editors

Machine Learning Paradigms

Advances in Data Analytics

Editors
George A. Tsihrintzis
University of Piraeus
Piraeus
Greece

Dionisios N. Sotiropoulos
University of Piraeus
Piraeus
Greece

Lakhmi C. Jain
Faculty of Engineering and Information Technology, Centre for Artificial Intelligence
University of Technology
Sydney, NSW
Australia

and

Faculty of Science, Technology and Mathematics
University of Canberra
Canberra, ACT
Australia

and

KES International
Shoreham-by-Sea
UK

ISSN 1868-4394 ISSN 1868-4408 (electronic)
Intelligent Systems Reference Library
ISBN 978-3-030-06777-9 ISBN 978-3-319-94030-4 (eBook)
https://doi.org/10.1007/978-3-319-94030-4

Softcover re-print of the Hardcover 1st edition 2019

Printed on acid-free paper

This Springer imprint is published by the registered company Springer International Publishing AG part of Springer Nature
The registered company address is: Gewerbestrasse 11, 6330 Cham, Switzerland

To my wife and colleague, Prof.-Dr. Maria Virvou, and to our daughters, Evina, Konstantina and Andreani

George A. Tsihrintzis

To my beloved family and friends

Dionisios N. Sotiropoulos

To my beloved family

Lakhmi C. Jain

Foreword

In 1959, Arthur Samuel (1901–1990) published *Some Studies in Machine Learning Using the Game of Checkers* [1]. The paper was one of the earliest uses of the words "machine learning" [2]. He wrote, "As a result of these experiments one can say with some certainty that it is now possible to devise learning schemes which will greatly outperform an average person and that such learning schemes may eventually be economically feasible as applied to real-life problems" [1, p. 548]. His program together with IBM's first stored-program computer, the 701, demonstrated this statement by winning a game of checkers against a human expert in Connecticut. Since that time, games have provided fertile research ground for artificial intelligence, and in 1996, the Chinook Project for checkers was recognized by the Guinness Book of World Records as the first computer program to win a human world championship [3]. The day to apply machine learning to challenging real-world problems is here and now.

What is "machine learning"? Several suggested definitions are discussed on the IBM community site [4], including "The purpose of machine learning is to learn from training data in order to make as good as possible predictions on new, unseen, data." This definition suggests some of the challenges with machine learning. The program needs to build a model based on training data that includes the correct answer (i.e., supervised learning) and that minimizes the error in predicting new data. Alternatively, algorithms may look for structure in the data and group similar clusters (i.e., unsupervised learning). Too closely mirroring the training data results in overfitting and poor results with unknown data, and too little fitting results in unacceptable errors in predictions. In addition, as updated known data become available, the model may need to be re-adjusted to retain generalizability. Thus, the methods used in machine learning are constantly being researched and assessed against real-life data from various fields along with the computer technologies needed to implement them.

This book applies and assesses machine learning for classes of important real-life problems, an area often referred to as "data analytics." Authors have contributed leading research in the areas of medical, biological and signal sciences; social studies and social interactions; traffic, computer and power networks; and digital forensics. The book also looks to the future for research areas that may yield theoretical advances. The editors have provided a valuable and much-needed collection of leading research in machine learning and data analytics that will increasingly impact each of us in our everyday lives. New and experienced researchers, practitioners, and those interested in machine learning will be inspired by the innovative ideas contained in its pages.

Baltimore, USA

Gloria Phillips-Wren, Ph.D.
Professor, Loyola University Maryland

References

1. Samuel, A.: Some studies in machine learning using the game of checkers. IBM J. 3(3), 210–229 (1959)
2. McCarthy, J., Feigenbaum, E.: In memoriam—Arthur Samuel: Pioneer in machine learning. AI Mag. 11(3), 10–11 (1990)
3. Chinook,: Arthur Samuel's legacy. Accessed on 28 May 2018, from https://webdocs.cs.ualberta.ca/~chinook/project/legacy.html (2018)
4. Puget, J-C.: What is machine learning? IBM Community, May 18 from https://www.ibm.com/developerworks/community/blogs/jfp/entry/What_Is_Machine_Learning?lang=en (2016). Accessed on 28 May 2018

Preface

At the dawn of the fourth Industrial Revolution, *data analytics* is emerging as a force that drives towards dramatic changes in our daily lives, the workplace and human relations. Synergies between physical, digital, biological and energy sciences and technologies, sewn together by *non-traditional data collection and analysis*, drive the digital economy at all levels and offer new, previously unavailable opportunities.

The need for data analytics arises in most modern scientific disciplines, including engineering, natural, computer and information sciences, economics, business, commerce, environment, healthcare and life sciences. The book at hand explores some of the emerging scientific and technological areas in which data analytics arises as a need and, thus, may play a significant role in the years to come.

Coming as the third volume under the general title *Machine Learning Paradigms* and following two related monographs, the book includes an editorial note (Chap. 1) and an additional twelve (12) chapters and is divided into five parts, namely: (1) *Data Analytics in the Medical, Biological and Signal Sciences*, (2) *Data Analytics in Social Studies and Social Interactions*, (3) *Data Analytics in Traffic, Computer and Power Networks*, (4) *Data Analytics for Digital Forensics* and (5) *Theoretical Advances and Tools for Data Analytics*.

This research book is directed towards professors, researchers, scientists, engineers and students of all disciplines. We hope that they all will find it useful in their works and researches.

We are grateful to the authors and the reviewers for their excellent contributions and visionary ideas. We are also thankful to Springer for agreeing to publish this book. Last, but not least, we are grateful to the Springer staff for their excellent work in producing this book.

Piraeus, Greece — George A. Tsihrintzis
Piraeus, Greece — Dionisios N. Sotiropoulos
Sydney, Australia/Canberra, Australia — Lakhmi C. Jain

Contents

Chapter 1
Machine Learning Paradigms: Advances in Data Analytics

George A. Tsihrintzis, Dionisios N. Sotiropoulos and Lakhmi C. Jain

Abstract At the dawn of the 4th Industrial Revolution, data analytics is emerging as a force that drives towards dramatic changes in our daily lives, the workplace and human relationships. Synergies between physical, digital, biological and energy sciences and technologies, brought together by non-traditional data collection and analysis, drive the digital economy at all levels and offer new, previously-unavailable opportunities. The need for data analytics arises in most modern scientific disciplines, including engineering; natural-, computer- and information sciences; economics; business; commerce; environment; healthcare; and life sciences. Coming as the third volume under the general title MACHINE LEARNING PARADIGMS, the book includes an editorial note (Chapter 1) and an additional 12 chapters, and is divided into five parts: (1) Data Analytics in the Medical, Biological and Signal Sciences, (2) Data Analytics in Social Studies and Social Interactions, (3) Data Analytics in Traffic, Computer and Power Networks, (4) Data Analytics for Digital Forensics, and (5) Theoretical Advances and Tools for Data Analytics. This research book is intended for both experts/researchers in the field of data analytics, and readers working in the fields of artificial and computational intelligence as well as computer science in general who wish to learn more about the field of data analytics and its applications. An extensive list of bibliographic references at the end of each chapter guides readers to probe further into the application areas of interest to them.

We are facing the dawn of the 4th Industrial Revolution (IR), which is expected to have a dramatic impact on our daily lives, the work place and human relations [1]. Its pace is unprecedented, its breadth of applications is almost universal and its impact will transform our societies radically.

G. A. Tsihrintzis (✉) · D. N. Sotiropoulos
University of Piraeus, Piraeus, Greece
e-mail: geoatsi@unipi.gr

L. C. Jain
University of Technology Sydney, Broadway, Australia

L. C. Jain
University of Canberra, Canberra, Australia

G. A. Tsihrintzis et al. (eds.), *Machine Learning Paradigms*, Intelligent Systems Reference Library 149, https://doi.org/10.1007/978-3-319-94030-4_1

Unlike the previous IRs, which have depended on coal and water to drive steam engines and oil to drive internal combustion engines (1st IR), the Mass Production Line (2nd IR) and advances in Electronics and Information Technology (3rd IR), the 4th IR is characterized by synergies between physical, digital, biological and energy sciences and technologies. The common thread that unites all these different scientific disciplines comes under the term *data*.

As has been eloquently stated, *Data in the* 21st *Century is like Oil in the* 18th *Century* [2, 3]. This is because good data drive the digital economy at all levels, offering new, previously-unavailable opportunities. However, traditionally-collected data, i.e. economic figures, are not sufficient. All sorts of additional data need to be collected, besides traditional economic figures, and this data needs to be efficiently stored, transmitted, processed and converted into information, knowledge and, eventually, wisdom [4].

Data Analytics is the term devised to describe specialized processing techniques, software and systems aiming at extracting information from extensive data sets and enabling their users to draw conclusions, to make informed decisions, to support scientific theories and to manage hypotheses [5, 6].

The need for Data Analytics arises in most modern scientific disciplines, including engineering, natural, computer and information sciences, economics, business, commerce, environment, healthcare, and life sciences. The book at hand explores some of the emerging scientific and technological areas in which Data Analytics arises as a need and, thus, may play a significant role in the years to come. The book comes as the *third volume under the general title MACHINE LEARNING PARADIGMS and follows two related monographs* [7, 8].

More specifically, the book at hand consists of an editorial chapter (Chap. 1) and an additional twelve (12) chapters. All chapters in the book were invited from authors who work in the corresponding area of Data Analytics and are recognized for their research contributions. In more detail, the chapters in the book are organized into five parts, as follows.

The first part of the book consists of five chapters devoted to *Data Analytics in the Medical, Biological and Signal Sciences*.

Specifically, Chap. 2, by Dessi, Recupero, Fenu and Consoli, is on "*A Recommender System of Medical Reports Leveraging Cognitive Computing and Frame Semantics*." The authors design and implement a medical recommender system to cluster a collection of medical reports and, subsequently, given a medical report for a specific patient as input, to recommend similar medical reports from patients who had similar symptoms.

Chapter 3, by Amelio and Amelio, is entitled "*Classification Methods in Image Analysis with a Special Focus on Medical Analytics*." The authors present and discuss supervised and unsupervised classification methods for used in medical analytics and outline future related methodologies.

Chapter 4, by Rjeily, Badr, El Hassani and Andres, is on "*Medical Data Mining for Heart Diseases and the Future of Sequential Mining in Medical Field*." The authors present an overview of various approaches in predicting heart failure and classifying heart disease.

Chapter 5, by Stąpor, Roterman-Konieczna, and Fabian, is on "*Machine Learning Methods for the Protein Fold Recognition Problem*." The authors present methodologies for addressing the problem of protein fold recognition, which is characterised by a high number of data classes, imbalance of the available data sets and presence of outliers.

Chapter 6, by Korvel, Kurowski, Kostek and Czyzewski, is on "*Speech Analytics based on Machine Learning*." The authors present methodologies to prepare speech data for machine learning-based processing.

The second part of the book consists of two chapters devoted to *Data Analytics in Social Studies and Social Interactions*.

Specifically, Chap. 7, by Troussas, Krouska and Virvou, is on "*Trends on Sentiment Analysis over Social Networks: Pre-processing Ramifications, Stand-Alone Classifiers and Ensemble Averaging*." The authors provide a guideline for the decision of optimal pre-processing techniques and classifiers for sentiment analysis over Twitter.

Chapter 8, by Sidorova, Rosander, Skold, Grahn and Lundberg, is on "*Finding a Healthy Equilibrium of Geo-demographic Segments for a Telecom Business: Who are Malicious Hotspotters*." The authors present a data-driven analytic strategy based on combinatorial optimization and analysis of the historical mobility designed to quantify the desirability of different geo-demographic segments in the telecommunications business.

The third part of the book consists of three chapters devoted to *Data Analytics in Traffic, Computer and Power Networks*.

Specifically, Chap. 9, by Gravvanis, Salamanis and Filelis-Papadopoulos, is on "*Advanced Parametric Methods for Short-Term Traffic Forecasting in the Era of Big Data*." The authors present several state-of-the-art methods used in all aspects of the traffic forecasting problems, with particular emphasis given on both the algorithmic and the efficiency aspects of the problem, in the light of the large amounts of available traffic data.

Chapter 10, by Leros and Andreatos, is on "*Network Traffic Analytics for Internet Service Providers—Application in Early Prediction of DDoS Attacks*." The authors model intra-values forecasts of a time-series Network Traffic using a mean reverting stochastic process and show that proposed algorithm was able to identify successfully unusual activities contained in test datasets and to produce proper warnings.

Chapter 11, by Androvitsaneas, Boulas and Dounias, is on "*Intelligent Data Analysis in Electric Power Engineering Applications*." The authors various intelligent approaches for modelling, generalization and knowledge extraction from data, which are applied in different electric power engineering domains of the real world.

The fourth part of the book contains one chapter on *Data Analytics for Digital Forensics*.

Specifically, Chap. 12, authored by Karampidis, Deligiannis and Papadourakis, is on "*Combining Genetic Algorithms and Neural Networks for File Forgery Detection*." The authors propose a digital forensic examiner which uses specialized forensic software to accurately identify the various file types to determine which of them may contain potential evidence.

Finally, the fifth part of the book contains one chapter on new *Theoretical Advances and Tools for Data Analytics*.

Specifically, Chap. 13, authored by Passalis and Tefas, is on "*Deep Learning Analytics.*" The authors present various architectures of (Deep) Neural Networks, from simple Multilayer Perceptrons to Convolutional Neural Networks and Recurrent Neural Networks, and also discuss their advanced training and optimization techniques.

In this book, we have presented some of the emerging scientific and technological areas in which Data Analytics arises as a need and, thus, may play a significant role in the years to come. The book has come as the *third volume under the general title MACHINE LEARNING PARADIGMS, following two related monographs* [7, 8]. Societal demand continues to pose challenging problems, which require ever more efficient tools, methodologies, and systems to de devised to address them. Thus, the reader may expect that additional volumes on other aspects of Machine Learning Paradigms and their application areas will appear in the future.

Bibliography

1. Schwabd, K.: The fourth industrial revolution—what it means and how to respond. Foreign Aff. (2015) https://www.foreignaffairs.com/articles/2015-12-12/fourth-industrial-revolution
2. Toonders, J.: Data is the new oil of the digital economy. Wired https://www.wired.com/insights/2014/07/data-new-oil-digital-economy/
3. https://www.economist.com/news/leaders/21721656-data-economy-demands-new-approach-antitrust-rules-worlds-most-valuable-resource
4. https://en.wikipedia.org/wiki/Data
5. https://en.wikipedia.org/wiki/Data_analysis
6. https://searchdatamanagement.techtarget.com/definition/data-analytics
7. Lampropoulos, A.S., Tsihrintzis, G.A.: Machine learning paradigms—applications in recommender systems. In: Intelligent Systems Reference Library Book Series, vol. 92 Springer (2015)
8. Sotiropoulos, D.N., Tsihrintzis, G.A.: Machine Learning paradigms—artificial immune systems and their applications in software personalization. In: Intelligent Systems Reference Library Book Series, vol. 118. Springer (2017)

Part I
Data Analytics in the Medical, Biological and Signal Sciences

Chapter 2
A Recommender System of Medical Reports Leveraging Cognitive Computing and Frame Semantics

Danilo Dessì, Diego Reforgiato Recupero, Gianni Fenu and Sergio Consoli

Abstract During the last decades, a huge amount of data have been collected in clinical databases in the form of medical reports, laboratory results, treatment plans, etc., representing patients health status. Hence, digital information available for patient-oriented decision making has increased drastically but it is often not mined and analyzed in depth since: (i) medical documents are often unstructured and therefore difficult to analyze automatically, (ii) doctors traditionally rely on their experience to recognize an illness, give a diagnosis, and prescribe medications. However doctors experience can be limited by the cases they are treated so far and medication errors can occur frequently. In addition, it is generally hard and time-consuming inferring information for comparing unstructured data and evaluating similarities between heterogeneous resources. Technologies as Data Mining, Natural Language Processing, and Machine Learning can provide possibilities to explore and exploit potential knowledge from diagnosis history records and help doctors to prescribe medication correctly to decrease medication error effectively. In this paper, we design and implement a medical recommender system that is able to cluster a collection of medical reports on features detected by IBM Watson and Framester, two emerging tools from, respectively, Cognitive Computing and Frame Semantics, and then, giving a medical report from a specific patient as input, to recommend similar other medical reports from patients who had analogues symptoms. Experiments and results have proved the quality of the resulting clustering and recommendations, and the key role that these innovative services can play on the biomedical sector. The proposed system is

D. Dessì (✉) · D. Reforgiato Recupero · G. Fenu
Mathematics and Computer Science Department, University of Cagliari,
Via Ospedale 72, 09124 Cagliari, Italy
e-mail: danilo_dessi@unica.it

D. Reforgiato Recupero
e-mail: diego.reforgiato@unica.it

G. Fenu
e-mail: fenu@unica.it

S. Consoli
Philips Research, Data Science Department, High Tech Campus 34,
5656 AE Eindhoven, The Netherlands
e-mail: sergio.consoli@philips.com

G. A. Tsihrintzis et al. (eds.), *Machine Learning Paradigms*, Intelligent Systems Reference Library 149, https://doi.org/10.1007/978-3-319-94030-4_2

able to classify new medical cases thus supporting physicians to take more correct and reliable actions about specific diagnosis and cares.

Keywords Health recommender systems · Data mining · Cognitive computation Personal health records · Clustering · Knowledge inference · Personalized medicine · Relevance computation · Biomedical text-mining

2.1 Introduction

During the last decades a lot of data have been collected in textual clinical datasets representing patients' health states (e.g. medical reports, treatment plans, laboratory results, clinical records, surgical transcriptions, researches results etc.). Hence, digital data available for patient-oriented decision making has extremely grown but is not often mined and analyzed. Therefore, efficient access to information becomes hard for end-users [1]. In order to overcome text data overload and transform the text into useful and understandable source of knowledge, automated processing methods are required. Undoubtedly, this data can be exploited for figuring out relevant insights in the healthcare industry through data mining and machine learning techniques. These can work as a potential base for developing recommender systems which employ documents as items, and try to suggest diagnosis for new patients who present a clinical state similar to those that have been previously evaluated.

Recommender systems can be divided into three main categories: collaborative, content-based, and hybrid systems. Collaborative recommender systems work on experience gathered from previous user experiences, i.e. exploiting items which have been previously chosen by other users similar to a target in order to predict similar needs. Content-based recommender systems focus on the characteristics of items, e.g. when searching for a car, the recommendation output could be based on its price, brand, and color. Finally, hybrid recommender systems combine features of context-based and collaborative systems [2]. Hereafter, we focus on content-based recommender systems.

Content-based recommender systems usually rely on descriptions of people and items to build models which can be exploited for suggesting items similar to those a target person already had in the past [3]. They often employ retrieval approaches as a Vector Space Model (VSM), e.g. *bag-of-words*, as in [4]. A VSM is a model where each item is represented in a N-dimensional space and each dimension is related to a word of the documents collection. Many times, word-based approaches have not been able to figure out features for good results raising problems of accuracy. Therefore, data should be more deeply analyzed to yield a better understanding of users' state. One main challenge with medical reports is that a lot of information is stored by using the natural language, which suffers from the classical problem of ambiguity. Polysemy, troponymy, metonymy, n-grams expressions, entity recognition and disambiguation are common inherent problems of traditional methods largely employed in literature for dealing with textual resources. They make hard to elaborate the contained information by means of machines, preventing the storing and sharing between different agents, processes and systems. As a consequence, recent studies have started to employ Semantic Web and knowledge based resources for obtaining better results.

New researches have introduced Semantic Web techniques combining ontologies and knowledge-based resources for shifting from word-based to concept-based representations of textual resources. This implies an increasing adoption of Semantic Web resources, tools and best practices for discovering the best features which play significant roles into unstructured texts, enabling high level categorization of contents. New systems, usually named Cognitive Computing systems, have earned a lot of attention for figuring out relevant insights from textual data. One system is IBM Watson[1] which can understand concepts, entities, sentiments, keywords, etc. from unstructured text through its Natural Language Understanding[2] service.

WordNet [5] and FrameNet [6], among others, are two of the most important linguistic open data resources that have been illustrated several times. WordNet is a lexical database that defines synsets as groups of synonyms. Each synset represents a unique meaning, which is semantically related to other meanings through derivation, hyponym/hypernymy, meronymy/holonymy, antonymy, entailment, etc. relations. FrameNet contains frames, which contextualize a general situation or state. Each frame includes semantic roles known as *frame elements* which are activated by lexical units of the speech (e.g. different verbs evoke different frames). However, its limited coverage and non-standard semantics are two major barriers for its wide adoption on natural language data analysis. To overcome these issues, a novel frame semantic tool, Framester [7], has been recently proposed. Framester works as a graph-linked data hub between open data systems as FrameNet, BabelNet [8] and WordNet, providing a dense interlinking between existing resources and enabling a novel formal semantics for frames. Framester can perform semantic frames and BabelNet synsets detection which may improve matchings between meanings of data expressed by different words. It is public available through an online interface[3] and an API.[4]

Technologies as Data Mining, Natural Language Processing, and Machine Learning can provide novel alternatives to explore and exploit potential retrieved knowledge from historical medical records, and help doctors to prescribe medication correctly to decrease medication errors effectively. In fact, text and data mining approaches have been already employed in healthcare for saving time, money and life [9–11].

Knowledge based techniques and tools, if reliable, can support medical staff in diagnosis, prevention and treatment of diseases, providing suggestions based on past medical cases. This chapter shows how to build a content-based recommender system within the healthcare domain leveraging Semantic Web technologies and cognitive computing tools.

---

[1]https://www.ibm.com/watson/.

[2]https://www.ibm.com/watson/services/natural-language-understanding/.

[3]https://lipn.univ-paris13.fr/framester/en/wfd_html/.

[4]https://github.com/framester/Framester/wiki/Framester-Documentation.

Moreover, we performed tests on a real dataset showing enhancements in embedding Semantic Web and Cognitive Computing tools. We examined which features better detect distinct characteristics from texts, and result suitable to cluster medical documents in order to provide high quality recommendations.

The chapter is organized as follows. First, we present the research on biomedical text analysis in Sect. 2.2. Then, we describe our recommender system in Sect. 2.3. In Sect. 2.4, we show our experiments and discuss the results we obtained. Finally, Sect. 2.5 proposes future development of our system and directions where we are headed.

2.2 State of the Art

It would be impossible to enumerate the numerous medical questions dealt with computational approaches for clinical enhancements. Here, we focus on an overview of the most interesting and promising text, data mining and machine learning methods, and their applications, to discover insightful information from textual data in order to support the development of a novel content-based recommender system.

2.2.1 Biomedical Information Retrieval

In recent years, many retrieval tools have appeared and have been used on textual resources for extracting relevant and insightful semantics [12, 13]. These tools usually exploit statistical techniques, even though there have been recently based on open linked data and machine learning techniques. Medical text processing is not a new question, but extracting biomedical data into a well-defined structural storage still remains a complex task [14]. Dealing with various medical domains does not help the development of systems to support medical activity. Because biomedical information is continuously being created in textual form more than ever before, there have been a lot of efforts for coding information into databases, and developing automatic processes which aim at finding useful ways to represent and organize data [15]. Medical text processing on medical domain, in particular using Natural Language Processing (NLP) approaches, has been explored into many other works [10, 11]. In general, researchers have usually tried to overcome text-depending issues focusing on classic entity recognition and text disambiguation techniques to create a domain-specific semantic content for the analysis of medical reports [14, 16, 17].

To alleviate textual inherit issues, some proposals have started to adopt Semantic Web practices in the medical system development. The first competition [18] on medical text-mining was run in 2002 during the Knowledge Discovery in Databases (KDD) Challenge Cup. Participants faced with a curation problem for assessing medical documents from the FlyBase dataset in order to determine whether a document should be curated based on the presence of experimental evidence of Drosophila gene products. Exploiting Part-of-Speech (POS) tagging and semantic controls determined by examining the training documents and by focusing on figures captions, a collection of manually constructed rules obtained best results on the presence of experimental evidence for the document clustering [19]. In [20] the authors used a Support Vector Machine which was trained on MEDLINE abstracts to distinguish abstracts containing information on protein-protein interactions, prior to curate this information into their BIND database. They used a bag-of-words model with classification techniques and discovered that classifiers could minimize the number of abstracts that the practitioners employed to read by about two-thirds.

Authors in [21] have proposed a new concept-based model which exploits various text mining approaches and their combinations for improving text clustering. They propose a labeler which evaluates the semantic contribute of each word in sentences, outperforming traditional methods and discovering that the semantics is less sensitive to noise. More recent approaches are based on semantic analysis which enables learning more accurate features defined by means of external knowledge bases. In [22] authors make able systems to face with challenges by exploiting cultural and linguistic background knowledge for better interpreting unstructured documents and reasoning on their content. In [23] an item recommender system has been provided for recommendation tasks of various resources (e.g. movies and books) exploiting Word Sense Disambiguation techniques based on WordNet lexical ontology for mapping contents by means of synsets. Similar techniques are studied today in medical domain.

2.2.2 *Biomedical Classification*

In this section, we present classification methods which have been adopted for dealing with unstructured clinical notes over past years.

Classification is a fundamental component in the biomedical domain due to its widespread utility in applications such as medical diagnosis and identification of genetic causes of disease. In [20] authors exploit various classification techniques as described in Sect. 2.2.1. One more approach on MEDLINE documents was proposed by [24] where authors applied a semi-supervised spectral approach technique for clustering contents over two types of constraint: must-link constraints on document pairs with high (MeSH)-semantic or global-content similarities, and cannot-link constraints on those with low similarities. The authors proved the good performance of their new method on MEDLINE documents, improving performance of linear combination methods and several well-known semisupervised clustering methods.

Authors in [25] experiment multi-label classification techniques by means of combinations of bag-of-words models, and adopt time series and dimensionality reduction approaches on the MIMIC II dataset. In [26], authors implemented a Support Vector Machine classifier on n-gram features retrieved from clinical notes of the Beth Israel Deaconess Medical Center to identify the mechanical ventilation and diagnosis of neonatal and adult patients. A Convolutional Neural Network classification approach has been proposed by [27] to build models which enable to generate context based representation of health related information at sentence level. Predefined disease labels have been adopted by [28] to classify free text clinical notes. They propose two techniques Sampled Classifier Chains (SCC) and Ensemble of Sampled Classifier Chains (ESCC), which extend their dataset with selected labels in order to obtain a relationship between disease and classification.

Performances of some classification methods applied on clinical notes have been recently evaluated in [29]. Authors focused on feature selection techniques investigating different approaches of transformation methods in order to improve the multi-label classification task. They report advantages of using filtering techniques and hybrid feature selection methods. One more recent work where classification methods have been evaluated is [30]. The best results have been obtained when a hierarchical approach to tag a document by identifying the relevant sentences for each label has been exploited.

2.2.3 Biomedical Clustering

In this section, we describe clustering methods applied to biomedical texts, and discuss recent works.

The clustering is the unsupervised task of finding groups of similar items by segmenting a collection into partitions called clusters, where items in the same cluster are more similar to each other than those in other clusters. In our work, biomedical text clustering items are medical reports. In general, document clustering can show various insights considering different levels of granularity of texts (i.e. clusters can be composed by whole documents, paragraphs, sentences or terms). In this case, the clustering can be employed as a tool for organizing and browsing documents in order to enhance the retrieval of information [31]. In biomedical domain, it could be essential to investigate patterns of a set of medical reports on features of different stuff so that similar patients can be treated concurrently in similar way.

An interesting medical document clustering has been proposed by [32] where authors exploited an ontology-based term similarity to index terms in a set of medical documents. They used a spherical k-means clustering algorithm on PubMed documents sets in order to evaluate the proposed similarity technique.

In [33] authors employed the KNN clustering method for evaluating a new similarity measure based on the semantic connection between words of a electronic medical report set. Authors in [34] performed cluster analysis on medical posts of online health communities for recognizing various types of content. They found that clusters can be associated to common categories as treatments, procedures, medications and so on. A framework based on clustering analysis has been developed by [35] for exploring health related topic automatically in online communities integrating data with medical domain specific knowledge.

Features as biomedical concepts and semantic relationships were identified with the help of ad-hoc ontologies for building a graph representation in order to enhance the recognition of categories by means of clustering techniques in [9].

2.2.4 Biomedical Recommendation

In literature, several systems refer to medicine for identifying active relations of new patients states with past ones, but few of them exploit natural language or text mining for accomplishing recommendation tasks. In [36], authors describe a recommendation procedure which uses similarity measures for finding relations between online users' health data and medical information of Wikipedia to increase patients' autonomy in their personal health. The task to predict future health risks by means of a recommendation technique has been proposed by [37], where authors developed an engine called CARE in order to predict the future diseases risks of patients. To provide more accurate and personalized doctor recommendations, authors in [38] mined emotions from previous users' ratings adopting a topic model technique for developing a system named iDoctor. To engender advances on health recommender systems, the ACM Conference on Recommender Systems hosted a workshop in years 2016 and 2017 where specific-purpose health recommender systems have been presented, but no one focused on textual resources as narrative medical reports. In addition, these systems deal with clinical data in order to provide specific online services which target patients as end-users, but there are not systems which exploit data for supporting diagnosis fruition and physicians' work.

2.2.5 Cognitive Computing and IBM Watson

With the terms Cognitive Computing systems we refer to those smart systems that learn at scale, can learn with purpose, and recently have modules for interacting directly with humans. They are being developed to reduce costs, increase efficiency, accelerate discovery, make essential connections in large amounts of data. With the rapid growth of the availability of massive amounts of data, Cognitive Computing systems provide new opportunities for augmenting human expertise in a broad range of domains. Embedding Cognitive Computing services in novel systems results fundamental for dealing with previous unmanageable issues. In medical domain, Cognitive Computing systems can play a relevant role for supporting activities of practitioners, providing understandable access to clinical data and enhancing the precision of the medicine. In our research, we have employed the most promising Cognitive Computing system, IBM Watson[5] which provides a cloud suite of services by means of the

---

[5] https://www.ibm.com/watson/.

IBM Cloud[6] platform for dealing with huge amount of data, and returns interesting features that can capture medical insights. More specifically, we employed the outcomes of the Natural Language Understanding service[7] which has been developed for analyzing textual data and, therefore, it is suitable for managing unstructured and narrative contents of medical reports.

2.2.6 Frame Semantics and Framester

Frame semantics is a linguistic theory that defines a meaning as a coherent structure of related concepts [39]. To relate various concepts, knowledge-based resources are usually employed as corner stones of semantic technological approaches. A content-based recommender system aware of semantics has the ability to interpret natural language texts and makes conclusions on their content. In order to embed frame semantics in our recommender system we exploits Framester, a novel data linked resource that works as a hub between linked open data systems as FrameNet, BabelNet and WordNet. It is a new frame-based ontological resource that leverages an inter-operable predicate space formalized according to frame semantics [6] and semiotics [40].

2.3 Architecture of Our System

In this section, we describe the modules of our system which needs proper techniques for representing items and comparing new and old users' states. An overview of our system is depicted in Fig. 2.1. The reader can see:

- *Medical Reports Collection*. This is the set of reports on which the system can learn about past clinical historical cases.
- *Content Analyzer Module*. The module takes as an input the collection of reports and the new report that the user (e.g. a physician) wants to evaluate. It embeds various resources for mining features from textual components of medical reports.
- *Represented Medical Reports*. This is the output given back by the Content Analyzer Module. The output is formatted so that machine learning algorithms can be easily applied.
- *Machine Learning Module*. This module implements a set of classification and clustering algorithms that are used for building models which describe patients' profiles.
- *Clinical Patients Profiles*. They are profiles that have been built by algorithms that had been employed in the Machine Learning Module.

---

[6]https://www.ibm.com/cloud/.

[7]https://www.ibm.com/watson/services/natural-language-understanding/.

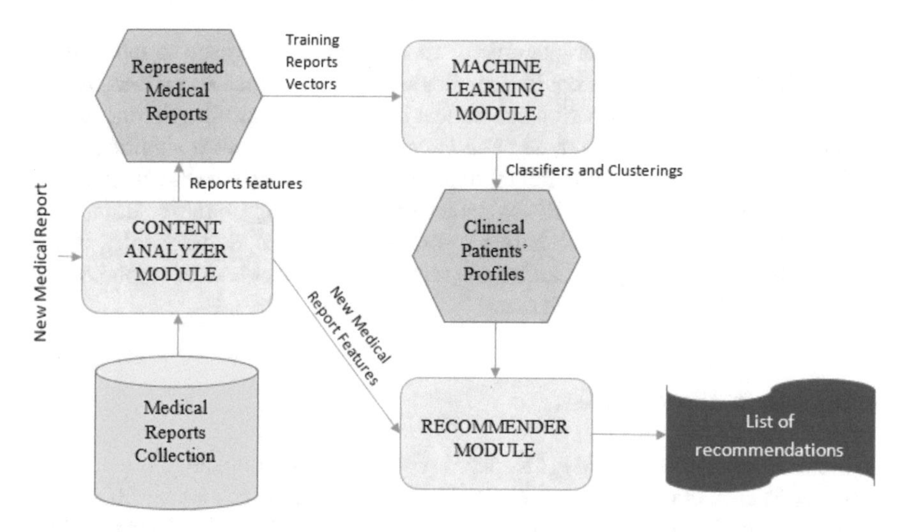

Fig. 2.1 Architecture of the content-based recommender system

- *Recommender Module.* The Recommender Module matches the new medical report features with the known patients' profile in order to make a list of recommendations.

2.3.1 Content Analyzer Module

The Content Analyzer Module takes as an input the collection of unstructured medical reports and produces a structured documents representation which enables the automatic computation of machine learning techniques performed by the Machine Learning Module. In addition, it must mine new unknown medical reports. In this section, the description of the model, the features and their characteristics are described.

2.3.1.1 Item Representation

For applying machine learning algorithms, data must be represented by sets of features usually called attributes. For example, to recommend books, attributes adopted to describe a book can be authors, editor, genre etc. When items are described by the same set of attributes and there are known values of these attributes, they are represented in structured data that can be employed for automatic computations. In case of biomedical textual documents there are not well-defined attributes, and textual features can raise difficulties when the system learns about patients. The main problem is that traditional term-based method can fail to capture the semantics of clinical states of patients. For example, if more words can be used to indicate the same pathology (e.g. *tumors* could be indicated with the names *neoplasms, malignancies* etc.) relevant information can be lost if two clinical profiles do not contain the same word. In this context, semantic analysis of data plays a significant role and promises surprising results for solving these issues. More specifically, we have employed words coming from IBM Watson which is a leading Cognitive Computing tool, and Framester a novel hub between semantic resources. Subsequently, in this section we show features that can be extracted from medical textual resources and discuss about advantages of each one.

2.3.1.2 Vector Space Models

A Vector Space Model (VSM) is a spatial representation. For example, in a word-based VSM each document is represented over a N-dimensional space, where each dimension corresponds to a word that belongs to the whole set of terms of the given collection of documents. Let $D = \{d_1, d_2, \ldots, d_k\}$ be a collection of medical reports and $A = \{a_1, a_2, \ldots, a_n\}$ the set of attributes employed for representing them. A can be built by means of a natural language process or semantic content exploration pipeline which applies methods (e.g. the English stop word and stemming steps) for representing D. Each medical report d_i is represented by a vector of values $d_i = \{v_{1i}, v_{2i}, \ldots, v_{ni}\}$ where each value v_{ki} indicates the degree of relation between the attribute a_k and the document d_i. Attributes can have various natures such as words, n-grams, semantic features which describe contents, and so on. In our recommender system we have employed 6 different types of attributes: Term Frequency-Inverse Document Frequency, Concepts, Keywords, Entities, BabelNet Synsets, and Frames.

2.3.1.3 Term Frequency-Inverse Document Frequency

This is a *bag-of-word* model where attributes are words within the collection. For assigning a value to each word w, we have employed the Term Frequency-Inverse Document Frequency (TF-IDF) technique in which (i) uncommon words are not less relevant from frequent ones, (ii) a word that occurs many times in a document is not less relevant than a single one, and (iii) the length of documents does not play a significant role for the comparison of documents. To put it more simply, words that frequently occur within a document, but rarely in the whole collection, have more probability to be relevant in the document. The TF-IDF formula is showed in (2.1) where w_{ki} is the number of occurrences of the word w_k in the document d_i, $|d_i|$ is the size of the document expressed as number of words, N is the number of documents in the collection, and n_k is the number of documents where the word w_k occurs at least once.

$$TF - IDF(w_k, d_i) = \frac{w_{ki}}{|d_i|} \cdot log \frac{N}{n_k} \tag{2.1}$$

In order to prevent that longer texts have higher probability to be chosen by a recommender system, TF-IDF values are usually normalized in a range [0,1].

To avoid that frequent and no-relevant data (e.g. words that do not carry any meaning for the medical purpose as articles *the, a, an*, preposition *about, therefore, at*, etc.) appear in the TF-IDF features, the module performs some cleaning steps on the input texts. It precisely removes numeric data, punctuation, and stop-words. In fact, they are considered unnecessary and their remotion serves for (i) reducing the size of the VSM and (ii) for the subsequent efficiency of using a smaller space of features. All terms are taken in their lower case shape, avoiding to consider more times different representations of the same word (e.g. *Cardiac* and *cardiac*).

2.3.1.4 Concepts

Concepts can be defined as cognitive units which model perceived abstract subjects. They depend on the ability to process domain dependent knowledge and efficiently learn insights which become fundamental keys in the meaning of contents. Concepts can embody structures and representation of real words discovered in text, hence, they enable capturing high level abstraction reducing the complexity of the computation space. Moreover, they enable the specialization of employed attributes for representing documents in the VSM. IBM Watson can be employed for discovering automatically concepts related to the medical domain from natural language texts. It assigns a weight to each concept we have used for building the VSM. More precisely, given a collection of medical reports we use as set of attributes A the union of almost fifty concepts returned by IBM Watson from each medical report.

2.3.1.5 Keywords

Keywords are words of texts that enable listing the content of a report, releasing information about which words result relevant for describing the content of a document. Keywords are automatically detected by IBM Watson which provides a weight for each one. The VSM model is built as in the case of concepts.

2.3.1.6 Entities

Entities are actors that make actions in a text. Specifically to the medical domain, they can be people (e.g. physicians or nurses), illnesses (e.g. tumor), medicine names and so on. By capturing entities, it is possible to find relations between different documents if they share similar actors, especially when they are specific (for example if in a subset of documents D' physicians are cardiologists and in another subset D'' they are physiotherapists, the entities are distinct and enable better separation of the document subsets in different topics). As with the previous IBM Watson features, a weight is returned for each entity and indicates its influence in a document.

2.3.1.7 BabelNet Synsets

BabelNet synsets are unique unambiguous identifiers of sets of words which share the same meaning. We have chosen these synsets because (i) they are the result of the integration of various linguistic and semantic resources as WordNet, Wikipedia, FrameNet, among others, and (ii) they are directly provided by Framester. Differently from IBM Watson features, we do not have weights, hence, only the presence of BabelNet synsets have been considered by means of boolean flags into the Content Analyzer Module.

2.3.1.8 Semantic Frames

A semantic frame is a coherent group of concepts such that complete knowledge of one concept depends on the knowledge of all them in a context. Given a text, they are activated by nouns and verbs. Each frame can have multiple hierarchical levels that indicate its abstractions. For example, in Fig. 2.2, the word *cardiology* is abstracted by frames *Medical_specialties* and *Cure*. It should be underlined that frames are different from IBM Watson concepts because they do not depend on the application domain, but on relations that words have into linguistic and knowledge resources Framester adopts. As with the BabelNet synsets, we use the frames presence in the VSM.

2.3.1.9 The Course of Dimensionality Problem

The course of dimensionality problem refers to the issue that regards the great size of the number of attributes required to describe the target collection. The VSM suffers of this problem, hence, it needs to be managed into content-based applications as our

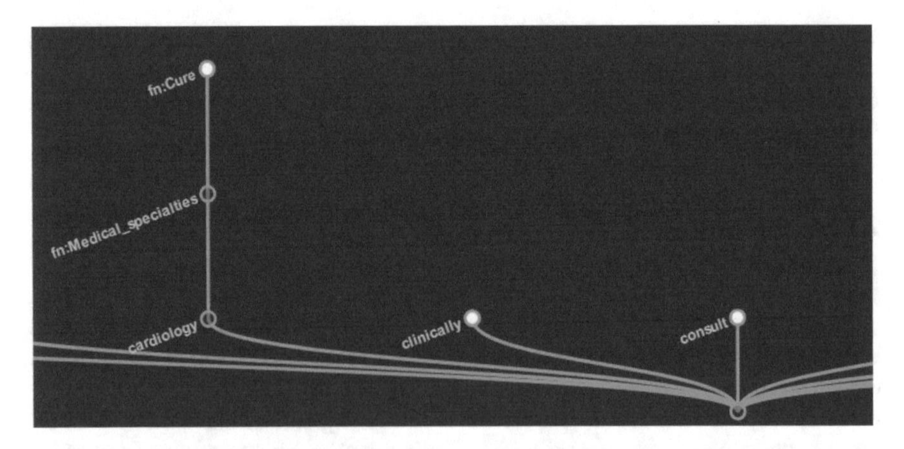

Fig. 2.2 Part of framester result on the sentence "*consider cardiology consult and further evaluation if clinically indicated*"

recommender system. One common method intensively applied in order to solve the issue is the Singular Value Decomposition (SVD). Let $A = \{a_1, a_2, \ldots, a_n\}$ be the set of attributes and $D = \{d_1, d_2, \ldots, d_i\}$ be the collection of our documents. The VSM is usually represented by a matrix M of size $|D| \times |A|$. M can be disjointed in three components $M = USV^T$ where S is a diagonal matrix containing the largest singular values, U is a matrix where columns are left singular vectors, and V is a matrix where columns define right singular values. In order to reduce complexity of data, the module applies a truncation which consists in holding only the largest k singular values, removing others which can be considered less relevant. This technique is known in literature as Truncated-SVD (TSVD). The module adopts the matrix $M' = U \times S$ which has a number of rows equivalent to the number of considered documents with a smaller number of attributes (columns) than the original matrix M. Besides decreasing the overall computational costs, an advantage of using the TSVD is deleting noise elements that might deteriorate the list of final recommendations. We want to point out that the value of k requires a trade-off between the amount of remaining and neglecting data to avoid the loss of information. Its value depends on the set of attributes A which characterizes the used collection.

2.3.2 *Machine Learning Module*

The Machine Learning Module receives a VSM as an input and returns a model which recognizes clinical patients' states. Its current version includes two clustering algorithms which are applied on all VSMs. The clustering techniques the module implements enable to deal with unsupervised data. They are (i) Hierarchical clustering algorithm and (ii) K-means clustering algorithm.

In this section we explain in depth how the chosen clustering algorithms work, and discuss about advantages they enable. Moreover, we show which machine learning algorithms might be employed in our system underlying which are requirements for an enhanced recommendation.

2.3.2.1 Hierarchical Clustering

Hierarchical Clustering builds a clusters hierarchy, or in other words, a tree of clusters which is usually called *dendrogram*. Each cluster contains children that are clusters as well, unless for the leafs of the tree. Sibling clusters split documents that are contained in the common parent cluster. A hierarchical clustering algorithm can be either *agglomerative* or *divisive*. In its agglomerative version, the algorithm starts with single elements of the collection, then it merges elements together based on a chosen measure (e.g. *Euclidean distance*). The agglomerative process is iterated as long as a unique cluster that covers all collection documents is obtained. The divisive variant of the algorithm starts with one cluster of all documents and recursively splits the most appropriate clusters according to a given criteria (e.g. splitting the largest cluster in each iteration.). The method continues its execution until a stop criterion (e.g a given number of clusters) is achieved. Our recommender system implements an agglomerative clustering, since we are interested in building groups looking for similarities starting from pairs of documents.

The hierarchical clustering algorithms are easily applicable on each kind of data, enable a manageable granularity of clusters and can be applied with any type of similarity measures. For these reasons, we felt that this type of clustering approach can lead good results on medical domain.

2.3.2.2 K-means Clustering

The K-means Clustering is a partition method. It builds a set of clusters minimizing the sum of squared distance between elements of a cluster and its center. The results is a single partition of data without any structure and, hence, can have advantages on applications which involve large sets of data for which the construction of a hierarchical structure can be onerous. The algorithm requires the number of clusters k as an input. This number is used to allocate k random centers which will be employed to build clusters. At beginning, it assigns each element to the cluster with the nearest center. Iteratively, centers are updated based on the built clusters and elements are moved into the cluster with their nearest center.

2.3.2.3 Similarity Measures

Precise clustering requires an accurate definition of the closeness between documents represented in the VSM. The closeness can be measured by either the pair-wised similarity or distance. A variety of similarity or distance measures have been proposed and discussed in literature. Our Machine Learning Module adopts the Cosine and Euclidean measures. The Cosine similarity quantifies the angle between two documents expressed by vectors. Its formula applied on two vectors v_p and v_q can be observed in (2.2).

$$CosS(v_p, v_q) = \frac{v_p \, v_q}{\parallel v_p \parallel \parallel v_q \parallel} \tag{2.2}$$

$CosS$ values 1 when v_p and v_q are completely similar, and 0 otherwise.

The Euclidean distance $EucD$ between two vectors v_p and v_q is defined as usual in (2.3).

$$EucD(v_p, v_q) = \sqrt{\sum_i (v_p(i) - v_q(i))^2} \tag{2.3}$$

Differently from the Cosine similarity, the Euclidean distance has not a limited range of values, therefore, it needs to be scaled before used for the similarity evaluation. For such reason the module adopts the formula (2.4) for scaling Euclidean-based values.

$$EucS(v_p, v_q) = \frac{1}{1 + EucD(v_p, v_q)} \tag{2.4}$$

2.3.3 *Recommendation Module*

This module uses the clinical patients' profiles for suggesting possible past medical cases that are similar to the new one by matching the new medical case against clinical profiles' clusterings of medical reports to be recommended. More specifically, the Recommendation Module takes a new medical report representation r and predicts whether there are clinical patients' profiles $p_1, \ldots, p_n$ that are interesting according to the relevance with r. It performs strategies to rank documents, and top-ranked ones are included in the list of recommendations that are provided to the final user. For doing so, the module computes the closeness between a new medical reports and clusters. In detail, given a new patients' medical report r and a clustering $C = \{c_1, \ldots, c_n\}$, the module finds the cluster c_i which has the closest center to r. Then elements within c_i are ranked from the most to the least similar to r. The produced ranking is used for finding the closest k medical reports as the final recommendation list.

2.4 Experiments

2.4.1 *The Test Dataset*

The employed dataset is a collection of no-labeled medical reports. It is freely available from the open-source iDASH repository.[8] In the dataset there are 2362 reports written in English. On the average, each report contains 400 words (ranging from 138 words for the shortest document to 1048 words for the longest one). There are singleton medical reports which might not have similarities with others, hence, for avoiding making unclear clustering groups they should be placed in one-element clusters.

Reports can be medical transcription samples including clinical notes, care plans, medical examinations, radiology reports etc. In the dataset, categories, their amount and distribution across reports, are not explicitly reported, although categories can be deduced from reports content (e.g. there are reports concerning heart issues whose can be placed in a category *heart*). The lack of predefined schema and the wide vocabulary of used terms make hard the categorization. Moreover, file names refer

[8] https://idash-data.ucsd.edu/.

to specific diseases or body parts issues that could be exploited to classify directly contents, but there could be ambiguous terms which may or may not refer to the same disease. Examples of medical reports names which can involve the same topic and discuss about the same issue are *cardiac-catheterization* and *hearth-catheterization*. As a consequence, these medical reports might be inappropriate to test supervised approaches, but they are suitable to test our unsupervised system which can manage unlabeled data.

2.4.2 Experiment Setup

2.4.2.1 Data Cleaning

Data cleaning is necessary in order to provide the same valid English text to the Content Analyzer Module services. First of all, we have cleaned all medical reports from HTML tags, removed all tables and structured format styles in order to obtain simple plain texts. Then we have matched reports words against those provided by WordNet, sending the word w' and getting the word w'' which has been placed in the text. At the end, only English text with correct grammar and punctuation composes the collection of medical reports.

2.4.2.2 Content Analyzer Module Setup

The Content Analyzer Module has been configured for providing more VSMs models which have been built on various features as described in Sect. 2.3.1. More precisely, let r_i be the i-th medical report and f_j be the j-th feature of a selected type. The outcomes of the module are:

- **5 Binary VSMs**: they include a matrix representation for the Concepts, Keywords, Entities, BabelNet Synsets and Semantic Frames features. Binary means that if f_j occurs within the inferred set of features of the medical reports r_i, in the VSM model M their relation is indicated by $M[i, j] = 1$, otherwise $M[i, j] = 0$;
- **4 Weighted VSMs**: they include a matrix representation for the Concepts, Keywords, Entities, and TD-IDF features. Weighted means that $M[i, j] = weight$, where $weight$ has been calculated exploiting the Natural Language Understanding service of IBM Watson or the TF-IDF approach as described above, and represents how strong is the relation between the medical report r_i and the feature f_j, otherwise $M[i, j] = 0$;
- **5 Counted VSMs**: they include a matrix representation for the Concepts, Keywords, Entities, BabelNet Synsets and Semantic Frames features. Counted means that $M[i, j] = count$ where $count$ is the number of times that a feature f_j occurs within the set of features of the medical reports r_i, otherwise $M[i, j] = 0$;

(a)

Report Name	Myocardial infarction	Heart	Atherosclerosis	Obesity	Cardiology	Cardiovascular system	Atheroma	Hypertension
heart-catheterization-ventriculography-angiography	1	1	1	0	1	1	0	0
cardiac-catheterization	1	1	1	0	1	0	1	0
cardiovascular-letter	1	0	1	1	0	0	0	1

(b)

Report Name	Myocardial infarction	Heart	Atherosclerosis	Obesity	Cardiology	Cardiovascular system	Atheroma	Hypertension
heart-catheterization-ventriculography-angiography	0.97	0.95	0.87	0	0.68	0.60	0	0
cardiac-catheterization	0.96	0.62	0.51	0	0.50	0	0.53	0
cardiovascular-letter	0.96	0	0.85	0.25	0	0	0	0.94

(c)

Report Name	Myocardial infarction	Heart	Atherosclerosis	Obesity	Cardiology	Cardiovascular system	Atheroma	Hypertension
heart-catheterization-ventriculography-angiography	4	6	4	0	4	2	0	0
cardiac-catheterization	3	3	2	0	2	2	2	0
cardiovascular-letter	4	0	3	3	0	0	0	3

Fig. 2.3 Samples of VSMs built on concepts extracted from three medical reports. Samples are related to **a** binary **b** weighted and **c** counted VSM

For more details on the three mentioned distances, the reader can look at examples of VSMs built on concepts extracted from three medical reports of the test dataset in Fig. 2.3. In the first row of each VSM, there are concepts that form the N-dimensional space. In the other rows, there are the names of reports on the first columns followed by values that indicate the degree of relation between the medical report and the i-th concept. The reader notices that (a) is built using the binary relation, (b) is built using the weighted relation and (c) is built using the counted relation.

2.4.2.3 Machine Learning Module Setup

The Machine Learning Module applied both clustering methods on all VSMs. In order to obtain high quality clusters, we set the module for exploiting the Silhouette width measure. Given a cluster c, its Silhouette width value $s(c)$ is computed as showed in Eq. (2.5) where $w(c)$ is the average dissimilarity within c and $o(c)$ is the lowest average dissimilarity of c to any other cluster.

$$s(c) = \frac{o(c) - w(c)}{max\{o(c), w(c)\}} \tag{2.5}$$

Values of Silhouette width range from -1 to 1. When the value is closer to 1, it means that the clusters are well separated; when the value is closer to 0, it might be difficult to detect the decision boundary; when the value is closer to -1, it means that elements assigned to a cluster might have been erroneously assigned. In general, we can consider good clusterings those that have high average values of Silhouette width. Unsurprisingly, the value of the Silhouette width depends on the type of features of the VSM under processing.

Hierarchical clustering. After the hierarchical clustering has been computed, the resulting dendrogram has been iteratively cut starting from its head, in order to increase the number of clusters for each iteration. In doing so, various clusterings obtained with different cut values have been produced. As a reminder, in our dataset we do not know how many groups can be formed. Therefore, we have exploited the highest value of average Silhouette width values in order to cut dendrogram where the clustering showed the best separation between medical reports.

K-means clustering. K-means Clustering has been performed with different values of k as number of clusters. For each value of k, the average Silhouette width measure has been computed similarly to hierarchical clustering setup. Then the clustering with the highest average value of Silhouette width has been hold as the output of the module.

2.4.3 Recommendation Module Setup

The recommendation module has been setup to receive an unknown medical report and a number k which represents the number of recommendations. In our experiments the adopted value of k is 10.

2.4.4 Results

At the current state, the quality of results of our recommender system mainly depends on the Content Analyzer Module and Machine Learning Module. In fact, a good quality of clusters means that medical reports similar to a new one can be correctly detected in the test dataset. First, for obtaining good clustering the features must allow a good separation of reports, and second the clustering algorithm must recognize which the best divisions are. Therefore, in this section we discuss about the most representative features of our dataset and the clustering algorithms performance. Results of clusterings quality can be observed in Figs. 2.4, 2.5, 2.6, 2.7 and 2.8.

The sets of features which have formed the best division of medical reports into clusters are those that have been computed using IBM Watson. In fact, they reach good levels of silhouette width. In more details, concepts and entities in their weighted and binary mappings have showed good performances in capturing medical information from medical reports of the test dataset. This fact suggests that the relevance of an entity or a concept into a medical report does not depend on the number of times that it appears. We can say that their role depends on the relations they have into reports, and

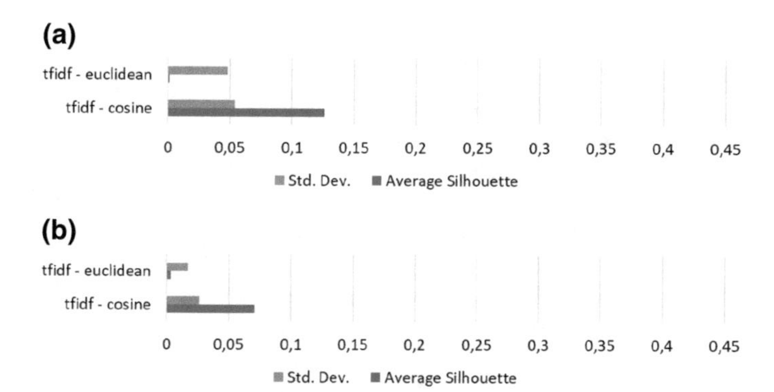

Fig. 2.4 The average and standard deviation values of the silhouette width measure of the clusterings computed on the TF-IDF measure. **a** Hierarchical clustering. **b** K-means clustering

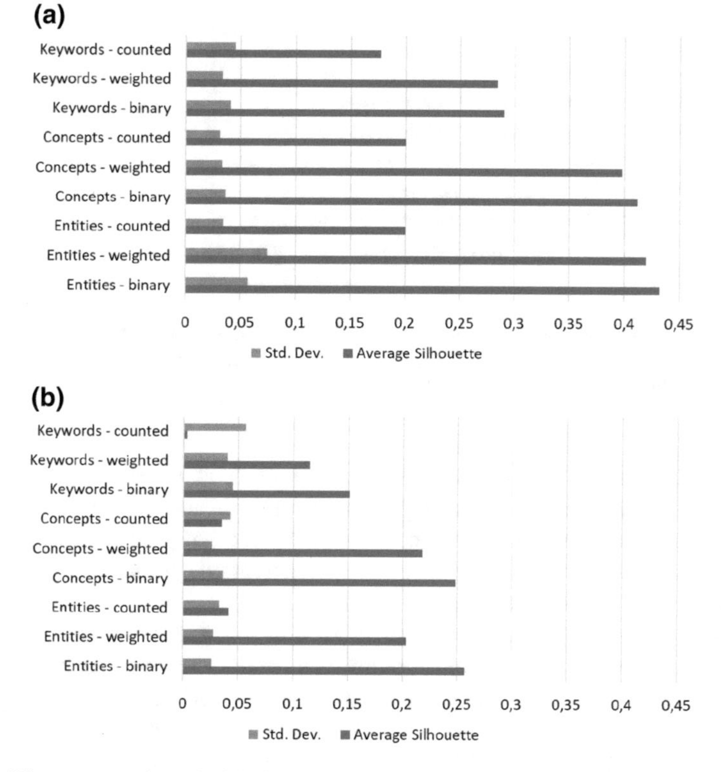

Fig. 2.5 The average and standard deviation values of the silhouette width measure of the clusterings computed on IBM Watson features. **a** Hierarchical clustering on cosine distance. **b** Hierarchical clustering on Euclidean distance

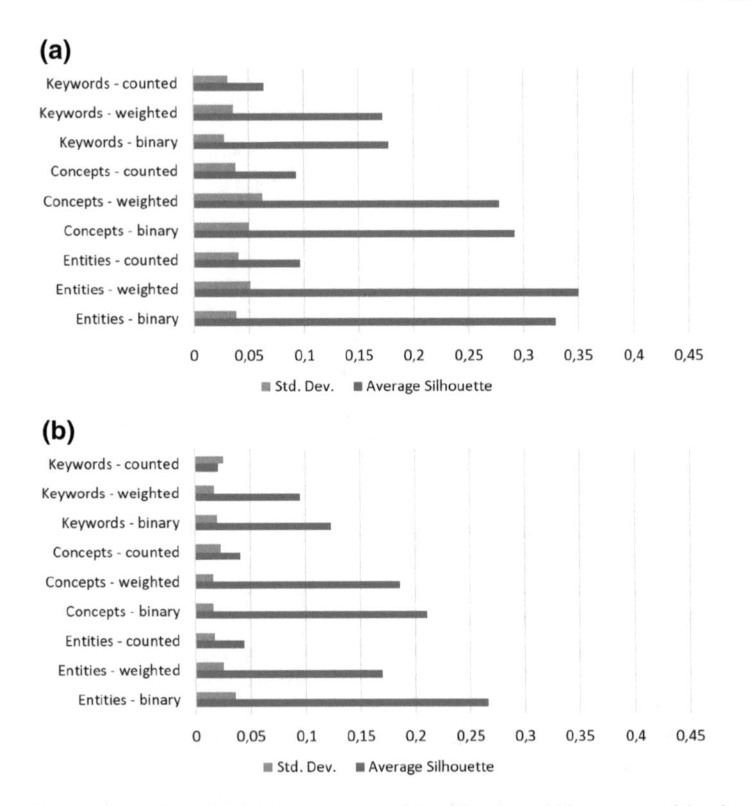

Fig. 2.6 The average and standard deviation values of the silhouette width measure of the clusterings computed on IBM Watson features. **a** K-means clustering on cosine distance. **b** K-means clustering on Euclidean distance

more influent their actions are, stronger their relevance is. Considering the number of times that a concept or an entity appears we do not add any additional information in our representative VSM. Keywords do not have showed good performances like entities and concepts but they could be considered as good alternatives in those cases where detecting entities and concepts can be hard.

Framester features do not have reached good results in the clusterings. This can depend on the fact that they are more abstract and not directly connected to the medical domain. Moreover, our test dataset could negatively influence this types of features since medical reports are strongly specific on patients' medical states. By contrast, they can result useful for those medical reports that describe the state of patients more in general without too clinical details (e.g. a starting examination visit). As for Framester features, the TF-IDF does not have showed good performances and same motivations can be observed.

One more point to consider is how the distance between two medical reports is computed. Results suggest that the cosine distance is more reliable than the Euclidean distance. Nevertheless, it is important underlying how they seem keeping a similar

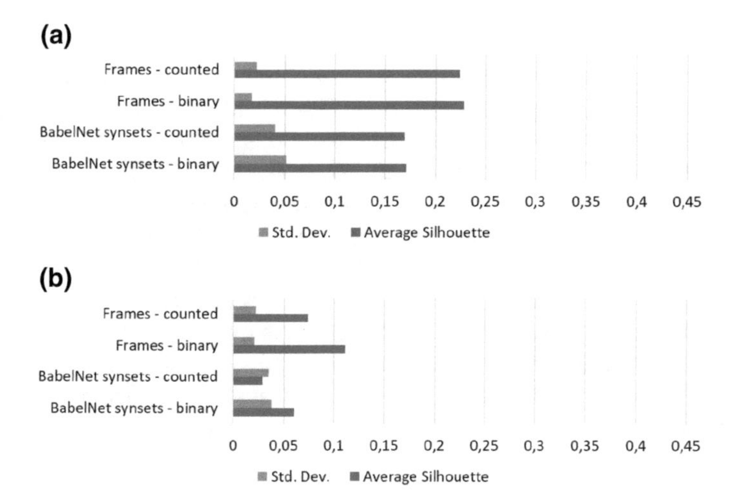

Fig. 2.7 The average and standard deviation values of the silhouette width measure of the clusterings computed on Framester features. **a** Hierarchical clustering on cosine distance. **b** Hierarchical clustering on Euclidean distance

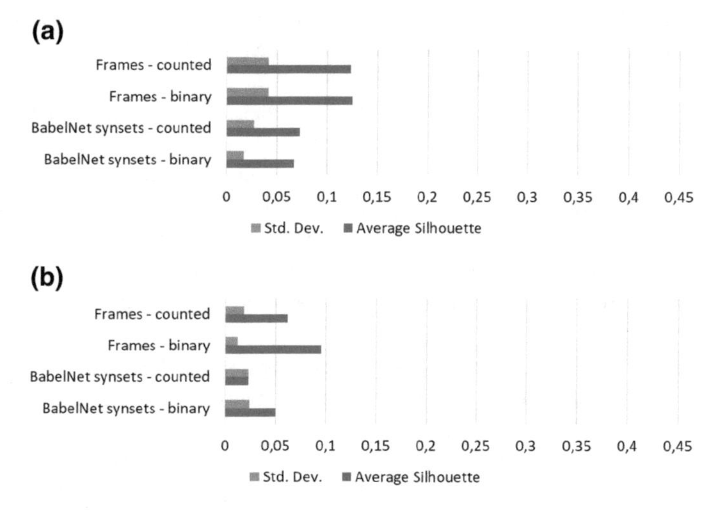

Fig. 2.8 The average and standard deviation values of the silhouette width measure of the clusterings computed on Framester features. **a** K-means clustering on cosine distance. **b** K-means clustering on Euclidean distance

Fig. 2.9 An example of list of recommendations built by our recommender system using as new report that called *heart-catheterization-angiography-1*

```
heart-catheterization-angiography-2
coronary-ct-angiography-ccta-2
heart-catheterization-ventriculography-angiography-6
stenting
cardiac-catheterization-8
heart-cath-coronary-angiography
nuclear-cardiac-stress-report
gen-med-consult-4
heart-catheterization-ventriculography-angiography-4
cardiac-catheterization-1
```

behavior on different features. To name an example, *entities-binary* and *entities-weighted* show a similar behavior both for cosine and for Euclidean distance.

Hierarchical clustering have outperformed the results of the K-means clustering, hence, if the recommender system would have been employed on a real medical case, the hierarchical clustering should be used. The agglomerative approach seems to be more suitable for finding medical cases similar to a new one.

Finally, to show how recommendation module has worked the reader can look at example in Fig. 2.9. The figure lists 10 medical reports of our test dataset that are returned by our recommender system when the report called *heart-catheterization-angiography-1* has been adopted for the evaluation of a new clinical state of a patient. The example shows how returned recommendations are correlated to heart issues and, hence, that our approach in building a recommender system can effectively recognize medical contents in order to suggest relevant past clinical cases.

2.5 Conclusion and Future Trends

Recommender systems are employed in many fields to help users to find important products and services for them. Similar approaches can be headed for providing diagnosis, thus supporting physicians in their work. In this chapter we presented a content-based recommender system within the medical domain, by providing an overview of recent information retrieval and semantic enrichment tools we employed. Our work addressed the challenge to find out which types of information can be directly processed by machines on large collections of medical reports, combining emergent cognitive computing systems in order to return reliable recommendation results. We discussed about the quality of features related to the representation of the medical reports content, underlying how they can capture the semantics from unstructured texts.

Subsequently, we discuss about two clustering approaches our recommender system currently implements. We used them to handle with various VSMs and explained their advantages and uses based on the type of dataset. In order to deal with classes of reports, the Machine Learning Module can be integrated with classification approaches and we aim at finishing this improvement in the immediate future. At the moment, we have not considered this evolution since it results hard to find datasets for a significant validation of classification tasks. We would like to underlie how recommender systems are substantial opportunities to progress in data science for the health-care field. For doing so, new resources and open datasets are required to enable further improvement of methods and designation of algorithms in clinical context.

Acknowledgements Danilo Dessì gratefully acknowledges Sardinia Regional Government for the financial support of his PhD scholarship (P.O.R. Sardegna F.S.E. Operational Programme of the Autonomous Region of Sardinia, European Social Fund 2014–2020—Axis III Education and training, Thematic goal 10, Priority of investment 10ii, Specific goal 10.5).

References

1. Mishra, R., Bian, J., Fiszman, M., Weir, C.R., Jonnalagadda, S., Mostafa, J., Del Fiol, G.: Text summarization in the biomedical domain: a systematic review of recent research. J. biomed. Inform. **52**, 457–467 (2014)
2. Sezgin, E., Ozkan, S.: A systematic literature review on health recommender systems. In: IEEE E-Health and Bioengineering Conference (EHB), pp. 1–4 (2013)
3. de Gemmis, M., Lops, P., Musto, C., Narducci, F., Semeraro, G.: Semantics-aware content-based recommender systems. In: Recommender Systems Handbook, pp. 119–159. Springer (2015)
4. Capelle, M., Hogenboom, F., Hogenboom, A., Frasincar, F.: Semantic news recommendation using wordnet and bing similarities. In: Proceedings of the 28th Annual ACM Symposium on Applied Computing, pp. 296–302. ACM (2013)
5. Lin, D.: Review of "WordNet: an electronic lexical database" by Christiane Fellbaum. The MIT Press 1998. Comput. Linguist. **25**(2), 292–296 (1999)
6. Baker, F.C., Fillmore, C.J., Lowe, J.B.: The berkeley framenet project. In: Proceedings of the 36th Annual Meeting of the Association for Computational Linguistics, ACL '98 and 17th International Conference on Computational Linguistics, vol. 1, pp. 86–90. Association for Computational Linguistics, Stroudsburg, PA, USA (1998)
7. Gangemi, A., Alam, M., Asprino, L., Presutti, V., Recupero, D.R.: Framester: a wide coverage linguistic linked data hub. In: 2016 20th International Conference on Proceedings of Knowledge Engineering and Knowledge Management, EKAW, pp. 239–254. Springer (2016)
8. Navigli, R., Ponzetto, S.P.: Babelnet: the automatic construction, evaluation and application of a wide-coverage multilingual semantic network. Artif. Intell. **193**, 217–250 (2012)
9. Bleik, S., Mishra, M., Huan, J., Song, M.: Text categorization of biomedical data sets using graph kernels and a controlled vocabulary. IEEE/ACM Trans. Comput. Biol. Bioinform. **10**(5), 1211–1217 (2013)
10. Cohen, A.M., Hersh, W.R.: A survey of current work in biomedical text mining. Brief. bioinform. **6**(1), 57–71 (2005)
11. Toor, R., Chana, I.: Application of IT in healthcare: a systematic review. ACM SIGBioinform. Rec. **6**(2), 1–8 (2016)
12. Presutti, V., Consoli, S., Nuzzolese, A.G., Recupero, D.R., Gangemi, A., Bannour, I., Zargayouna, H.: Uncovering the semantics of wikipedia pagelinks. In: Lecture Notes in Computer Science, vol. 8876, pp. 413–428 (2014)
13. Presutti, V., Nuzzolese, A.G., Consoli, S., Gangemi, A., Recupero, D.R.: From hyperlinks to semantic web properties using open knowledge extraction. Semant. Web **7**(4), 351–378 (2016)
14. Lushnov, M., Safin, T., Lapaev, M., Zhukova, N.: Medical text processing for SMDA project. In: EMSA-RMed@ESWC (2016)
15. Consoli, S., Stilianakis, N.I.: A quartet method based on variable neighbourhood search for biomedical literature extraction and clustering. Int. Trans. Oper. Res. **24**(3), 537–558 (2017)
16. Chernyshevich, M., Stankevitch, V.: IHS-RD-BELARUS: clinical named entities identification in French medical texts. Physiology **279**, 291 (2015)
17. Dessì, D., Recupero, D.R., Fenu, G., Consoli, S.: Exploiting cognitive computing and frame semantic features for biomedical document clustering. In: Proceedings of the Workshop on Semantic Web Solutions for Large-scale Biomedical Data Analytics co-located with 14th Extended Semantic Web Conference, SeWeBMeDA@ESWC 2017, pp. 20–34 (2017)
18. Yeh, A.S., Hirschman, L., Morgan, A.A.: Evaluation of text data mining for database curation: lessons learned from the KDD challenge cup. Bioinformatics **19**(Suppl. 1), 331–339 (2003)
19. Regev, Y., Finkelstein-Landau, M., Feldman, R.: Rule-based extraction of experimental evidence in the biomedical domain: the KDD cup 2002 (task 1). ACM SIGKDD Explor. Newslett. **4**(2), 90–92 (2002)
20. Donaldson, I., Martin, J., de Bruijn, B., Wolting, C., Lay, V., Tuekam, B., Zhang, S., Baskin, B., Bader, G.D., Michalickova, K., Pawson, T., Hogue, C.W.V.: PreBIND and textomy—mining

the biomedical literature for protein-protein interactions using a support vector machine. BMC Bioinform. **4**(1), 11 (2003)
21. Shehata, S., Karray, F., Kamel, M.: An efficient concept-based mining model for enhancing text clustering. IEEE Trans. Knowl. Data Eng. **22**(10), 1360–1371 (2010)
22. Lops, P., De Gemmis, M., Semeraro, G.: Content-based recommender systems: state of the art and trends. In: Recommender Systems Handbook, pp. 73–105. Springer (2011)
23. Degemmis, M., Lops, P., Semeraro, G.: A content-collaborative recommender that exploits wordnet-based user profiles for neighborhood formation. User Model. User-Adapt. Interact. **17**(3), 217–255 (2007)
24. Gu, J., Feng, W., Zeng, J., Mamitsuka, H., Zhu, S.: Efficient semisupervised MEDLINE document clustering with MeSH-semantic and global-content constraints. IEEE Trans. Cybern. **43**(4), 1265–1276 (2013)
25. Bromuri, S., Zufferey, D., Hennebert, J., Schumacher, M.: Multi-label classification of chronically ill patients with bag of words and supervised dimensionality reduction algorithms. J. Biomed. Inform. **51**, 165–175 (2014)
26. Marafino, B.J., Davies, J.M., Bardach, N.S., Dean, M.L., Dudley, R.A., Boscardin, J.: N-gram support vector machines for scalable procedure and diagnosis classification, with applications to clinical free text data from the intensive care unit. J. Am. Med. Inform. Assoc. **21**(5), 871–875 (2014)
27. Hughes, M., Li, I., Kotoulas, S., Suzumura, T.: Medical text classification using convolutional neural networks. Stud. Health Technol. Inform. **235**, 246–50 (2017)
28. Zhao, R.W., Li, G.Z., Liu, J.M., Wang, X.: Clinical multi-label free text classification by exploiting disease label relation. In: 2013 IEEE International Conference on Bioinformatics and Biomedicine (BIBM), pp. 311–315. IEEE (2013)
29. Glinka, K., Woźniak, R., Zakrzewska, D.: Improving multi-label medical text classification by feature selection. In: 2017 IEEE 26th International Conference on Enabling Technologies: Infrastructure for Collaborative Enterprises (WETICE), pp. 176–181. IEEE (2017)
30. Baumel, T., Nassour-Kassis, J., Elhadad, M., Elhadad, N.: Multi-label classification of patient notes a case study on icd code assignment. CoRR abs/1709.09587 (2017)
31. Allahyari, M., Pouriyeh, S., Assefi, M., Safaei, S., Trippe, E.D., Gutierrez, J.B., Kochut, K.: A Brief Survey of Text Mining: classification, clustering and extraction techniques. arXiv:1707.02919 (2017)
32. Zhang, X., Jing, L., Hu, X., Ng, M., Xia, J., Zhou, X.: Medical Document Clustering using Ontology-based Term Similarity Measures (2008)
33. Zhang, Y., He, Z., Yang, J.J., Wang, Q., Li, J.: Re-structuring and specific similarity computation of electronic medical records. In: 2017 IEEE 41st Annual Computer Software and Applications Conference (COMPSAC), vol. 2, pp. 230–235. IEEE (2017)
34. Chen, A.T.: Exploring online support spaces: using cluster analysis to examine breast cancer, diabetes and fibromyalgia support groups. Patient Educ. Couns. **87**(2), 250–257 (2012)
35. Lu, Y., Zhang, P., Deng, S.: Exploring health-related topics in online health community using cluster analysis. In: 2013 46th Hawaii International Conference on System Sciences (HICSS), pp. 802–811. IEEE (2013)
36. Wiesner, M., Pfeifer, D.: Adapting recommender systems to the requirements of personal health record systems. In: Proceedings of the 1st ACM International Health Informatics Symposium, pp. 410–414. ACM (2010)
37. Davis, D.A., Chawla, N.V., Blumm, N., Christakis, N., Barabási, A.L.: Predicting individual disease risk based on medical history. In: Proceedings of the 17th ACM Conference On Information and Knowledge Management, pp. 769–778. ACM (2008)
38. Zhang, Y., Chen, M., Huang, D., Wu, D., Li, Y.: iDoctor: personalized and professionalized medical recommendations based on hybrid matrix factorization. Future Gener. Comput. Syst. **66**, 30–35 (2017)
39. Fillmore, C.: Frame semantics. In: Linguistics in the Morning Calm, pp. 111–137 (1982)
40. Gangemi, A.: What's in a Schema? pp. 144–182, Cambridge University Press, Cambridge (2010)

Chapter 3
Classification Methods in Image Analysis with a Special Focus on Medical Analytics

Lucio Amelio and Alessia Amelio

Abstract This paper describes the design and application of classification methods for image analysis and processing. Accordingly, the main trends and challenges of the machine learning are presented in multiple contexts where the image analysis plays a very important role, including security and biometrics, aerospace and satellite monitoring, document analysis, natural language understanding, and information retrieval. This is accomplished by introducing a categorisation of the most challenging classification methods according to the thematic context and classification typology. Hence, supervised and unsupervised classification methods are presented and discussed. It is followed by a special focus on the medical context, where the classification methods for image analysis are of prior importance in supporting the medical diagnosis process. Accordingly, the second part of the paper surveys the recent and current research in medical analytics where the image classification is a key aspect, and tracks the horizon of the research for future challenges in the field.

Keywords Classification · Clustering · Image analysis · Medical analytics
Pattern recognition

3.1 Introduction

One of the first applications of digital imaging dates back in 1920s in the newspaper industry with the Bartlane cable picture transfer service, where images were

---

L. Amelio (✉)
Faculty of Medicine and Surgery, University of Bologna, Via Massarenti 9, 40138 Bologna, Italy
e-mail: lucio.amelio@studio.unibo.it

A. Amelio
DIMES University of Calabria, Via Pietro Bucci 44, 87036 Rende, CS, Italy
e-mail: aamelio@dimes.unical.it

G. A. Tsihrintzis et al. (eds.), *Machine Learning Paradigms*, Intelligent Systems Reference Library 149, https://doi.org/10.1007/978-3-319-94030-4_3

coded and transferred by the submarine cable between London and New York and reconstructed at the receiver on a telegraph printer [71]. From that time, different attempts were performed to improve the quality of the Bartlane system introducing new reproduction methods based on photographic strategies and increasing the number of tones of the images. However, the plenty of techniques that characterise the digital image processing were introduced starting from 1960s at the Jet Propulsion Laboratory, Massachusetts Institute of Technology, Bell Laboratories, University of Maryland, and some other research centers, in different fields of interest, including satellite imagery, character recognition, medical imaging and picture enhancement [98]. In particular, in 1964 computers were adopted for improving the quality of images in different space missions, including the Apollo landing. However, processing an image was hard and expensive at that time due to the limited hardware resources in terms of computer storage and CPU power. Starting from the 1970s, with the increase of the computer power and the availability of dedicated hardware, the techniques of image processing have become more accessible, because images could be also processed in real-time. In particular, the digital image processing has started to be used in the medical context, with the invention of the tomography in 1979. Because computers increased their speed and power, they could be used for processing images as a dedicated hardware. However, they still were not able to manage intensive and specialised image processing operations. Lately, starting from 2000s with the proliferation of last generation high-power fast and performant computers, the digital image processing has become the most common form of operating on images with fast and cheap methods.

Over the years, the digital image processing has been enriched with complex techniques, which can be categorised as: (i) classification, (ii) multi-scale signal processing, (iii) feature extraction, (iv) pattern recognition, and (v) projection [98].

In this paper, we focus our analysis on classification approaches in digital image processing, which is one of the most important applications of statistical classification in data mining. They consist of giving a category to an image or elements of an image using statistical and machine learning methods. Usually, these methods require a low level processing of the image before their application, whose description is out the scope of this paper, which can be noise removal and/or image sharpening for making more visible the contours inside the image. Reader can refer to [47] for a depth explanation about these pre-processing techniques.

The paper is organised as follows. Section 3.2 provides a general background of the basic concepts underlying the image classification. Section 3.3 describes the main approaches of feature representation for image classification. Sections 3.4–3.6 present some relevant image classification methods respectively in the contexts of security and biometrics, aerospace and satellite monitoring, document analysis and language understanding. Section 3.7 describes some important methods of image classification in information retrieval. Section 3.8 introduces a categorisation of the medical imaging from a clinical point of view and describes the most important image classification methods for each category. Then, the horizon of the research is described together with a few challenges for future work in this direction. At the end, Sect. 3.9 draws conclusions.

3.2 Background

Image classification in its most general meaning is the process of inferring one or more categories for one or a set of images, the regions or the pixels composing the images. Basically, the image classification can be: (i) supervised, or (ii) unsupervised, commonly referred as clustering. Accordingly, the supervised image classification can be at: (i) image, (ii) object, or (iii) pixel-level. On the contrary, the unsupervised image classification or clustering can be at: (i) image, or (ii) pixel-level. Figure 3.1 shows an overview of the hierarchy of image classification methods.

When unsupervised classification or clustering at image-level is performed, we are given a set of images whose categories are unknown. The aim is to find groups of images with similar characteristics which define the categories. Figure 3.2 depicts the process of unsupervised classification or clustering of an image set composed of twenty elements. Images of the same semantic class belong to the same cluster, for a total of five clusters. On the contrary, the unsupervised classification at pixel-level is employed for grouping the image pixels based on similar characteristics, in order to find uniform image regions. This is an important step for the image segmentation. Figure 3.3 shows how the unsupervised image classification or clustering at pixel-level determines the image segmentation. Pixels which are characterized by homogeneous color and texture are grouped in the same cluster or image region.

In the supervised classification at image-level, we are given a set of images each belonging to a specific category corresponding to the training set. The classifier learns the image categories by using the training set. Then, it is able to generalize the classification by categorizing unseen images which are not included in the training

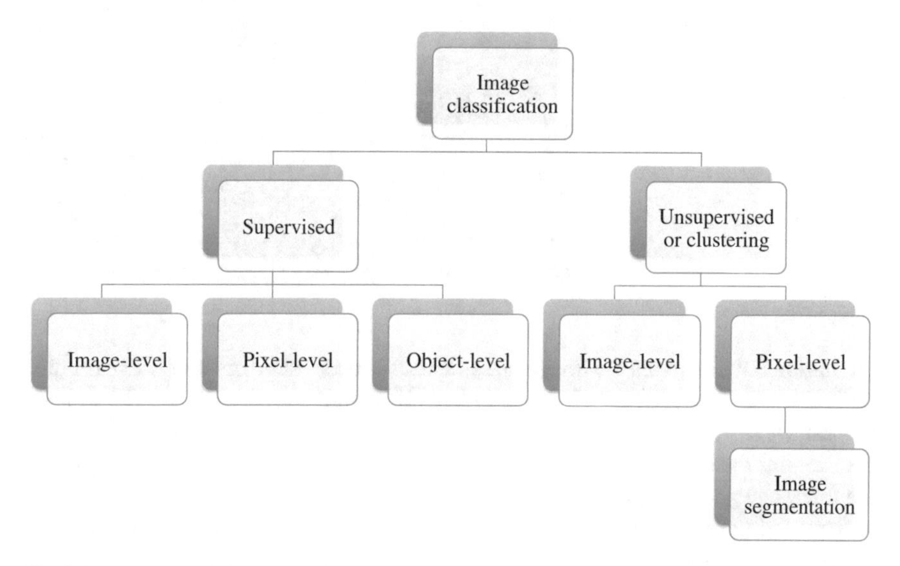

Fig. 3.1 Overview of the image classification methods

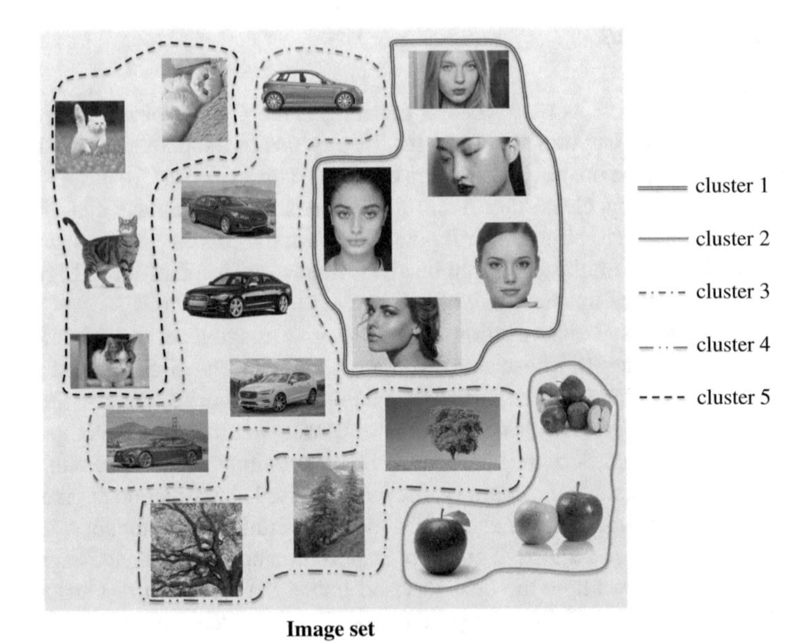

Fig. 3.2 Unsupervised classification or clustering of an image set composed of twenty images. The five clusters of images belonging to the same semantic class are differently colored and marked

Fig. 3.3 Image unsupervised classification or clustering at pixel-level. The four image regions with uniform pixels in terms of color and texture are bounded by different contours

set. Figure 3.4 illustrates the image classification process at image-level. The training set is composed of images belonging to five different categories: (i) cat, (ii) woman, (iii) car, (iv) apple, and (v) tree. After the training phase for learning the model from the training data, the test phase is performed on three unseen images belonging to the three categories: (i) car, (ii) tree, and (iii) woman. These images do not belong to the training set. At pixel-level, the training set is composed of pixels of different categories belonging to a given image. The aim is to classify the pixels of unknown category as belonging to a given category. This is performed after a training phase

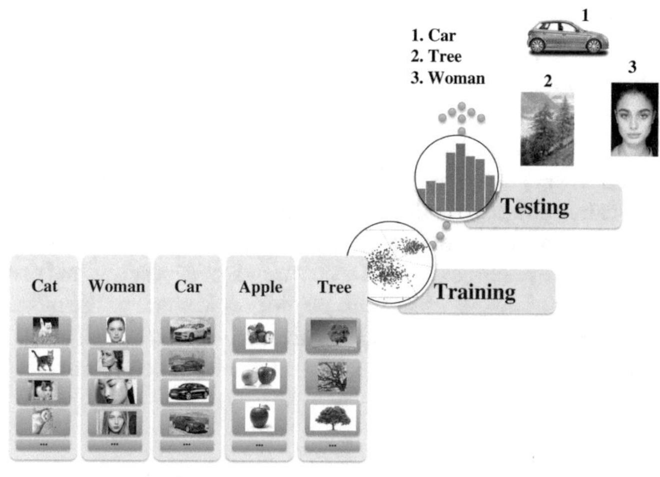

Fig. 3.4 Supervised image classification at image-level. The image training set is divided into five categories: (i) cat, (ii) woman, (iii) car, (iv) apple, and (v) tree. After the training phase, the classifier is ready to classify three unseen images belonging to the following categories: (i) car, (ii) tree, and (iii) woman

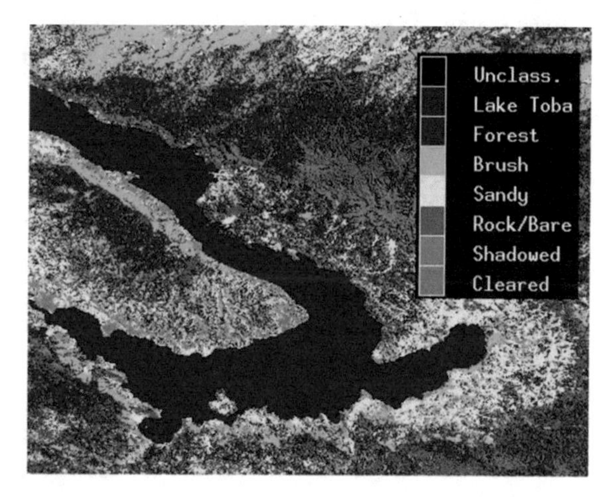

Fig. 3.5 Supervised image classification at pixel-level. The image pixels are classified as belonging to one of the seven different classes corresponding to different land cover types

where the classifier learns the pixel categories from the training set. Figure 3.5 shows the process of supervised classification at pixel-level. In particular, the image pixels are categorised in seven different classes corresponding to different land cover types. Finally, the supervised classification at object-level employs unsupervised classification at pixel-level for the identification of representative regions inside the image. Then, it performs the supervised classification on the extracted regions. A complete description of the concept of classification can be found in [99].

An essential pre-processing step in image classification is the transformation of the input images, objects or pixels in the feature domain. Basically, each image, object or pixel element is represented by one or more feature vectors characterising the main aspects by which the classification process is performed. Accordingly, a good feature representation can achieve a more accurate image classification. In the following, we provide a broad categorisation of the most general features employed in image classification and describe some relevant feature representations (summarised in Table 3.1). The reader may refer to [53, 62] for further analysis of different image descriptors. Then, we describe the most recent trends of supervised and unsupervised image classification at image, object and pixel-level in some of the most important contexts of security and biometrics, aerospace and satellite monitoring, document analysis and natural language understanding, and information retrieval.

3.3 Feature Representation for Image Classification

Prior of any classification task at image-level, the images can be represented by different feature models, including:

- global features,
- local features [53],
- a combination of global and local features [63], and
- Bag of visual words (BoVW) [43].

For the classification at pixel-level, every image pixel can be characterised by different features, such as brightness, colour, and texture.

3.3.1 Global Features

The global features represent the image by using a single feature vector embedding information about colour, texture or shape derived by all pixels. Sometimes the feature vector combines the different aspects of colour, texture and shape. In the last years, multiple global features have been introduced in the literature for image classification.

Two well-known feature representations based on textural content for image classification are: (i) grey-level co-occurrence matrix (GLCM), and (ii) run-length statistics (RL). In recent time, other global feature representations have also been proposed.

3.3.1.1 Grey-Level Co-occurrence Matrix

The GLCM features, also called Haralick features [52], were computed from the grey-level co-occurrence matrix embedding information about the frequency of

Table 3.1 Overview of the different feature types described in Sect. 3.3

Feature	Method	Image-level		Pixel-level	
		Supervised	Unsupervised	Supervised	Unsupervised
Global	GLCM [52]	×			
	RL [32, 34, 45]	×			
	Güld et al. [51]	×			
	Amelio and Pizzuti [12]		×		
	Miao et al. [75]	×			
	Zhang et al. [105]	×			
	Brodić et al. [25, 28, 29]		×		
Local	SIFT [68]	×			
	SURF [13]	×			
	LBP [82]	×			
	Dawood et al. [35]	×			
	Brodić et al. [18, 26]		×		
	Margolin et al. [73]	×			
	Morales et al. [77]	×			
	Gragnaniello et al. [49]	×			
	Yang et al. [103]			×	
	Shi and Malik [89]				×
	Amelio and Pizzuti [11]				×
	Amelio and Pizzuti [9, 10]				×
	Vandenbroucke et al. [95]			×	
	Lu and Weng [69]			×	
Global-Local	Kuric et al. [63]	×			
BoVW	Fei-Fei and Perna [43]	×			

co-occurrence of the grey levels in the image. Haralick features were originally used for the supervised classification of photomicrograph, aerial photographic and satellite images [52].

3.3.1.2 Run-Length Statistics

The RL statistics [32, 34, 45] were extracted from the grey-level run-length matrix. It counted the frequency of pixel runs at the different image grey levels in a given texture direction. A run is a sequence of consecutive pixels in the image. RL features were originally tested for the supervised classification of different textural images, including terrain images [45].

3.3.1.3 Other Global Features

Gueld et al. [51] evaluated different types of feature representations, i.e. texture, structure and down-scaled features, in characterising medical images for supervised classification. For the texture, multiple feature representations, including three-dimensional histograms of coarsness, contrast and directionality computed at pixel-level, were used. For the structure, specific properties of the image edges were extracted and adopted as features. Finally, down-scaled images were also employed as feature representation.

Also, Amelio and Pizzuti [12] used a combination of the Haralick features and colour centiles for representing the texture and colour image content in a process of clustering natural images.

In order to classify farmland images, Miao et al. [75] introduced a feature representation based on hue saturation value, hue saturation lightness and hue saturation intensity colour space models.

Furthermore, Zhang et al. [105] proposed a texture feature vector based on the grey-level co-occurrence matrix and a weighting factor, derived from a measure of directionality of the image texture. The feature representation was tested for the supervised classification of high-resolution remote sensing images.

Finally, Brodić et al. [25, 28] employed a variant of the grey-level co-occurrence matrix for a 1-D document image coding in order to cluster documents in multiple languages, including English, French and Serbian, and German historical documents in different scripts, including Latin and Fraktur. A variant of the run-length matrix was also proposed for clustering documents in different historical Croatian scripts [29].

3.3.2 *Local Features*

Differently from the global features, the local features are descriptors extracted from salient image regions which correspond to patches localised in the neighbourhood of points of interest, also called keypoints. In particular, the local features should be invariant to rotation, illumination and viewpoint changes.

Three widespread local image descriptors for image classification are [42]: (i) SIFT, (ii) SURF, and (iii) local binary patterns (LBP). Also, alternative approaches have been recently proposed in the literature.

Local descriptors are based on the concept of image gradient representing intensity changes over the image. In particular, the image gradient is characterised by its magnitude (intensity) and direction (orientation).

3.3.2.1 SIFT

The SIFT approach [68] finds locations and scales that are identifiable from different views of the same object. A scale represents the image with a given smoothing effect. Then, it discards keypoints with low contrast or poorly localised on an edge. After that, an orientation is assigned to the remaining keypoints according to local image properties. Finally, this data is used to create the keypoint descriptors. These are histograms computed in a window centered on the keypoints.

3.3.2.2 SURF

Similarly to the SIFT, in the SURF approach [13] the keypoints of a given image are represented as salient features from a scale-invariant modelling. This multiple-scale analysis is based on the convolution of the image with discrete masks at different scales (box filters). In the second step, orientation invariant descriptors are detected by using local gradient statistics (intensity and orientation). Both approaches were originally experimented for supervised object classification and recognition in real-life object images.

3.3.2.3 Local Binary Pattern

LBP features [82] are obtained by concatenating frequency histograms of digits extracted from the image. Each digit is derived from the comparison of the center pixel with the eight neighbouring pixels in a 8×8 window sliding over the image. Considering the co-occurrence of adjacent LBPs in the image plan extends the LBP feature representation to the adjacent local binary pattern (ALBP) [80]. LBP features were originally experimented for the supervised classification of various texture images [82].

3.3.2.4 Other Local Features

Dawood et al. [35] introduced an extension of the weber local descriptor (WLD) for the supervised classification of textural images. It was accomplished by dividing the image in small regions and computing an histogram for each region representing the orientation from the image gradient.

Also, Brodić et al. proposed a variant of the ALBP feature representation and a combination of ALBP and RL statistics for clustering document image codings in different languages and scripts, e.g. Macedonian cyrillic, Slovakian latin, Serbian latin and Serbian cyrillic [18], and in modern and vulgar Italian language [26].

In order to classify scene images, Margolin et al. [73] introduced the oriented texture curves (OTC) descriptor. Basically, the texture of a patch was represented by the shape properties of the curves defining colour variations of that patch along multiple orientations.

Furthermore, Morales et al. [77] experimented the use of SIFT descriptors for supervised ear image classification and recognition in biometric systems, showing promising results under controlled conditions.

Finally, Gragnaniello et al. [49] introduced a local descriptor for the fingerprint liveness detection. It captured the spatial and frequency information from the image in order to create a bi-dimensional histogram representing the image feature vector.

3.3.3 Bag of Visual Words

The BoVW model [43] is based on the generation of a visual vocabulary which is obtained by clustering of a large collection of local features. In particular, the collection of local features is extracted from the training images and clustered. Each obtained cluster divides the local feature space in representative regions which define the visual words. The set of visual words characterises the visual vocabulary. According to this model, each image is represented by the histogram of the visual vocabulary. It quantifies the frequency of the visual words extracted from that image.

3.3.4 Pixel-Level Features

Pixel-level features aim at generating a feature vector for each image pixel which can include colour, texture and/or shape information retrieved at pixel-level. Different works have been introduced in the literature employing these features for supervised or unsupervised image classification at pixel-level.

Yang et al. [103] introduced a descriptor including colour and texture content in terms of edge saliency, colour saliency, local maximum energy, and multiresolution texture gradient, for supervised classification of natural images at pixel-level.

Also, Shi and Malik [89] proposed a graph representation of the image pixels, where nodes corresponded to the pixels and weighted edges characterised the similarity between pixels in terms of brightness and spatial closeness. This representation was employed for solving the image segmentation problem on different types of natural images.

A similar representation was used by Amelio and Pizzuti in [11] which extended the Shi and Malik's method for segmentation of natural and medical images. Another variant embedding the colour, texture and brightness pixel characteristics at multiple scales and orientations was introduced by Amelio and Pizzuti in [9] for the segmentation of natural images, and in [10] for the segmentation of medical images of skin lesions.

Also, Vandenbroucke et al. [95] proposed a colour-texture pixel descriptor which considered the best subset of texture features, such as mean, median, mode and skewness, and colour features derived by multiple colour representation systems, computed in a neighbourhood of the reference pixel. The pixel descriptor was used for the supervised classification of pixels in soccer images.

Finally, Lu and Weng [69] surveyed the process of supervised image classification for remotely sensed images. In particular, they indicated different pixel-level features which could be used for the identification of land-cover types or vegetation classes, such as textural or contextual information, vegetation indices, spectral signatures, multi-temporal, transformed and multi-sensor images, and ancillary data.

3.4 Security and Biometrics

The image classification characterises different aspects of a biometric system, whose aim is securely recognising the identity of a person by measuring the anatomical as well as behavioural features. In fact, usual tasks include the identification of images like fingerprint, palm, face and iris of the eye [2]. A biometric image-based system is composed of a decision-making module where the biometric traits extracted from the input image are recognised as belonging to a given person from a database of different persons' biometric traits. It requires the comparison of the biometric traits of the unknown person with a large database of biometric traits, which can be time-consuming in large scale applications. In this context, the classification can be very important to categorise the elements of the database into predefined classes. Hence, the biometric traits of the unknown person can be only compared with the subset of elements from the database belonging to the same corresponding class. In some other cases, the classification can be used to train a model from the database of biometric traits in order to recognise the identity of unknown human traits. This type of classification is usually supervised at image and object-level.

In the following, we describe some important image classification methods for biometric systems and refer to [104] for further details about the topic.

3.4.1 Supervised Classification

In the context of fingerprint recognition, four categories of supervised image classification methods can be identified [4]: (i) heuristic-based, (ii) structure-based, (iii) neural-based, and (iv) statistical-based. Also, different methods have been introduced for the identification of other human traits, e.g. face and iris.

3.4.1.1 Heuristic-Based Methods

The heuristic-based methods consist of extracting singularity and ridge features as landmarks to be used for the classification process. In particular, the singularity features aim at extracting the number and position of core and delta points which represent accurate features for the identification task. The ridge features can be global features. Accordingly, a rule-based classifier using the number of singularities and global ridge features obtains accurate classification performances. A rule-based classifier is characterised by a set of rules of type: *if* {condition} *then* {conclusion}, where the condition is stated over the data, and the conclusion is a class label. These rules are extracted from the training set and used to classify unknown instances.

3.4.1.2 Structure-Based Methods

The structure-based methods may be: (i) syntactic, or (ii) graph matching. Syntactic methods represent each class like a grammar characterising the fingerprint. Each fingerprint is modelled as a pattern like a string or phrase. Then, the fingerprint is syntactically analysed and associated to that grammar of which the fingerprint follows the rules, corresponding to its class [96]. The graph matching methods represent each fingerprint by a relational graph codifying the segmented regions of homogeneous ridge and valley orientation and their adjacency. The minimum edit distance between the fingerprint and each class is computed. Then, posterior probability vectors are calculated from each edit distance, and the class is selected corresponding to the maximum probability [88].

3.4.1.3 Neural-Based Methods

Neural-based methods employ multilayer perceptron (MLP) artificial neural networks which are trained for identifying the class of the fingerprints. An MLP is characterised by a layer of input neurons defining the size of the input, one or more layers of hidden neurons, and a layer of output neurons, where the classification is returned. Each layer is connected with the adjacent layer by a set of connections. Each connection is equipped by a weight. An MLP can be considered as a logistic regression classifier. It learns a non linear-transformation by adjusting the weights in order to map the input data into another space where it becomes linearly separable.

3.4.1.4 Statistical-Based Methods

Statistical-based methods use a statistical classifier for the categorisation of fixed-size feature vectors representing the fingerprints. They include: (i) K-nearest neighbour (KNN), and (ii) support vector machine (SVM). The KNN [7] consists in finding a fixed number of elements from the training set which is the nearest to the test sample according to a given distance function. Then, the test sample is classified according to this subset by adopting a majority voting strategy. The SVM [33] aims at learning the parameters of an hyperplane in the feature space optimally separating the elements of the training set in the different classes. The hyperplane divides the feature space in parts corresponding to the different classes. Then, the classification of a test sample is performed according to its location versus the obtained hyperplane.

3.4.1.5 Other Methods

Ramesha et al. [84] introduced a biometric identification system based on face, gender and age recognition. It employed a posterior class probability model based on face shape features for the gender supervised classification, and an artificial neural network using textural face features for the age supervised classification. In the posterior class probability model, the posterior probability of observing an age class given the features was computed according to the probability of observing the features given the age class and the prior probability of observing the age class. The probability values were computed from the training set. Then, a test feature vector was classified with the label obtaining the maximum posterior probability. Also, Ali et al. [6] proposed an iris recognition and identification system based on object-level classification using an SVM approach. After a phase of iris segmentation and normalisation from which the Gabor wavelet-based features were extracted, the SVM classifier, learnt from the iris database, was used to recognise the identity of unknown iris.

3.5 Aerospace and Satellite Monitoring

Classification is also employed in the satellite imaging, with the main objective to determine the land cover class of each image pixel or region. In the supervised classification at pixel-level, the land cover classes characterising the area must be detected. Then, a set of representative pixels for each land cover class is selected as the training set by manual inspection of the image terrain. After that, a classifier is learned from the training set in order to identify the land cover class of each image pixel. An automatic procedure for selecting the land cover classes prior of any land use identification is the unsupervised classification or clustering at pixel-level. It partitions the satellite image in regions which are spectrally homogeneous inside, while there is a meaningful spectral difference between them [39].

Another type of widespread supervised classification for the satellite images is the object-level one [15]. It applies clustering on the satellite image for the identification of representative cover type regions. Then, it performs the supervised classification of the extracted regions.

All these tasks are of prior importance for detecting and quantifying the land cover change for conservation [16], e.g. forest loss and conservation.

Next, we investigate the main approaches currently used for satellite image classification. Readings about alternative techniques can be found in [94].

3.5.1 *Supervised Classification*

Relevant supervised image classification algorithms for satellite images are [58]: (i) maximum likelihood classifier (MLC), (ii) minimum distance classifier (MDC), (iii) parallelepiped classifier (PC), (iv) random forest (RF), and (v) artificial neural network. In this context, the input of the classifiers is the feature vector representing the spectral signature of a given pixel.

3.5.1.1 Maximum Likelihood Classifier

MLC computes the probability that a pixel belongs to a given class by following the Bayes theorem. It requires the probability of observing the pixel given the class (likelihood function), expressed in terms of covariance matrix and mean vector of the class. Under the assumption of multivariate Gaussian data distribution, the unknown pixels are assigned to the class with the highest value of log-likelihood monotone function, which is computed from the training data.

3.5.1.2 Minimum Distance Classifier

MDC computes a mean vector of instances from each prototype class of the training set. In order to classify a test pixel, the method measures the distance (e.g. Euclidean) between the unknown pixel and each mean vector. At the end, the pixel is assigned to that class corresponding to the lowest spectral distance.

3.5.1.3 Parallelepiped Classifier

PC represents the pixels in the training set in their multidimensional spectral space. Accordingly, each prototype class of the training set describes a multidimensional structure according to the value of its pixels. In order to classify a test pixel, it is checked if the unknown pixel lies in a given multidimensional structure of the spectral space. Then, it is assigned to its corresponding class.

3.5.1.4 Random Forest

Ma et al. [70] presented a meta-analysis of different studies proposed in the recent literature for the identification of land cover classes. According to this analysis, the RF classifier showed the best performances in object-level classification. RF creates a set of tree-based classifiers, trained by random subsets of the training objects. A tree-based classifier is a prediction model where leaves correspond to class labels and branches correspond to feature combinations determining those class labels. The classification of unknown objects is performed by majority voting over the set of the classifiers.

3.5.1.5 Artificial Neural Network

Recently, Geng et al. [46] introduced a deep supervised and contractive neural network for the classification of SAR images at pixel-level. The network was composed of four layers of supervised and contractive autoencoders, which determined a modelling of the features and provided the prediction of the class of tested pixels. An autoencoder was characterised by an encoder and a decoder. The encoder provided a coding of the input. The decoder aimed at reconstructing the input from its coding. Hence, the autoencoder aimed at minimising the reconstruction error between the input of the encoder and the output of the decoder. Penalty factors were included in the autoencoders in order to improve the robustness of the classifier agains the image noise.

3.5.2 Unsupervised Classification

Two unsupervised classification methods obtaining better performances in clustering satellite image pixels are [39]: (i) fuzzy K-means, and (ii) fuzzy maximum likelihood (ML). Both classifiers consider that every pixel is associated with the different clusters by a value of membership.

3.5.2.1 Fuzzy K-means

The fuzzy K-means algorithm finds the K clusters of pixels which minimise the objective function. This is the total Euclidean distance between the pixels in each cluster and their centroid weighted by the degree of membership of that pixel to that cluster. The centroid of a given cluster is the mean vector of all pixels inside that cluster weighted by the membership coefficients of each pixel to that cluster. The algorithm for detecting the fuzzy partitioning iteratively updates the membership coefficients and the cluster centroids until the objective function is optimised. The first step of the algorithm computes the centroids from initial membership coefficients. The obtained centroids are used for updating the membership coefficients. The procedure is iterated until there are small variations in the membership coefficients.

3.5.2.2 Fuzzy Maximum Likelihood

In the fuzzy ML clustering, the pixels are assumed to be realisations of random vectors independently selected from K multivariate normally distributed populations which correspond to the pixel clusters. Accordingly, the random vectors are expressed in terms of covariance matrix and expected value. The aim of the method is the maximisation of the posterior probability of observing the clustering given the pixel data. It can be expressed as the minimisation of the log-likelihood of observing the pixel data given the clustering. This function represented in terms of covariance matrices and expected values of the pixel categories defines the cost function weighted by the membership coefficients of a given pixel to a given cluster. In the fuzzy ML clustering, these membership coefficients are computed as the posterior probabilities of observing the cluster given the pixel data in that cluster.

3.6 Document Analysis and Language Understanding

The supervised classification at image-level in document analysis aims at identifying the category of a given document. Some categories correspond to different document types, such as letter, newspaper, journal paper, form, etc. Other categories may define language classes, such as Italian and English, script classes, such as Latin and Fraktur, orthography classes, dialect or sub-dialect classes.

On the contrary, the unsupervised classification at image-level is adopted for detecting classes of documents with similar characteristics according to different criteria, e.g. similar content, language or script. In the handwriting context, it can be used for grouping documents from the same writer (writer identification).

Finally, the image segmentation is used for the discrimination of different objects inside the document, e.g. text, stamp and logo, resulting in faster document identification and retrieval.

The input of the classifier at image-level is a fixed-size feature vector representing the document image instance or the pixel matrix of the document image.

In the following, we describe some relevant approaches introduced in the literature for the classification of document images, and refer to [81] for further details about the topic.

3.6.1 Supervised Classification

The supervised document image classification methods at image-level can be categorised as [30]: (i) statistical methods, (ii) structural methods, (iii) knowledge-based methods, (iv) combination of multiple classifiers, and (v) multi-stage classification.

3.6.1.1 Statistical Methods

Statistical methods include: (i) KNN classifier, (ii) decision tree, (iii) MLP artificial neural network, (iv) SVM classifier, (v) Naive Bayes, and (vi) convolutional neural network (CNN).

A decision tree is a tree-based classifier built from the training set, where leaf nodes correspond to the classes and internal nodes are feature condition points which divide instances with different characteristics. The class of a test instance is the leaf node along the path in the tree tracked by the feature values of the instance.

The Naive Bayes method assumes the independence of the features on a given class. Accordingly, the posterior probability of the class given the feature values of an instance can be defined in terms of prior probabilities of the class, and product of likelihood of each feature value given the class. The probability values are computed from the training set. The class determining the highest posterior probability on the test instance corresponds to the predicted value. The Naive Bayes classifier was adopted in [17, 20, 22] for the identification of documents in different orthography styles, i.e. old and new Glagolitic, and Slavonic-Serbian and Serbian languages. Also, the Naive Bayes and SVM classifiers were employed in [23] for the recognition of documents in the evolved Slavonic-Serbian into modern Serbian language, and in [21] for the recognition of documents in different pronunciations of the Shtokavian dialect. Also, the KNN and Naive Bayes classifiers were adopted in [27] for the identification of a minority language like Serbian among widespread in the world languages, i.e. German, Spanish, English and French.

A CNN is an artificial neural network characterised by a set of convolutional layers, by which the input matrix of pixels is processed. The output of these layers is classified by adopting two or three fully-connected layers. Kang et al. [60] trained a CNN for the identification of documents of different type, e.g. letter, news, note, email, and tax-forms.

3.6.1.2 Structural Methods

The structural methods generate models, represented by objects like trees or graphs, for each document class from the training set. Then, an unknown document is classified according to the minimum distance computed with the generated document class models. Different methods can be employed for generating class models.

A first method is a decision tree which can be extended to manage document tree-based representations in order to classify documents. Accordingly, the leaves of the classifier are labeled trees and the internal nodes correspond to shared sub-trees. The classifier is generated using operations of insertion, descending and splitting, according to sub-tree similarity matching.

A second method extends the KNN classifier for managing tree-based representations of documents in order to classify pages. The adopted distance for the classifier is a tree version of the edit distance.

3.6.1.3 Knowledge-Based Methods

The knowledge-based methods [86] generate a set of rules, a hierarchy of frames or similar representations from training document data. They model the expert knowledge about the categorisation of the documents in the different classes. Unknown documents are classified according to the generated rules or other representations. A method for generating the set of rules is using Inductive Logic Programming on the training set of labeled documents [40]. In particular, a set of definite clauses optimally covering the training set instances is generated. Each clause is composed of a head representing the class label, and a body which is the combination of different feature values. The algorithm starts by selecting a seed instance and generating a set of clauses covering that instance. Then, a preference criterion is used for selecting the best clause from the set. Finally, all instances covered by the selected clause are discarded from the set. The procedure is iterated until all instances are covered by a clause.

3.6.1.4 Combination of Multiple Classifiers

The combined classifiers consist in different classification methods which are combined in different modalities, e.g. hierarchical, in order to improve the final document classification result. Accordingly, three different classifiers, KNN, MLP and tree matching, can be employed for form classification [55]. The tree matching is a structural classifier which represents the form content like a tree structure. Document image features are used as input to KNN and MLP. On the contrary, structural features based on layout are used as input to the tree matching classifier. The combination of the different classifiers can be performed by adopting a hierarchical or parallel strategy with majority voting. In the first case, the classifiers are applied for reducing the number of candidate labels. Then, the structural classifier is employed on the data labeled by the other classifiers in order to detect the final labelling.

3.6.1.5 Multi-stage Classification

Finally, the multi-stage classification methods produce an increasingly refined classification of the document from an initial classification with a small number of coarse classes. Accordingly, these methods require different class models and classification approaches to be used. A similar method could provide a two-phase classifier where in the first phase the documents are categorised as journal papers or business letters, or business letters from different senders, and in the second phase the business letters are categorised in more refined classes or different message types are identified.

3.6.2 Unsupervised Classification

Different approaches for the unsupervised classification of document images include: (i) evolutionary computation methods, and (ii) statistical methods.

3.6.2.1 Evolutionary Computation Methods

Brodić et al. [28] proposed *Genetic Algorithms Image Clustering for Document Analysis* (GA-ICDA), a graph-based genetic algorithm for document image database clustering. The algorithm generated a weighted undirected graph, where nodes were the document images, and edges connected the n most similar and spatially close nodes. Then, a genetic algorithm, which optimised the weighted modularity function, was employed for detecting the graph communities, corresponding to document image clusters. A high weighted modularity corresponded to a solution where each node community was well-connected with edges of high weight, while connections among communities were low-weighted and sparse. This realised clusters of similar documents, which were dissimilar each others. The genetic algorithm was based on the generation of an initial population of chromosomes, each representing a possible partitioning of the graph in node communities. After that, the weighted modularity function was computed on the individuals of the chromosome population. Then, operators of mutation and crossover were applied on the chromosomes for changing the node membership to communities. The mutation randomly changed one gene of the chromosome. The uniform crossover applied on a pair of parent chromosomes generated a child randomly selecting the genes from the first or the second parent. The last two steps were iterated until a termination condition was satisfied.

The traditional algorithm used the Manhattan distance for the discrimination of documents given in different languages [28], i.e. English, French, Serbian and Slovenian. A variant of the algorithm used the Euclidean distance for the discrimination of documents given in closely related languages [24], i.e. Serbian and Croatian. An extension of GA-ICDA, called *Genetic Algorithms Image Clustering for Document Analysis-Plus* (GA-ICDA$^+$) [26], was adopted for clustering documents given in languages evolved over time, i.e. Italian language evolved from Italian vulgar to modern Italian. GA-ICDA$^+$ introduced a more flexible parameterisation in the similarity computation, and the managing of singleton clusters in the genetic procedure.

GA-ICDA was also employed for the discrimination of German historical documents given in different scripts [25], i.e. Latin and Fraktur, the discrimination of documents given in South-Slavic scripts [29], i.e. Cyrillic, Latin and Glagolitic, and the discrimination of documents given in multiple languages with different scripts [18], e.g. Macedonian Cyrillic, Serbian Cyrillic and Slovakian Latin. Finally, a reduced version of GA-ICDA$^+$, only considering the parameterisation in the similarity computation, was used in [19] for clustering historical documents in old Cyrillic, angular and round Glagolitic as well as Antiqua and Fraktur scripts.

3.6.2.2 Statistical Methods

Diem et al. [38] proposed a semi-automatic document image clustering method based on multiple features. They represented layout characteristics, such as: (i) supporting material, e.g. paper colour and texture, (ii) structure, e.g. font height and font slant, and (iii) writing, e.g. writing colour and writer identification. The user selected the number of desired clusters. Then, a hierarchical procedure based on the K-means algorithm was employed for clustering the documents at multiple levels according to the different feature subsets. The K-means algorithm detects the clustering which minimises the total Euclidean distance between the instances in each cluster and their centroid. The number K of clusters is an input to the algorithm. The first step randomly selects K initial centroids. Then, each instance is assigned to the nearest centroid in terms of Euclidean distance. A centroid is computed as the mean value on the data belonging to that cluster. After that, new K centroids are computed according to the new assignment of the instances to the clusters. The last two steps of the algorithm are iterated until the K centroids do not meaningfully change their location.

Also, Dey et al. [37] introduced a consensus-based segmentation method for document images. It extracted from the document image a set of foreground blocks, each characterised by different features. Then, similarity values were computed between the blocks according to each feature using a statistical test. After that, clustering of the blocks was performed according to the consensus of the similarity values for the different features. This clustering procedure was employed together with a classifier in order to recognise the block type, e.g. stamp, logo and noise.

3.7 Information Retrieval

The image classification approaches considered as "information retrieval" include methods for the classification of natural texture-colour images. Because of the high variability of these methods, we describe some examples of supervised and unsupervised classification at image and pixel-level. A critical overview of supervised and unsupervised image classification methods can be found in [3, 72].

3.7.1 Supervised Classification

Kanan and Cottrell [59] proposed a supervised classification method at image-level for object, faces and flowers images based on sequential fixation-based visual attention. A sequential fixation was a random sampling T times of the saliency map corresponding to a sort of image segmentation. Hence, the posterior probability of observing the class given the fixation feature vector acquired in a time period T from the image was computed in terms of prior probability of observing the class and the product over time of the probabilities of observing the feature vector given the class. The classification of a test image was performed by computing the probability terms according to the training set. Finally, the class obtaining the highest posterior probability was assigned to the test image.

Also, Hiremath and Bhusnurmath [56] employed the KNN classifier for supervised classification at image-level of colour and texture images. The input to the classifier was represented by features extracted from the RGB colour space and a variant of the local binary patters.

In the context of supervised classification at pixel-level, Zhu et al. [106] proposed a method for shadow identification in monochromatic natural images. The method employed a random forest of decision trees, each learned from a random sample of the training set, which provided a probability distribution for each image pixel over the shadow classes. Then, a sampling strategy from each distribution was used as input for estimating the marginal distribution over the class of each pixel, by which the likelihood of a pixel to be assigned to a given class was obtained.

3.7.2 Unsupervised Classification

Amelio and Pizzuti [12] introduced *Genetic Algorithms Image Clustering* (GA-IC), an unsupervised method at image-level for clustering databases of natural colour-texture images. The method was based on the construction of a weighted undirected graph, where nodes represented images, and edges connected the most similar images according to the Manhattan distance of the related feature vectors. Then, a genetic algorithm optimising the weighted modularity function was employed on the obtained graph for detecting the node communities corresponding to image clusters. The adopted genetic algorithm was the same introduced for GA-ICDA in Sect. 3.6.2, with the difference that node communities represented by the chromosome corresponded to groups of images.

Also, Amelio and Pizzuti [8] proposed an image segmentation method called *Genetic Normalized Cut* (GeNCut). It represented the image as a weighted undirected graph, where nodes were associated to the image pixels, and edges connected the nearest neighbour pixels. They were spatially close and similar pixels according to a criterion of brightness homogeneity. Then, a genetic algorithm, minimising the weighted normalised cut, was applied on the image graph in order to detect the node clusters which were pixel regions. The weighted normalised cut considered the total weight of the edges among node clusters. It was minimised when the pixel regions were homogeneous inside, and spatially distant and dissimilar in brightness each others. The employed genetic algorithm was the same proposed in GA-ICDA (see Sect. 3.6.2), except that node clusters represented by the chromosome were pixel regions, and the adopted objective function was the weighted normalised cut. The method was further tested on a variety of images, including satellite, medical, faces and natural images [11]. GeNCut was further extended to *Color Genetic Normalized Cut* (C-GeNCut) for including colour and texture features in the similarity computation of the pixels [9], and tested on different types of natural and human scenes.

3.8 Classification in Image-Based Medical Analytics

The image classification methods are also an essential part of the medical diagnostic process, assisting the physician during the achievement of the patient's general wellness. These methods can represent a huge help in the global treatment of the disease from the aspects of secondary prevention to the management of the main consequences and prognostic perspective. In particular, the methods of medical image classification can be used to quickly identify a specific disease, or signs of it, in an apparently healthy person visualising and recognising elements to deeply study with further accurate exams (secondary prevention). Also, techniques of image classification can help the physician during the accurate diagnosis and/or the treatment of a known disease. A last important utility can be found in the follow-up process of diseases, giving a visualisation of patient's conditions during future time, predicting eventual symptoms return and addressing the patient to the best quality of life.

In the following, a categorisation of medical images is introduced according to a clinical and general point of view. It includes a first section of diagnostic inspective acquisition, covering two of the most common fields of pure inspective imaging which are: (i) dermatology, and (ii) ophthalmology. Then, a second section is characterised by the main different fields of nuclear medicine imaging: (i) scintigraphy, (ii) single-photon emission computed tomography (SPECT), and (iii) positron emission tomography (PET). At the end, a third and last section is related to the general fields of clinical-sided radiology: (i) ultrasonography, (ii) computed tomography (CT), (iii)

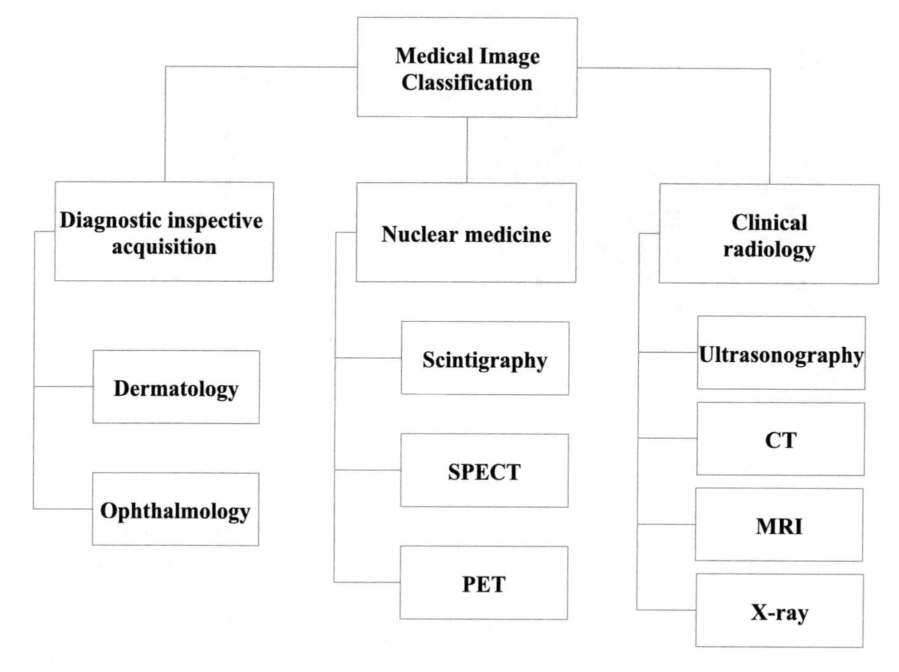

Fig. 3.6 Overview of the main fields of medical imaging

magnetic resonance imaging (MRI), and (iv) X-ray. According to this categorisation, shown in Fig. 3.6, some relevant techniques of image classification are described and discussed.

Readers can refer to [36, 100] for further details about classification methods in medical imaging.

3.8.1 *Diagnostic Inspective Acquisition Imaging*

The diagnostic inspective acquisition imaging, including dermatology and ophthalmology, consists of a visualisation of direct images that can be obtained during the inspective part of the medical examination. It is performed through the utilisation of specific medical instruments, such as fundus camera for the digital capturing of retinal fundus images and generic cameras for dermatological images. Figure 3.7 shows a sample of retinal image captured by a fundus camera and a dermatological image captured by a generic camera.

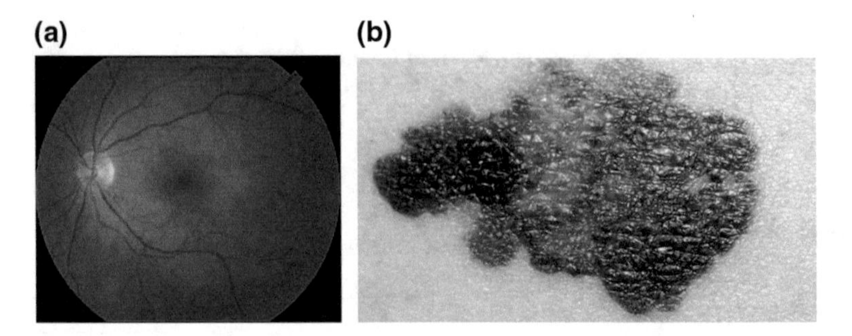

Fig. 3.7 Diagnostic inspective acquisition imaging: **a** retinal fundus image, and **b** melanoma dermatological image

3.8.1.1 Dermatology

In dermatology, supervised image classification at object-level is used to determine the eventual progression of skin lesions, focusing on the different inspective appearance and diversity from no pathological lesions. Also, some works are finalised to the comparison of different skin lesion types, e.g. trying to differentiate melanoma cancerous lesions from other cancer types such as squamous cell carcinoma (differential diagnosis). Furthermore, image segmentation is used to extract lesion's borders which can be later analysed by supervised image classification algorithms.

In particular, KNN and SVM classifiers were used for the identification of skin lesions from images [91]. First, a region growing algorithm was adopted for segmenting the skin lesions in the image. Region growing randomly selected seed pixels as initial regions. Then, it increased the region size by clustering neighbour pixels to each seed until the pixels were similar. After that, the extracted lesion segments represented by colour and texture features were classified by using KNN and SVM. The SVM classifier was also employed for lesion severity classification from skin images [1]. In the first step, image segmentation was performed for extracting the lesion regions. Then, multiple features were computed from the extracted regions, e.g. central shape asymmetry, colour asymmetry and border irregularity. In the second step, a BoVW model of concepts was created by classifying the extracted features in semantic categories using SVM. The lesion severity was computed as the sum of the scores associated with the semantic concepts. Furthermore, CNN was used for the identification of skin lesions [41]. Input to the network was the skin lesion image. The network was trained for recognising fine-grained classes of skin lesions, which determined a probability distribution over the classes. In order to compute the probability on coarser-level classes of skin lesions, the probability of their descendants was summed. The algorithm was tested for the discrimination of keratinocyte carcinomas and benign seborrheic keratoses, and for the distinction between malignant melanomas and benign nevi.

In the context of image segmentation, Amelio and Pizzuti [10] proposed the application of C-GeNCut on skin lesion images for the identification of the lesion borders. In particular, the algorithm was tested on images representing different types of melanoma.

A wider review of classification algorithms for skin lesion images can be found in [44].

3.8.1.2 Ophthalmology

On the contrary, in ophthalmology, supervised and unsupervised image classification methods can be used to check retinal vessels, state to monitor the progression and the eventual diagnosis of diabetic retinopathy. Eventually, these methods can be applied to study macular diseases in different expressions or can be used to recognise different types of no retinal pathologies such as cataract and glaucoma.

In this context, K-means clustering algorithm was employed for segmenting retinal blood vessels from fundus images [54]. It could be used for analysing the status of the retinal blood vessels in order to capture the advancement of the diabetes.

Also, CNN was used for the supervised classification at image-level of retinal images derived from the fundus inspection [31]. In particular, the CNN was capable of classifying the retinal images in multiple categories of normal retina and retina affected by different diseases. Also, the SVM classifier was employed for the identification of ophthalmic images in order to recognise the pediatric cataract [97].

A detailed survey about fundus image analysis for the diagnosis of glaucoma is reported in [74].

3.8.2 Nuclear Medicine Imaging

Nuclear medicine imaging, including scintigraphy, SPECT and PET, is based on the acquisition of radiation emitted from inside the patient instead of using external radiation sources. In particular, small amounts of radioactive material called radiotracers are injected in the patient's body. When the radiotracers reach the interested area, they emit radiation which is captured by a specific instrument in order to generate the image.

In the scintigraphy and SPECT, radiotracers based on emission of gamma rays are injected in the patient's body. On the contrary, in PET imaging the radiotracers emit positrons. Among the different types of imaging, PET and SPECT imaging give a 3D view, while the scintigraphy provides a 2D view.

Figure 3.8 shows a sample of scintigraphy, SPECT and PET images. Further details about nuclear medicine imaging and instrumentation can be found in [61].

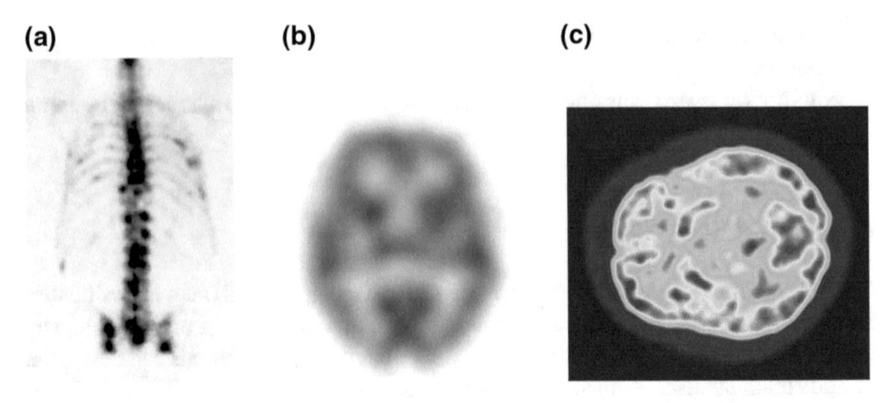

Fig. 3.8 Nuclear medicine imaging: **a** scintigraphy of prostatic cancer, **b** brain SPECT image, and **c** brain PET image

3.8.2.1 Scintigraphy

The classification of scintigraphy images mostly involves object-level supervised algorithms whose aim is the recognition of multiple bone illnesses from features extracted from the scintigraphy.

Among the others, a knowledge-based algorithm was proposed for the identification of different pathologies from bone scintigraphy [87]. First, an image segmentation method was introduced for segmenting the whole-body bone scintigraphy for further diagnosis on single bones. Then, a set of parameterised knowledge-based rules was adopted for supporting image processing algorithms in detecting reference points of the anterior and posterior whole-body skeletal regions, e.g. shoulders, head, pelvis, thorax. It determined parameterised bone images according to the bone reference points, which were the input to an SVM classifier. The output of the SVM was a class label corresponding to some bone pathology, e.g. lesion, malignom, metastasis, degenerative changes, inflammation, or no pathologies.

3.8.2.2 Single-Photon Emission Computed Tomography

Differently from the scintigraphy, the classification methods for SPECT images aim at recognising specific regions of brain and body tissues in order to identify benign versus malignant pathologies, e.g. tumour pathologies, neurodegenerative pathologies (Alzheimer, Parkinson disease), or other types of diseases.

Among the others, the SVM classifier was adopted for the supervised classification at object-level of SPECT images in order to identify the Alzheimer disease [48]. First, a gaussian mixture model (GMM) was used for approximating the intensity profile of the image. In the GMM, the data was realisation of a probability distribution as the sum of k gaussians, with parameters given by the expectation values μ_i and covariance matrices Σ_i, $i = 1, ..., k$. The aim was to estimate the parameters which maximised the likelihood of a GMM with k elements using an expectation maximisation (EM) algorithm. EM was based on a hidden variable, which was added to simplify the maximisation of the likelihood. The procedure was characterised by a first step of expectation, where it was estimated the distribution of the hidden variable given the data and the value of the parameters. Then, a step of maximisation used the hidden variable for changing the parameters' value in order to maximise the likelihood of the data and of the hidden variable. The output of this procedure was a set of k regions characterising the image. From each region, mean intensities were computed which determined a final feature vector of the image. At the end, the SVM was trained using the obtained feature vectors in order to classify unknown brain images as normal or affected by the Alzheimer disease. Another supervised classification system at object-level used the random forest classifier for the recognition of brain SPECT images affected by the Alzheimer disease [85]. A first step of identification of meaningful image regions was employed for detecting the most discriminant features in order to train the classifier. At the end, unknown brain images were classified by the random forest as normal or abnormal.

3.8.2.3 Positron Emission Tomography

The aim of the classification of PET images is relatively similar to the SPECT image classification. Also in this context, the SVM classifier was used for the supervised classification at image-level of brain images for the automatic diagnosis of the Alzheimer disease [76]. The classifier was trained by images represented by a set of texture feature vectors. Textural features were extracted by an extension in 3D of the LBP operators. At the end, unknown images were classified as normal or affected by the Alzheimer disease. At object-level, the random forest classifier was employed on PET images for the automatic identification of the lymphoma area [50]. The PET image was hierarchically decomposed in meaningful regions embedding the spatial and spectral information of the image. This was represented as a tree data structure where each node corresponded to a given region, with associated features representing shape, texture or intensity of the region. Then, the random forest classifier was trained on these features in order to classify each node of the tree with three possible labels: lesions, organs and non-relevant parts.

Finally, in the context of the image segmentation, an extension of the fuzzy K-means was introduced for the segmentation of tumour masses from PET images, e.g. laryngeal squamous cell carcinoma [14]. In particular, spatial information and a method for managing the segmentation of heterogeneous lesions were integrated in the clustering algorithm.

3.8.3 Clinical Radiology Imaging

Clinical radiology imaging consists of a visualisation of images that can be obtained during the routinary general radiological approach to the patient. The different techniques comprehend various types of images, not based on the same physical laws but grouped here from a diagnostic point of view. Figure 3.9 shows a sample of breast ultrasonic image, liver CT image, head MRI image, and chest X-ray image.

3.8.3.1 Ultrasonography

Ultrasonography imaging is based on images obtained and scanned from a sound wave source instrument (ultrasound probe) that provides the visualisation of different types of tissues. In medical clinical diagnosis, ecography is usually essential and fast,

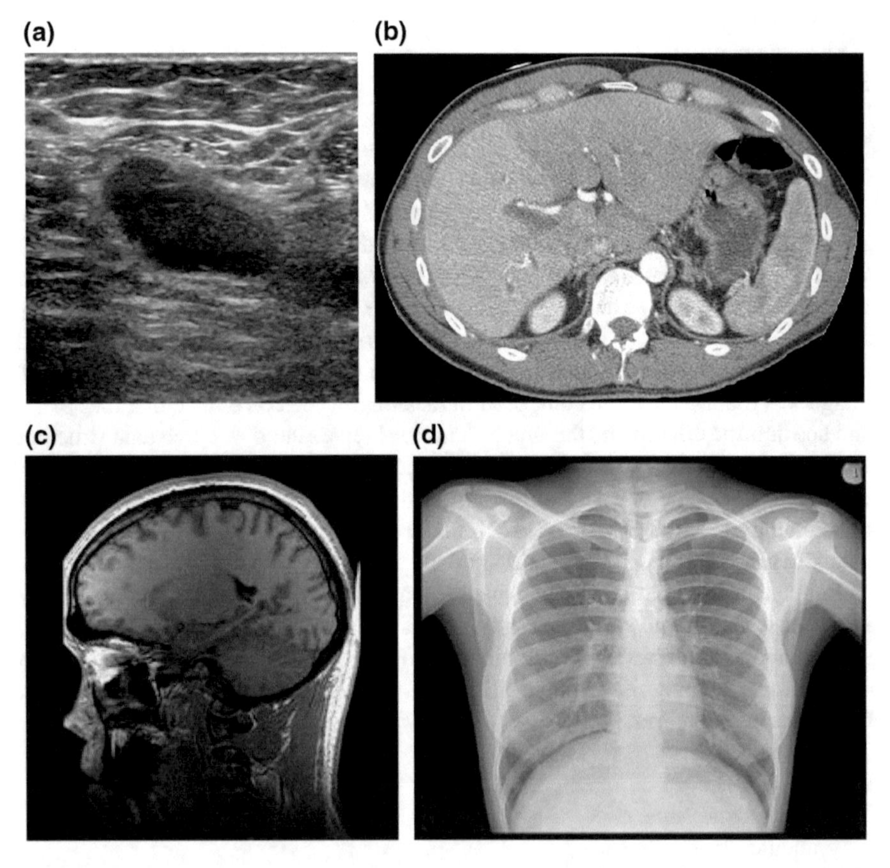

Fig. 3.9 Clinical radiology imaging: **a** breast ultrasonic image, **b** liver CT image, **c** head MRI image, and **d** chest X-ray image

letting to the physic a quick and substantial comprehension of a specific group of symptoms and associated disease. The reader can refer to [93] for a wider description of the ultrasonography imaging.

Among the image segmentation methods, a bayesian non-parametric clustering was employed for segmenting breast ultrasound images in order to discriminate a breast cancer lesion from the background [83]. The feature vectors characterising the images were considered to be the realisation of a mixture of distributions, whose parameters followed a Dirichlet process. The objective was the estimation of the set of parameters and set of mixture components such that the probability of observing the parameters given the data was maximised. Also, the K-means clustering was used in the segmentation of ultrasonic images for supporting detection and classification of different stages of chronic kidney disease [57]. After a pre-processing step of cleaning and noise removal and a step of manual selection of regions of interest (ROIs), the K-means segmented each ROI into five local regions, in order to detect the borders of renal pelvis and parenchyma. According to the distribution of black and white pixels in the different regions, the stages of the patients with chronic kidney disease could be detected. Further methods of image segmentation for ultrasonic images can be found in [79].

In the context of supervised classification at image-level, CNN and SVM methods were adopted inside a system for the diagnosis of the cirrhosis from ultrasonic images [67]. A pre-processing step included the extraction of liver capsules from the images. Then, a CNN was trained according to the liver capsules. It was used for computing features from patches composing the capsules. At the end, an SVM was trained from labeled features and used for the classification of unknown images into normal or abnormal cases.

3.8.3.2 Computed Tomography Scan

On the contrary, CT imaging is based on images obtained with X-ray from a 3D point of view using a tomograph. CT scan is today a needful part of medical diagnosis, irreplaceable and fast. It lets to the physic a global understanding of the patient's state, giving the possibility to analyse even regions not directly affected by the pathology but associated with the pathology progression.

In this context, the CNN method was employed for the supervised classification at image-level of optical coherence tomography (OCT) images. The aim was the identification of normal OCT macular images as well as OCT images of patients affected by age-based macular degeneration [65]. The input of the CNN was the OCT macular image. After a training phase of the neural network, an unknown OCT macular image could be classified as normal or abnormal. Deep learning in terms of CNN, deep neural network (DNN) and stacked autoencoder (SAE), was also used for the supervised classification at image-level of benign and malignant lung nodules from CT images [90]. The DNN was composed of an input layer, an output layer and a set of fully connected hidden layers with increasing number of neurons. Each hidden layer was a nonlinear transformation of the output layer.

The SAE was characterised by multiple autoencoders which were combined in a network with different hidden layers, an input layer and an output layer. In the training phase, labeled CT lung nodule images were used as input for training the networks modelling the images at different size for each hidden layer. In the first training step, all weights of the networks were randomly set. Then, the output value was forward propagated according to the functions defined at neuron-level. After that, the weights were re-computed backward in order to minimise the error. Given an unknown image as input, the output of the networks was the classification of the lung nodule as benign or malignant. Results demonstrated that CNN obtained the highest accuracy in classification of lung nodules.

In the context of the image segmentation, fuzzy K-means was adopted for the segmentation of the liver region from CT images [66]. This is of prior importance for building 3D models of the liver in order to automatically detect liver diseases.

3.8.3.3 Magnetic Resonance Imaging

Differently from the previous images, MRI is a type of medical image based on the spin orientation variation of hydrogen atoms exposed to a variable magnetic field. Such variation is employed by the MRI scanner for generating the image. These types of images are used for a huge spectrum of diseases, usually when CT scan imaging is not so appropriate, or directly in the study of specific pathologies.

In this context, deep learning CNN was widely adopted for the supervised classification of brain tumour magnetic resonance images [5]. In particular, three architectures of CNN were used in the state-of-the-art for solving this task: (i) Patch-Wise CNN, (ii) Semantic-Wise CNN, and (iii) Cascaded CNN. A Patch-Wise CNN identified a $N \times N$ patch in the neighbourhood of each image pixel and trained the network from these patches and their class labels in order to identify normal or abnormal (tumour) part. In some cases, the CNN employed multiple pathways, each trained from patches of different size. The final output was given by a neural network which learned a model from the single outputs in order to detect the final classification. Semantic-Wise CNN realised a supervised classification at pixel-level using autoencoders. Finally, cascaded CNN was characterised by a first network which provided a first classification, and a second network receiving as input the outputs of the first network in order to refine the classification. Also, the SVM classifier was used for the supervised classification of brain MR images in order to identify patients with normal condition, Alzheimer's disease or mild cognitive impairment [101]. Textural features as well as region-based local features were extracted from the image and given as input to the SVM classifier for the patient's condition identification.

In the context of the image segmentation, an extended version of the fuzzy K-means was introduced for the segmentation of brain MRI images [92]. It is of prior importance for the identification and further analysis of brain structures, e.g. cerebrospinal fluid and white matter. In particular, the optimisation function of the fuzzy K-means was modified for considering also intensity inhomogeneity, making the segmentation more robust to noise.

3.8.3.4 X-ray

Finally, X-ray images are the most common and direct type of medical images. They allow a rapid analysis of specific anatomical districts such as chest, abdomen or bones. An X-ray image is based on the exposure of the patient to a flow of photons. It provides a rapid realisation of a clinical condition, helping identifying regions to check with more accurate radiological approaches.

The CNN classifier was also used in the X-ray imaging for the supervised classification at image-level of images showing the presence/absence and position (low/normal) of an endotracheal tube on chest/abdominal radiographs, which is an important task for the radiologist [64]. Obtained results showed that CNN performed well for the differentiation of chest and abdominal radiographs. On the contrary, more training images were needed for obtaining high accuracy in more difficult cases, i.e. differentiation of presence and absence, and low and normal position of an endotracheal tube. Also, CNN was adopted for the identification of vessel regions in contrast to the background in angiography images [78], which is of prior importance for the diagnosis of the coronary artery disease. The input image was scanned by a fixed-size window which progressively extracted patches as input to the CNN. The output indicated if the central pixel of the patch was inside a vessel region or outside. At the end, the trained CNN was employed for identifying the vessels locations inside the patches of the input image, and dividing the image in two regions of vessel/background. Finally, SVM and KNN classifiers were employed for the supervised classification of esophageal X-ray images in order to identify normal or different cancerous images [102]. After a pre-processing step where the image was resized to a small region of interest and enhanced, a feature vector was extracted from the image and the most discriminative features selected. After that, SVM and KNN were trained using the selected features for classifying different types of esophageal cancer images.

3.8.4 Horizon of the Research and Future Challenges

The utility offered by these techniques is a great tool supposed to guide the sanitary operator through the entire diagnostic process. The direction taken from these studies is clearly projected to the future where the several ideas, day after day, paint a universe rich of innovation and technology.

From a clinical point of view, the possibility (economical and time related) to access to imaging instrumentation that can apply and utilise these algorithms as fast as possible, is crucial to let these methods concretely enter into the medical common approach. Today, there are a lot of horizons that can be discussed, going from the emergency medicine to the pure radiology. A first future challenge could be a simple camera system, taking pictures of the patient's face from different angles. Then, it can be used before the medical examination to auto-check the images and identify eventual cranial nerve paralysis or facial bone fractures. In fact, deviation of the muscular normal position or signs of traumatic wounds or fractures can be detected and analysed so that the physician, in a second time, can directly focus on these lesions. All this can be imagined as a support during the triage assessment even before the real medical intervention, aimed to create a first classification of the patient identifying regions of interest to treat or study deeper with different types of radiological exams (CT). Another challenge could be the utilisation of these image classification methods with portable ultrasound machines. This could let the physician to have an assisted and real-time oriented diagnosis using a portable device everywhere it is needed, for example in contexts of disaster medicine, or at home for fragile elderly patients or at risk ones. Another interesting idea could be the introduction of new image classification methods to be used in a smart monitoring system for stroke patients. This kind of patients very often require a serial CT monitoring to check the eventual progressions of central nervous system (CNS) lesions or even the appearance of new ones. The hypothetical existence of an imaging confrontation system for CT images could help the physician during the monitoring of the patient's neurological state, highlighting variations of the cerebral lesions during the post-acute phase. A last context for the application of image classification algorithms could be represented by the radioguided surgery. In fact, smart image recognition systems could be used to closely assist in real-time the surgeon during the different phases of the surgery that require a radiological assistance.

Some of the proposed ideas could be the object of our future research.

3.9 Conclusions

This paper surveyed the most recent trends in supervised and unsupervised image classification. This was accomplished by providing a general overview of the image classification problem in all main aspects. Then, the analysis was focused on the main applicative contexts where the image classification has played an essential role, including security and biometrics, aerospace and satellite monitoring, document image processing and language understanding, and information retrieval. At the end, the most relevant methods of image classification in the medical analytics were presented according to an introduced categorisation from a clinical perspective. It was the starting point for shedding light on the future challenges of the image classification in the medical context, which are of prior importance for improving the diagnosis process. Accordingly, this paper could be a valid support for newcomers as well as students who approach for the first time to the image classification problem. Also, it can be particularly useful to teachers and researchers in academia as well as industry in order to present the advancement of the state-of-the-art and promote the design and development of new methods in image classification.

3.10 Further Readings

Thomas M. Mitchell. 1997. Machine Learning (1 ed.). McGraw-Hill, Inc., New York, NY, USA.
Jiawei Han. 2005. Data Mining: Concepts and Techniques. Morgan Kaufmann Publishers Inc., San Francisco, CA, USA.
Waseem Rawat and Zenghui Wang. 2017. Deep Convolutional Neural Networks for Image Classification: A Comprehensive Review. Neural Computation, Volume 29, Issue 9, pp. 2352–2449.
Abdullah-Al Nahid and Yinan Kong. 2017. Involvement of Machine Learning for Breast Cancer Image Classification: A Survey. Computational and Mathematical Methods in Medicine, Volume 2017, Article ID 3781951.
Mark Nixon. 2013. Feature Extraction and Image Processing. Elsevier.
Anirban Mukhopadhyay, Ujjwal Maulik, and Sanghamitra Bandyopadhyay. 2015. A Survey of Multiobjective Evolutionary Clustering. ACM Comput. Surv., Volume 47, Issue 4, Article 61.
John A. Richards. 2013. Remote Sensing Digital Image Analysis—An introduction. Springer-Verlag Berlin Heidelberg.
Signal and Image Processing for Biometrics. 2012. Editor(s): Amine Nait-Ali, Régis Fournier, Wiley.
Handbook of Document Image Processing and Recognition. 2014. Editor(s): David Doermann, Karl Tombre, Springer-Verlag London.
Klaus D. Toennies. 2017. Guide to Medical Image Analysis - Methods and Algorithms. Springer-Verlag London.

References

1. Abbes, W., Sellami, D.: Automatic skin lesions classification using ontology-based semantic analysis of optical standard images. Procedia Comput. Sci. **112**, 2096–2105 (2017). In: Proceedings of the 21st International Conference on Knowledge-Based and Intelligent Information & Engineering Systems, KES-2017, 6–8 September 2017, Marseille, France
2. Adámek, M., Matýsek, M., Neumann, P.: Security of biometric systems. Procedia Eng. **100**(Suppl C), 169–176 (2015). In: 25th DAAAM International Symposium on Intelligent Manufacturing and Automation (2014)
3. Aggarwal, C., Reddy, C.: Data Clustering: Algorithms and Applications. Data Mining and Knowledge Discovery Series. CRC Press, Chapman & Hall/CRC (2016)
4. Ahmad, F., Mohamad, D.: A review on fingerprint classification techniques. In: 2009 International Conference on Computer Technology and Development, vol. 2, pp. 411–415, November 2009
5. Akkus, Z., Galimzianova, A., Hoogi, A., Rubin, D.L., Erickson, B.J.: Deep learning for brain MRI segmentation: state of the art and future directions. J. Digit. Imaging **30**(4), 449–459 (2017)
6. Ali, H., Salami, M.J.E., Wahyudi: Iris recognition system by using support vector machines. In: 2008 International Conference on Computer and Communication Engineering, pp. 516–521, May 2008

7. Altman, N.S.: An introduction to kernel and nearest-neighbor nonparametric regression. Am. Stat. **46**(3), 175–185 (1992)
8. Amelio, A., Pizzuti, C.: An evolutionary and graph-based method for image segmentation. In: 12th International Conference on Parallel Problem Solving from Nature—PPSN XII Taormina, Italy, 1–5 September 2012, Proceedings, Part I, pp. 143–152. Springer, Berlin, Heidelberg (2012)
9. Amelio, A., Pizzuti, C.: A genetic algorithm for color image segmentation. In: 16th European Conference on Applications of Evolutionary Computation, EvoApplications 2013, Vienna, Austria, 3–5 April 2013, Proceedings, pp. 314–323. Springer, Berlin, Heidelberg (2013)
10. Amelio, A., Pizzuti, C.: Skin lesion image segmentation using a color genetic algorithm. In: Proceedings of the 15th Annual Conference Companion on Genetic and Evolutionary Computation, GECCO '13 Companion, pp. 1471–1478. ACM, New York, NY, USA (2013)
11. Amelio, A., Pizzuti, C.: An evolutionary approach for image segmentation. Evolut. Comput. **22**(4), 525–557 (2014)
12. Amelio, A., Pizzuti, C.: A new evolutionary-based clustering framework for image databases. In: 6th International Conference on Image and Signal Processing, ICISP 2014, Cherbourg, France, 30 June–2 July 2014, Proceedings, pp. 322–331. Springer International Publishing (2014)
13. Bay, H., Tuytelaars, T., Van Gool, L.: SURF: speeded up robust features. In: 9th European Conference on Computer Vision, ECCV 2006, Graz, Austria, 7–13 May 2006, Proceedings, Part I, pp. 404–417. Springer, Berlin, Heidelberg (2006)
14. Belhassen, S., Zaidi, H.: A novel fuzzy c-means algorithm for unsupervised heterogeneous tumor quantification in PET. Med. Phys. **37**(3), 1309–1324 (2010)
15. Blaschke, T.: Object based image analysis for remote sensing. ISPRS J. Photogramm. Remote Sens. **65**(1), 2–16 (2010)
16. Boyle, S.A., Kennedy, C.M., Torres, J., Colman, K., Pérez-Estigarribia, P.E., de la Sancha, N.: High-resolution satellite imagery is an important yet underutilized resource in conservation biology. PLOS ONE **9**(1), 1–11 (01 2014)
17. Brodić, D., Amelio, A.: Classification of the hand-printed and printed medieval glagolitic documents using differentiation in orthography. In: 2017 40th International Convention on Information and Communication Technology, Electronics and Microelectronics (MIPRO), pp. 1110–1115, May 2017
18. Brodić, D., Amelio, A., Milivojević, Z.N.: An approach to the language discrimination in different scripts using adjacent local binary pattern. J. Exp. Theor. Artif. Intell. **29**(5), 929–947 (2017)
19. Brodić, D., Amelio, A., Milivojević, Z., Jevtić, M.: Document image coding and clustering for script discrimination. ICIC Express Lett. **10**(7), 1561–1566 (2016)
20. Brodić, D., Amelio, A.: Dating the historical documents from digitalized books by orthography recognition. In: Digital Libraries and Archives: 13th Italian Research Conference on Digital Libraries, IRCDL 2017, Modena, Italy, 26–27 January 2017, Revised Selected Papers, pp. 119–131. Springer International Publishing (2017)
21. Brodić, D., Amelio, A.: Discrimination of different Serbian pronunciations from Shtokavian dialect. Procedia Comput. Sci. **112**, 1935–1944 (2017). In: Proceedings of the 21st International Conference on Knowledge-Based and Intelligent Information & Engineering Systems, KES-2017, 6–8 September 2017, Marseille, France
22. Brodić, D., Amelio, A.: Recognizing the orthography changes for identifying the temporal origin on the example of the Balkan historical documents. Neural Comput. Appl. (2017)
23. Brodić, D., Amelio, A., Janković, R., Milivojević, Z.N.: Analysis of the reforming languages by image-based variations of LBP and NBP operators. In: 11th International Workshop on Multi-disciplinary Trends in Artificial Intelligence, MIWAI 2017, Gadong, Brunei, 20–22 November 2017, Proceedings, pp. 238–251. Springer International Publishing (2017)
24. Brodić, D., Amelio, A., Milivojević, Z.N.: Characterization and distinction between closely related south Slavic languages on the example of Serbian and Croatian. In: 16th International Conference on Computer Analysis of Images and Patterns, CAIP 2015, Valletta, Malta, 2–4 September 2015 Proceedings, Part I, pp. 654–666. Springer International Publishing (2015)

25. Brodić, D., Amelio, A., Milivojević, Z.N.: Identification of Fraktur and Latin scripts in German historical documents using image texture analysis. Appl. Artif. Intell. **30**(5), 379–395 (2016)
26. Brodić, D., Amelio, A., Milivojević, Z.N.: Clustering documents in evolving languages by image texture analysis. Appl. Intell. **46**(4), 916–933 (2017)
27. Brodić, D., Amelio, A., Milivojević, Z.N.: An image texture analysis method for minority language identification. In: 18th International Workshop on Combinatorial Image Analysis, IWCIA 2017, Plovdiv, Bulgaria, 19–21 June 2017, Proceedings, pp. 280–293. Springer International Publishing (2017)
28. Brodić, D., Amelio, A., Milivojević, Z.N.: Language discrimination by texture analysis of the image corresponding to the text. Neural Comput. Appl. **29**(6), 151–172 (2018)
29. Brodić, D., Milivojević, Z., Amelio, A.: Analysis of the south Slavic scripts by run-length features of the image texture. Elektronika ir Elektrotechnika 21(4) (2015)
30. Chen, N., Blostein, D.: A survey of document image classification: problem statement, classifier architecture and performance evaluation. Int. J. Doc. Anal. Recognit. (IJDAR) **10**(1), 1–16 (2007)
31. Choi, J., Yoo, T., Seo, J., Kwak, J., Um, T., Rim, T.: Multi-categorical deep learning neural network to classify retinal images: a pilot study employing small database. PLoS ONE **12**(11), e0187336 (2017)
32. Chu, A., Sehgal, C., Greenleaf, J.: Use of gray value distribution of run lengths for texture analysis. Pattern Recognit. Lett. **11**(6), 415–419 (1990)
33. Cortes, C., Vapnik, V.: Support-vector networks. Mach. Learn. **20**(3), 273–297 (1995)
34. Dasarathy, B.V., Holder, E.B.: Image characterizations based on joint gray level-run length distributions. Pattern Recognit. Lett. **12**(8), 497–502 (1991)
35. Dawood, H., Dawood, H., Guo, P.: Texture image classification with improved Weber local descriptor. In: 13th International Conference on Artificial Intelligence and Soft Computing, ICAISC 2014, Zakopane, Poland, 1–5 June 2014, Proceedings, Part I, pp. 684–692. Springer International Publishing (2014)
36. Dey, N.: Classification and clustering in biomedical signal processing. In: Advances in Medical Technologies and Clinical Practice. IGI Global (2016)
37. Dey, S., Mukherjee, J., Sural, S.: Consensus-based clustering for document image segmentation. Int. J. Doc. Anal. Recognit. (IJDAR) **19**(4), 351–368 (2016)
38. Diem, M., Kleber, F., Fiel, S., Sablatnig, R.: Semi-automated document image clustering and retrieval. In: DRR. SPIE Proceedings, vol. 9021, pp. 90210M–90210M–10. SPIE (2014)
39. Duda, T., Canty, M.: Unsupervised classification of satellite imagery: choosing a good algorithm. Int. J. Remote Sens. **23**(11), 2193–2212 (2002)
40. Esposito, F., Malerba, D., Lisi, F.A.: Machine learning for intelligent processing of printed documents. J. Intell. Inf. Syst. **14**(2), 175–198 (2000)
41. Esteva, A., Kuprel, B., Novoa, R.A., Ko, J., Swetter, S.M., Blau, H.M., Thrun, S.: Dermatologist-level classification of skin cancer with deep neural networks. Nature **542**(7639), 115–118 (2017)
42. Fan, B., Wang, Z., Wu, F.: Local image descriptor: modern approaches. In: Springer Briefs in Computer Science. Springer (2015)
43. Fei-Fei, L., Perona, P.: A Bayesian hierarchical model for learning natural scene categories. In: 2005 IEEE Computer Society Conference on Computer Vision and Pattern Recognition (CVPR'05), vol. 2, pp. 524–531, June 2005
44. Filho, M., Ma, Z., Tavares, J.M.R.S.: A review of the quantification and classification of pigmented skin lesions: from dedicated to hand-held devices. J. Med. Syst. **39**(11), 177 (2015)
45. Galloway, M.M.: Texture analysis using gray level run lengths. Comput. Gr. Image Process. **4**(2), 172–179 (1975)
46. Geng, J., Wang, H., Fan, J., Ma, X.: Deep supervised and contractive neural network for SAR image classification. IEEE Trans. Geosci. Remote Sens. **55**(4), 2442–2459 (2017)
47. Gonzalez, R.: Digital Image Processing. Pearson Education (2009). https://books.google.it/books?id=a62xQ2r_f8wC

48. Górriz, J., Segovia, F., Ramírez, J., Lassl, A., Salas-Gonzalez, D.: GMM based SPECT image classification for the diagnosis of Alzheimers disease. Appl. Soft Comput. **11**(2), 2313–2325 (2011). The Impact of Soft Computing for the Progress of Artificial Intelligence
49. Gragnaniello, D., Poggi, G., Sansone, C., Verdoliva, L.: Local contrast phase descriptor for fingerprint liveness detection. Pattern Recognit. **48**(4), 1050–1058 (2015)
50. Grossiord, E., Talbot, H., Passat, N., Meignan, M., Najman, L.: Automated 3D lymphoma lesion segmentation from PET/CT characteristics. In: 2017 IEEE 14th International Symposium on Biomedical Imaging (ISBI 2017), pp. 174–178, April 2017
51. Gueld, M.O., Keysers, D., Deselaers, T., Leisten, M., Schubert, H., Ney, H., Lehmann, T.M.: Comparison of global features for categorization of medical images. In: Ratib, O.M., Huang, H.K. (eds.) Medical Imaging 2004: PACS and Imaging Informatics. Proceedings of the SPIE, vol. 5371, pp. 211–222 (2004)
52. Haralick, R.M., Shanmugam, K., Dinstein, I.: Textural features for image classification. IEEE Trans. Syst. Man Cybern. **SMC-3**(6), 610–621 (1973)
53. Hassaballah, M., Abdelmgeid, A.A., Alshazly, H.A.: Image features detection, description and matching. In: Image Feature Detectors and Descriptors: Foundations and Applications, pp. 11–45. Springer International Publishing (2016)
54. Hassan, G., El-Bendary, N., Hassanien, A.E., Fahmy, A., Shoeb, A.M., Snasel, V.: Retinal blood vessel segmentation approach based on mathematical morphology. Procedia Comput. Sci. **65**, 612–622 (2015). In: International Conference on Communications, management, and Information technology (ICCMIT'2015)
55. Heroux, P., Diana, S., Ribert, A., Trupin, E.: Classification method study for automatic form class identification. In: Proceedings of the Fourteenth International Conference on Pattern Recognition (Cat. No. 98EX170), vol. 1, pp. 926–928, August 1998
56. Hiremath, P.S., Bhusnurmath, R.A.: RGB-based color texture image classification using anisotropic diffusion and LDBP. In: 8th International Workshop on Multi-disciplinary Trends in Artificial Intelligence, MIWAI 2014, Bangalore, India, 8–10 December 2014, Proceedings, pp. 101–111. Springer International Publishing (2014)
57. Ho, C.Y., Pai, T.W., Peng, Y.C., Lee, C.H., Chen, Y.C., Chen, Y.T., Chen, K.S.: Ultrasonography image analysis for detection and classification of chronic kidney disease. In: 2012 Sixth International Conference on Complex, Intelligent, and Software Intensive Systems, pp. 624–629, July 2012
58. Jog, S., Dixit, M.: Supervised classification of satellite images. In: 2016 Conference on Advances in Signal Processing (CASP), pp. 93–98, June 2016
59. Kanan, C., Cottrell, G.: Robust classification of objects, faces, and flowers using natural image statistics. In: 2010 IEEE Computer Society Conference on Computer Vision and Pattern Recognition, pp. 2472–2479, June 2010
60. Kang, L., Kumar, J., Ye, P., Li, Y., Doermann, D.: Convolutional neural networks for document image classification. In: 2014 22nd International Conference on Pattern Recognition, pp. 3168–3172, August 2014
61. Kharfi, F.: Principles and applications of nuclear medical imaging: a survey on recent developments. In: Kharfi, F. (ed.) Imaging and Radioanalytical Techniques in Interdisciplinary Research—Fundamentals and Cutting Edge Applications, Chap. 01. InTech, Rijeka (2013). http://dx.doi.org/10.5772/54884
62. Krig, S.: Computer Vision Metrics: Survey, Taxonomy, and Analysis. SpringerLink: Bücher, Apress (2014). https://books.google.it/books?id=ktKuAwAAQBAJ
63. Kuric, E., Bielikova, M.: ANNOR: efficient image annotation based on combining local and global features. Comput. Gr. **47**(Suppl C), 1–15 (2015)
64. Lakhani, P.: Deep convolutional neural networks for endotracheal tube position and X-ray image classification: challenges and opportunities. J. Digital Imaging **30**(4), 460–468 (2017)
65. Lee, C.S., Baughman, D.M., Lee, A.Y.: Deep learning is effective for classifying normal versus age-related macular degeneration OCT images. Ophthalmol. Retina **1**(4), 322–327 (2017)

66. Li, X., Luo, S., Li, J.: Liver segmentation from CT image using fuzzy clustering and level set. J. Signal Inf. Process. **4**, 36–42 (2013)
67. Liu, X., Song, J.L., Wang, S., Zhao, J., Chen, Y.Q.: Learning to diagnose cirrhosis with liver capsule guided ultrasound image classification. Sensors **17**(1), 149 (2017)
68. Lowe, D.G.: Object recognition from local scale-invariant features. In: Proceedings of the Seventh IEEE International Conference on Computer Vision, vol. 2, pp. 1150–1157 (1999)
69. Lu, D., Weng, Q.: A survey of image classification methods and techniques for improving classification performance. Int. J. Remote Sens. **28**(5), 823–870 (2007)
70. Ma, L., Li, M., Ma, X., Cheng, L., Du, P., Liu, Y.: A review of supervised object-based land-cover image classification. ISPRS J. Photogramm. Remote Sens. **130**, 277–293 (2017)
71. Mac Namee, B.: Digital image processing: introduction. http://www.imageprocessingplace.com/downloads_V3/root_downloads/tutorials/Image%20ProcessingIntroductionBryanMacNamee.pdf
72. Marée, R., Geurts, P., Visimberga, G., Piater, J., Wehenkel, L.: A comparison of generic machine learning algorithms for image classification. In: Coenen, F., Preece, A., Macintosh, A. (eds.) Research and Development in Intelligent Systems, vol. XX, pp. 169–182. Springer, London (2004)
73. Margolin, R., Zelnik-Manor, L., Tal, A.: OTC: a novel local descriptor for scene classification. In: 13th European Conference on Computer Vision—ECCV 2014, Zurich, Switzerland, 6–12 September 2014, Proceedings, Part VII, pp. 377–391. Springer International Publishing (2014)
74. Mary, M.C.V.S., Rajsingh, E.B., Naik, G.R.: Retinal fundus image analysis for diagnosis of glaucoma: a comprehensive survey. IEEE Access **4**, 4327–4354 (2016)
75. Miao, R.H., Tang, J.L., Chen, X.Q.: Classification of farmland images based on color features. J. Vis. Commun. Image Represent. **29**(Suppl C), 138–146 (2015)
76. Montagne, C., Kodewitz, A., Vigneron, V., Giraud, V., Lelandais, S.: 3D local binary pattern for PET image classification by SVM—application to early Alzheimer disease diagnosis. In: Alvarez, S., Sol-Casals, J., Fred, A.L.N., Gamboa, H. (eds.) Biosignals, pp. 145–150. SciTePress (2013)
77. Morales, A., Ferrer, M.A., Diaz-Cabrera, M., Gonzlez, E.: Analysis of local descriptors features and its robustness applied to ear recognition. In: 2013 47th International Carnahan Conference on Security Technology (ICCST), pp. 1–5, October 2013
78. Nasr-Esfahani, E., Samavi, S., Karimi, N., Soroushmehr, S.M.R., Ward, K., Jafari, M.H., Felfeliyan, B., Nallamothu, B., Najarian, K.: Vessel extraction in X-ray angiograms using deep learning. In: 2016 38th Annual International Conference of the IEEE Engineering in Medicine and Biology Society (EMBC), pp. 643–646, August 2016
79. Noble, J.A., Boukerroui, D.: Ultrasound image segmentation: a survey. IEEE Trans. Med. Imaging **25**(8), 987–1010 (2006)
80. Nosaka, R., Ohkawa, Y., Fukui, K.: Feature extraction based on co-occurrence of adjacent local binary patterns. In: Advances in Image and Video Technology: 5th Pacific Rim Symposium, PSIVT 2011, Gwangju, South Korea, 20–23 November 2011, Proceedings, Part II, pp. 82–91. Springer, Berlin, Heidelberg (2012)
81. O'Gorman, L.: Document Image Analysis: An Executive Briefing, 1st edn. IEEE Computer Society Press, Los Alamitos, CA, USA (1997)
82. Ojala, T., Pietikinen, M., Harwood, D.: A comparative study of texture measures with classification based on featured distributions. Pattern Recognit. **29**(1), 51–59 (1996)
83. Rahman, M.M.: An unsupervised segmentation algorithm for breast ultrasound images using local histogram features. In: 2016 International Conference on Medical Engineering, Health Informatics and Technology (MediTec), pp. 1–6, December 2016
84. Ramesha, K., Srikanth, N., Raja, K.B., Venugopal, K.R., Patnaik, L.M.: Advanced biometric identification on face, gender and age recognition. In: 2009 International Conference on Advances in Recent Technologies in Communication and Computing, pp. 23–27, October 2009

85. Ramirez, J., Gorriz, J.M., Chaves, R., Lopez, M., Salas-Gonzalez, D., Alvarez, I., Segovia, F.: Spect image classification using random forests. Electron. Lett. **45**(12), 604–605 (2009)
86. Romeo, S., Ienco, D., Tagarelli, A.: Knowledge-Based Representation for Transductive Multilingual Document Classification, pp. 92–103. Springer International Publishing (2015)
87. Šajn, L., Kukar, M., Kononenko, I., Milčinski, M.: Automatic segmentation of whole-body bone scintigrams as a preprocessing step for computer assisted diagnostics. In: Miksch, S., Hunter, J., Keravnou, E.T. (eds.) Artificial Intelligence in Medicine, pp. 363–372. Springer, Berlin, Heidelberg (2005)
88. Serrau, A., Marcialis, G.L., Bunke, H., Roli, F.: An experimental comparison of fingerprint classification methods using graphs. In: 5th IAPR International Workshop on Graph-Based Representations in Pattern Recognition, GbRPR 2005, Poitiers, France, 11–13 April 2005, Proceedings, pp. 281–290. Springer, Berlin, Heidelberg (2005)
89. Shi, J., Malik, J.: Normalized cuts and image segmentation. IEEE Trans. Pattern Anal. Mach. Intell. **22**(8), 888–905 (2000)
90. Song, Q., Zhao, L., Luo, X., Dou, X.: Using deep learning for classification of lung nodules on computed tomography images. J. Healthc. Eng. **2017** (1 2017)
91. Sumithra, R., Suhil, M., Guru, D.: Segmentation and classification of skin lesions for disease diagnosis. Procedia Comput. Sci. **45**, 76–85 (2015). In: International Conference on Advanced Computing Technologies and Applications (ICACTA)
92. Sun, S., Yan, S., Wang, Y., Li, Y.: Brain MRI image segmentation based on improved fuzzy c-means algorithm. In: 2016 International Conference on Smart City and Systems Engineering (ICSCSE), pp. 503–505, November 2016
93. Szabo, T.: Diagnostic ultrasound imaging: inside out. In: Biomedical Engineering, Elsevier Science (2013). https://books.google.it/books?id=wTYTAAAAQBAJ
94. Ünsalan, C., Boyer, K.: Multispectral satellite image understanding: from land classification to building and road detection. In: Advances in Computer Vision and Pattern Recognition. Springer, London (2011)
95. Vandenbroucke, N., Macaire, L., Postaire, J.G.: Color image segmentation by supervised pixel classification in a color texture feature space. Application to soccer image segmentation. In: Proceedings 15th International Conference on Pattern Recognition, ICPR-2000, vol. 3, pp. 621–624 (2000)
96. Wahab, A., Chin, S.H., Tan, E.C.: Novel approach to automated fingerprint recognition. IEE Proc. Vis. Image Signal Process. **145**(3), 160–166 (1998)
97. Wang, L., Zhang, K., Liu, X., Long, E., Jiang, J., An, Y., Zhang, J., Liu, Z., Lin, Z., Li, X., Chen, J., Cao, Q., Li, J., Wu, X., Wang, D., Li, W., Lin, H.: Comparative analysis of image classification methods for automatic diagnosis of ophthalmic images. Sci. Rep. **7**, 41545 (2017)
98. Wikipedia: Digital image processing. https://en.wikipedia.org/wiki/Digital_image_processing
99. Witten, I.H., Frank, E., Hall, M.A., Pal, C.J.: Data Mining, Fourth Edition: Practical Machine Learning Tools and Techniques, 4th edn. Morgan Kaufmann Publishers Inc., San Francisco, CA, USA (2016)
100. Wu, G., Shen, D., Sabuncu, M.: Machine Learning and Medical Imaging. Elsevier Science (2016)
101. Xiao, Z., Ding, Y., Lan, T., Zhang, C., Luo, C., Qin, Z.: Brain MR image classification for Alzheimer's disease diagnosis based on multifeature fusion. Comput. Math. Methods Med. **2017**, 1952373:1–1952373:13 (2017)
102. Yang, F., Hamit, M., Yan, C.B., Yao, J., Kutluk, A., Kong, X.M., Zhang, S.X.: Feature extraction and classification on esophageal X-ray images of Xinjiang Kazak Nationality. J. Healthc. Eng. **2017**(4620732), 11 (2017)
103. Yang, H.Y., Zhang, X.J., Wang, X.Y.: LS-SVM-based image segmentation using pixel color-texture descriptors. Pattern Anal. Appl. **17**(2), 341–359 (2014)
104. Zhang, D., Xu, Y., Zuo, W.: Discriminative Learning in Biometrics. Springer, Singapore (2016)

105. Zhang, X., Cui, J., Wang, W., Lin, C.: A study for texture feature extraction of high-resolution satellite images based on a direction measure and gray level co-occurrence matrix fusion algorithm. Sensors **17**(7, 1474) (2017)
106. Zhu, J., Samuel, K.G.G., Masood, S.Z., Tappen, M.F.: Learning to recognize shadows in monochromatic natural images. In: 2010 IEEE Computer Society Conference on Computer Vision and Pattern Recognition, pp. 223–230, June 2010

Lucio Amelio is receiving his M.Sc. from the Faculty of Medicine and Surgery, University of Bologna from 2011. Currently, he is a student at the Faculty of Medicine and Surgery, University of Bologna, Italy. His current research interests include different aspects of computerized medical analytics, radiology, internal medicine, and computerized diagnosis in oncology.

Alessia Amelio received her B.Sc. and M.Sc. magna cum laude in Computer Science Engineering from University of Calabria in 2005 and 2009, as well as Ph.D. in computer science engineering and systems from the Faculty of Engineering, University of Calabria in 2013. During her Ph.D., she was visiting research scholar at College of Computing, Georgia Institute of Technology. From 2011 to 2014 she was research fellow at the National Research Council of Italy. From 2015 to 2016 she was researcher at the National Research Council of Italy. Now she is research fellow and contract professor of computer science at the Department of Computer Science Engineering, Modeling, Electronics and Systems, University of Calabria, Italy. Her current research interests include different aspects of image processing, document classification, pattern recognition from sensor data, social network analysis, data mining and artificial intelligence methods for the web. She co-authored more than 20 journal papers indexed in Scopus and Web of Science, and more than 40 conference papers, book chapters and magazine papers. She was chair and co-organizer of different invited sessions as well as editorial board member in different international journals. She has also served as a reviewer as well as a member of program committee for leading journals and conferences in the fields of data mining, knowledge and data engineering, artificial intelligence, and physics.

Chapter 4
Medical Data Mining for Heart Diseases and the Future of Sequential Mining in Medical Field

Carine Bou Rjeily, Georges Badr, Amir Hajjarm El Hassani and Emmanuel Andres

Abstract Data Mining in general is the act of extracting interesting patterns and discovering non-trivial knowledge from a large amount of data. Medical data mining can be used to understand the events happened in the past, i.e. studying a patients vital signs to understand his complications and discover why he has died, or to predict the future by analyzing the events that had happened. In this chapter we are presenting an overview on studies that use data mining to predict heart failure and heart diseases classes. We will also focus on one of the trendiest data-mining field, namely the Sequential Mining, which is a very promising paradigm. Due to its important results in many fields, this chapter will also cover all its extensions from Sequential Pattern Mining, to Sequential Rule Mining and Sequence Prediction. Pattern Mining is the discovery of important and unexpected patterns or information and was introduced in 1990 with the well-known Apriori. Sequential Patterns Mining aims to extract and analyze frequent subsequences from sequences of events or items with time constraint. The importance of a sequence can be measured based on different factors such as the frequency of their occurrence, their length and their profit. In 1995, Agrawal et al. introduced a new Apriori algorithm supporting time constraints named AprioriAll. The algorithm studied the transactions through time, in order to extract frequent patterns from the sequences of products related to a customer. Time dimension is a very important factor in analyzing medical data, making it necessary to present a positioning of Sequential Mining in the medical domain.

C. Bou Rjeily · A. Hajjarm El Hassani
Nanomedicine Lab, Université de Bourgogne Franche - Comté, UTBM Belfort, Belfort, France
e-mail: carine.bourjeily@utbm.fr

A. Hajjarm El Hassani
e-mail: amir.hajjam-el-hassani@utbm.fr

G. Badr (✉)
TICKET Lab, Antonine University, Hadath, Baabda, Lebanon
e-mail: georges.badr@ua.edu.lb

E. Andres
Université de Strasbourg, Centre Hospitalier Universitaire, Strasbourg, France
e-mail: emmanuel.andres@chru-strasbourg.fr

G. A. Tsihrintzis et al. (eds.), *Machine Learning Paradigms*, Intelligent Systems Reference Library 149, https://doi.org/10.1007/978-3-319-94030-4_4

Keywords Data mining · Healthcare · Heart disease · Sequential pattern mining Algorithms

4.1 Introduction

The World Health Organization estimates that every year more than 12 million deaths are caused by heart diseases. The term Heart disease includes all the diseases related to the heart. Cardiovascular diseases cause half of deaths in the United States and other developed countries. It shows to be the primary reason of the death worldwide. Medical data mining is very important for exploring the hidden patterns in data. Discovering hidden patterns, which are the significant information from data according to the user, is very useful for clinical diagnosis. Heart failure (HF) is a complex clinical pathology [1–3]. It burdens or even disables the circulatory circuit of the body, because the ventricle whose main role is to distribute blood is handicapped. The symptoms that reveal themselves are as follows: breathlessness, ankle swelling and fatigue These symptoms may be accompanied by elevated jugular venous pressure, pulmonary crackles, and peripheral edema, due to structural and/or functional cardiac or non-cardiac abnormalities. Heart failure is a condition that contributes greatly in the rise of mortality rates. A study lead by the European Society of Cardiology (ESC)[1] shows that 26 million adults globally are diagnosed with HF, with 3.6 million new entrants every year. It is estimated that not less than 17–45% of the patients suffering from HF succumb within the first year and the remaining will not make it beyond 5 years. The expenses related to heart failure treatments represent around 1–2% of gross healthcare costs. Most of these costs are associated with regular re-entering to the hospital.

The phenomenon of heart failure has become so widespread with costs rising beyond belief, reduced quality of life and increased mortality. HF is now classified as an epidemic hitting the entire world and Europe in particular. The scale at which heart failure is being met highlights the need for regular check-ups in order to attain an early diagnosis and an immediate treatment. The clinical procedure involving the diagnosis includes a history and physical examination through ancillary tests (blood tests, radiography, electrocardiography, echocardiography). The analysis of the data obtained through the previously mentioned tests gives away several criteria (e.g. Framingham, Boston, the Gothenburg and the ESC criteria) that help detect any risk of HF [4]. The process of diagnosis is followed by a process of classification in which is determined the severity of HF exposure using either the New York Heart Association (NYHA) or the American College of Cardiology/American Heart Association (ACC/AHA)[2] Guidelines classification systems. These systems allow them to find the most appropriate treatment (medication treatment, guidelines regarding nutrition and physical activity exercising) to be followed [5].

---

[1] https://www.escardio.org/.

[2] http://www.acc.org/.

Many progresses were made in understanding the complexity of the pathological and physiological aspects of HF. However analyzing and interpreting the huge amount of complicated data into an appropriate therapeutic diagnosis with the right results is a quite challenging task. But the fact that it is possible to combine these factors up to a certain point and extract a usually successful treatment, prevention and recovery plan is a sign of the good things to come. Thanks to that, it is now possible to improve patients quality of life, prevent condition worsening while maintaining medical costs at the decrease. This explains the increasing popularity in the usage and application of machine learning techniques to analyze, predict and classify medical data. The classification methods are at the center of focus of data mining techniques in the eyes research groups. Accurate classification of disease stage or subtypes permits treatments and interventions to be executed in an efficient and targeted way and allows the patients progress to be assessed and controlled. The data of patients and hospitals need first to be collected and organized to form the hospital or medical information system. After that, the technics of data mining give the user the ability to find novel and important patterns in data.

In this chapter, we will present some studies about data mining and heart diseases, where data mining algorithms, classification and clustering algorithms where used to predict heart failure or to classify the heart disease of a patient. A light description of each algorithm is firstly presented. Later, Sequential Mining (SM) algorithms and their usage in medical domain will be covered. A description of the algorithms and their extensions is presented with a performance and memory consumption comparison for some of the existing Sequential Pattern Mining (SPM) algorithms. Before concluding, a discussion for new ideas for SM that can be useful for medical purposes and other values based applications

4.2 Classical Data Mining Technics and Heart Diseases

Data mining is the process of discovering interesting, meaningful and actionable patterns hidden in large amounts of data. Using data mining in medicine will reduce the cost of hospitalization and help doctors in the diagnosis of patients. The most used data mining techniques in medicine are classification and clustering. Lately SM has been introduced in this domain and showed to be efficient. SM in medicine will be discussed later. In this paragraph we will cover some existing studies that implement data mining in healthcare; in this chapter we are strictly interested in heart diseases. Data-mining techniques have two main tasks: predictive data mining and descriptive data mining.

1. Predictive Data Mining: is the fact of using some variables to predict unknown or future values of other variables. The most important practices in predictive data mining are:

- Classification: given a set of records with different attributes, the idea is to find a model to relate the class attribute in function of the others. The main goal is to assign a class as accurately as possible to previously unseen records.
- Regression: it is the function of estimating or predicting some continuous value called the target attribute. This value of the target is a function of the other attributes (predictors).

2. Descriptive Data Mining: Find human-interpretable patterns that describe the data

   - Clustering: it is the fact of discovering groups of similar instances in a way the instances in the same cluster should be more similar to one another. On the other hand, data of different clusters should be less similar to one another.
   - Association rules extraction: given a set of records of attributes, the idea is to discover dependency rules among attributes. This will help in predicting occurrence of an item based on occurrences of others.

4.2.1 Popular Data Mining Algorithms

Before presenting the studies that use data mining for detecting, predicting or classifying heart diseases, we will list some important algorithms used in this field e.g. Decision tree, Nave Bayes, k-means, Artificial Neural Network and others.

Decision Tree

Decision Trees (DT) is a way to display the data. It uses a tree-like graph as predictive model. The goal of DTs is to create a model to predict a result or a value based on input variables. The results are an important classification and widely used for decision-making. This technic is a popular tool in machine learning that help to find the appropriate strategy to reach the satisfying conclusions because it can be transformed to a set of important rules by matching the root nodes to the leaf nodes [6].

C4.5

is a decision tree based classifier method founded on information gain and pruning to detect important results. The advantage of C4.5 algorithm is that it is fast and gives a clear output easy to be studied [7].

ID3 Algorithm

Iterative Dichotomiser 3 builds a decision tree that classifies objects by testing the values of the properties. It uses the top down fashion using a set of objects and the specification of properties. The idea is to test the property at each node of the tree and based on the results, the objects are set. This process will recursively continue until all the sub trees contain homogeneous sets that respect the criteria of classification.

Those will become leaf nodes. The property is tested to divide the given candidate set to the most similar subsets [8].

Support Vector Machine (SVM)

This method classifies data into two classes. It is very similar to C4.5 but it does not use Decision trees. SVM aims to maximize the distance between the hyper plane and the closest two data points from each following class. This distance is called margin. This process decreases the risk of misclassification [9].

Naive Bayes (NB)

This technique is a probabilistic classifier that uses Bayes theorem. The NB theorem is the following: $P(C|X) = P(X|C) \times P(C)/P(X)$. Given the class variable, NB classifiers say that the value of any feature is independent of any other given feature. X is the data and C is the class. P(X) is constant or the same for all the classes. NB works well on large data set, knowing that is based on the condition that attributes value are conditionally independent which is unrealistic [10].

Artificial Neural Network (ANN)

A neural network is based on the idea of biological neural networks; it is performed on the computer to do some tasks like clustering, pattern reorganization and classification. ANN is a nonlinear statistical data model because complex relationships between inputs and outputs are modeled. The structure of ANN is affected by the flow of information because this structure changes and learns based on the traversing input and output in the neural network [11].

CART

CART algorithm is based on Classification And Regression Trees methodology. The classification tree is used to identify the class of the target categorical variable. The regression tree is used to predict the value of the target continuous value. The CART algorithm proceeds as follow: lets imagine a sequence of questions. The answer of each question leads to the next question if it exists. The answer of the questions are a tree structure where the terminal nodes indicate that there are no more questions or in computational term, queries [12].

Regression

This statistical concept determines the weight of the relationship between a dependent and a fix variable Y and other independent variables. The used regression techniques are linear and multiple linear regression but for complicated data studies non-linear regression methods are performed.

J48

The J48 algorithm is developed by WEKA[3] [13] project team. It is a decision tree implementation of ID3. The advantage of J48 is that it doesnt need a discretization for numeric attributes.

[3]https://www.cs.waikato.ac.nz/ml/weka/.

Fuzzy Logic

Was mainly developed for controlling issues but can also be used in data mining. Fuzzy logic tends to simulate human perception of the environment. In classical logic, the only accepted values are true or false. In fuzzy logic, the truth is a result of partial knowledge. It falls in a range between 0 and 1. It depends on mathematical methods to calculate the degree of truth of a numerical value, called membership functions [14].

K-means

K-means works very well in large datasets. The idea of K-mean Algorithm is the creation of k groups where a set of given input of objects is divided by groups, where each group contains the more similar objects. Those groups are called clusters. K-means is a semi supervised learning method, because it can learn the clusters without any information about which cluster a particular observation should belong to [15].

Association Rules

Association Rules are based on the well-known if/then statements and can be used to find the relationships between data in a warehouse that can appear to be unrelated. An association rule is composed from two parts, the antecedent (if) and a consequent (then). After analyzing data the set of frequent if/else patterns contains the Association rules. The support and confidence, defined later, define the most important relationships.

Random Forests

Random Forests are learning methods also known as Random Decision Forests [16, 17]. This method is used for classification, regression and other tasks. The idea behind these methods is building many decision trees at training time and outputting the class of the individual trees (classification) or their mean prediction (regression).

4.2.2 Data Mining and Heart Diseases

Many papers in the literature investigate in the study of heart diseases and especially heart failure. Researchers try to predict HF and its types, discover the causes, the symptoms and classes of heart diseases. Below, we are presenting some of the most important and recent related works that used data mining techniques. The results of these works showed the importance of data mining in medical field.

In 2016, a fuzzy-cart algorithm used to predict potential heart disease was developed by Suganya et al. [18]. The authors use an efficient approach for extracting significant patterns from the heart disease data warehouses for the efficient prediction of heart attack. The fuzziness is to remove uncertainty of the measured data. The CART-means clustering algorithm was used to cluster the warehouse containing the heart disease data. Frequent items were then extracted using the CART algorithm.

Thus, all the frequent patterns that occur more than the predefined threshold were taken as reference for the prediction of heart attack.

Another research in [19] explores the Random Forests algorithm on long-term ECG time series in order to congestive heart failure (CHF). Several databases were used to get ECG signals (check the databases in [20]). Two phases were needed for the study: feature extraction and classification. Compared to other algorithms (e.g C4.5, K-nearest neighbor, SVM, ANN,) the random forest algorithm showed 100% classification accuracy.

In [21], Pandey et al. used a J48 decision tree classifier over the medical dataset available in [22] to predict heart disease. The data was divided into a training set and a testing set with a 60–40% ratio. The obtained accuracy reached 75.73%.

Same dataset was processed by Bashir et al. [23] who proposed a framework that uses majority vote based novel classifier ensemble to combine different data mining classifiers. The classifier used three different concepts namely, Nave Bayes, Decision Tree and Support Vector Machine. They obtained 82% accuracy, 74% sensitivity and 93% specificity for heart disease dataset.

Again, data mining appeared to be efficient in the prediction of HF in [24] where a framework containing CART, ID3 and decision trees was used. This framework applies a tenfold cross validation on the given dataset and the accuracies obtained are the following: CART decision tree gave the highest accuracy with 83.49% followed by DT with 82.50% and finally the ID3 with 72.93%.

The study of Gharehchopogh et al. [25] was based on artificial neural networks and a set of 40 patient to detect HF. The used attributes were gender, age, blood pressure, and smoking habit. The normal values of these attributes were used to compare them with the ones taken from patients. Only two patients were not classified correctly with 95% True positive rate.

The authors, Uppin et al. [26], used 7 out of 13 attributes of [22] and implement them in a C4.5 decision tree classifier. Their main purpose was to reduce the number of attributes to avoid the redundant features. Results showed an accuracy of 85.96%.

The K-means cluster in [27] was used in order to diagnose heart disease. They obtained an accuracy of 83.9% by applying the inlier method with two clusters. An alternative for conventional decision tree was proposed by Bohacik et al. [28]. This method differs from the classic one by splitting each part of the decision tree multiple times while the classic method splits only leaf nodes. A dataset from Hull LifeLab [4] containing 9 parameters and 2 prediction classes was used. A tenfold cross validation was used resulting a 77.65% of accuracy.

Similarly to [19], ECG recordings were used in Melillo et al. [29] with CART decision tree. This method was able to classify patients according to their risk factors of heart disease and achieved 85.4% of accuracy.

In [2], risk level appeared again, but here with the C4.5 decision tree classifier. C4.5 was used with the dataset in [22] to predict and classify heart failure into 5 risk levels. The results were pretty satisfying with 86.5% sensitivity and 86.53% accuracy.

[4] http://www.hull.ac.uk/.

Sathish et al. in 2015 uses the Pruning Classification Association Rule (PCAR) data mining technique for the prediction of heart disease [30]. It is an efficient approach for mining association rules to predict heart attack. Authors end up by showing the validity of using association rules on medical data, for the association rules have combinatorial nature and can act with medical data records that contains categorical, numerical and time attributes.

Isler in 2016 [31] analyzed the heart rate variability to distinguish patients with systolic Congestive Heart Failure (CHF) from patients with diastolic CHF. Authors performed the classification using a multi-layer perception and the nearest neighbor. The study was performed on a total of 30 patients: 18 of them having systolic CHF and 12 having diastolic CHF. The maximum accuracy is obtained as 96.43% with MLP classifier.

The objective of Shah et al. 2015 [32] was to separate the heart failure with ejection fraction (HFpEF) subtypes. Their study was based on 397 patients. Data was collected from detailed clinical, laboratory, electrocardiographic phenotypes. 67 continuous variables were extracted and given as input to a phenomapping analysis algorithm. The study results in an improved classification of heterogeneous clinical syndromes.

The authors in [33] tried to find the association among frequent and infrequent attributes of a dataset using the Attribute Association. Their main goal was to find the strength or the relation between the symptoms and their frequencies and how their influence in Coronary Vascular Disease (CVD).

Far from the classification and clustering, the well-known Apriori algorithm, Predictive Apriori and Tertius were used on the UCI Cleveland dataset in [34]. First rules based on gender (male or female) were extracted, then other attributes were considered to classify the patients and indicates sick and healthy conditions.

In [35] the authors used both classification modeling techniques, and association classification technique to predict the risk to have a heart failure. For effective heart disease prediction K-means clustering with the decision tree method were applied. Again the Cleveland Clinic Foundation Heart Disease dataset with 13 attributes was used. The maximum prediction accuracy calculated was 83.9% after testing different combinations for the centroid.

13 attributes to different medical profiles were taken from the Cleveland Dataset. The authors in [36] analyzed the extracted data with many algorithms like Decision Trees, Nave Bayes and Neural Network. Moreover, and to find significant relationships in heart diseases, the authors used the Apriori algorithm, along with MAFIA algorithm. And finally the prediction process, for predicting and analyzing the type of heart disease for each patient was based on Decision Trees, Nave Bayes, and Association Rules.

Frequent item sets were used in [37] to predict the heart disease risk. The generation of the frequent itemsets was based on some symptoms and minimum support value given by the user. Experiments were performed on a simulated dataset with 1000 records of patients with 19 attributes. The proposed method showed efficient results in identifying risk level of heart disease in comparison of Apriori, Semi-Apriori, and Association rule mining algorithms.

New approach were adopted by Subramanian et al. [38] to predict Coronary Heart Disease but this time by extracting hidden knowledge from the text documents, using text mining. The collected data was mapped into a structured format and based on the rules generated by the Apriori Algorithm (defined in Sequential Ming section) the predictions were made with efficient and accurate results.

A scoring model that allows the detection of HF and its severity was proposed by Yang et al. [39]. This model is based on Support Vector Machine (SVM) model. Patients were classified into 3 groups according to their heart health: HF group, HF-prone group and healthy group. Results showed a maximum accuracy of 87.5%.

The literature is full of other researches that were conducted in order to predict heart disease and help medical agent to diagnose the severity. These methods include SVM, the C4.5 decision tree classifier, Fuzzy logic, CART algorithm, genetic algorithm, feature subset selection, swarm optimization algorithm, etc. In the current section, we presented a quick overview on some recent existing works that have explored the popular data mining technics to diagnose and predict heart diseases.

4.3 Sequential Mining in Medical Domain

In this section we will present some studies about using the sequential Mining paradigm in the medical field. Authors exploits this technic to help mining medical records and for better decision-making. Sequential mining and its extension will be discussed later on, so the readers will have a clear idea about the algorithms of SM and know which one is appropriate for their medical studies. To our knowledge, the usage of SM algorithms is still limited. Below some of the latest works on SM in the medical field.

Recently in [40, 41] the latest Sequence Prediction algorithm CPT+ was used to predict the existing or the absence of heart failure as well as its classes. The algorithm first analyzes the training sequences and then tries to predict the next element in a new given sequence. Authors had the idea to consider patients vital signs as a sequence of elements. Two datasets were downloaded from the UCI repository. Each record contains 13 input attributes (patients vital signs) and one class attribute. The first dataset was to indicate the presence or the absence of heart failure and was used in [41]. The accuracy achieved is 87.03%. This accuracy proves that a sequence prediction algorithm CPT+ can be useful to predict if a patient has or not a heart disease.

Other experiments were conducted on the second dataset [40]. The last attribute indicates the class of the patient depending to his HF severity. The class or the severity was defined by The New York Heart Association classifies. The heart disease is divided into four different classes based on patient symptoms. In this paper, CPT+, the Sequence prediction algorithm was used to predict to which of the 4 classes of heart disease the patient belongs. The used dataset is the Cleveland Clinic Foundation heart disease dataset. The accuracy of the prediction reaches 90.5%.

The SPM algorithm SPADE [42] was used in [43] in order to predict potential illness one can develop based in past or current data. A medical database (The Health Improvement Network (THIN)) containing medical records of UK was used. The idea was to find rules between age, gender and patients medical history to highlight a susceptible future illness. As a result, additional information is generated presenting the likelihood of re-infection. This knowledge would considerably improve health-care and reduce its costs.

In [44], authors relied on pattern mining approach to analyze electronic health records (EHRs) and classify temporal data. They used real-world clinical records of patients who potentially develop heparin-induced thrombocytopenia. Each record has multiple time series of collected variable for a specific patient e.g. laboratory tests, medication orders and physiological parameters. Additional information such as past diseases, surgical intervention, etc. are also available.

In 2014, the authors of [45] used SPM to predict the next prescribed medications. The CSPADE algorithm, a new version of the well-known SPADE, was used to mine the sequential patterns of diabetes medication prescriptions at different levels: drug class and generic drug level. CSPADE algorithm was used for a test set of patients to identify temporal patterns of medications prescribed for diabetes. The result is temporal relationships from the mined patterns. Resulted knowledge was use to generate rules to predict the next diabetes medication prescribed.

A short list for the usage of SPM in medical fields was presented. All related works and authors agree that it is a very promising approach. For that we are going to detail the SM and its algorithms, as well as the extensions in the next section.

4.4 Sequential Mining

In this part, we will adopt our classification of sequential mining algorithms that classify the algorithms into three main categories: SP Mining, Sequential Rule (SR) Mining, and Sequence Prediction [46]. Three new extensions of SPM will be added to this chapter to cover all the existing techniques and the most important algorithms in this field.

First of all, it is important to mention that all the SM algorithms belonging to the same category, e.g. all the sequential pattern-mining algorithms return the same sequences as result of mining. They only differ in term of performance and memory consumption. Similarly, for the sequence prediction algorithms, all the algorithms of this category return the same next element of a given sequence. In this part, we will discuss the ideas and the techniques on which are based the algorithms. Then, we will present a performance and memory consumption comparison for some important algorithms in the SPM category.

No matter where we want to use the SM approach, before choosing any algorithm, we have to ask those three questions: (1) What is the type of data to be analyzed, Numeric or symbols? (2) What is the type of patterns to be extracted? (3) How this patterns or knowledge will be used?

In the following paragraphs, the reader will be able to construct a general idea of SM algorithm, and will be able to choose the appropriate one for his studies if he is willing to work on SM.

4.4.1 Important Terms and Notations

Before presenting the algorithms and their classification, it is important to define some basic terms used in sequential pattern mining in order to understand the mining process. These terms are commonly used in data mining Processes and especially in Sequential patterns mining.

1. An item is an entity that can have multiple attributes: date, size, color, etc.
2. $I = \{i_1, \ldots, i_n\}$ is a non-empty set of items. A k-item set is an itemset with k items.
3. A sequence "S" is an ordered list of item sets. An itemset X_y in a sequence, with $1 \leqslant y \leqslant L$, is called a transaction. L denotes the length of the sequence, which refers to the number of its transactions. $S = \{(a, b); (b, c); (e, d)\}$ this means that the items a and b are occurring together in the same time, while the items b and c are occurring together although in the same time but after a and b happening together and so on.
4. A Sequential Database (SDB) is a list of sequences with a sequence ID (SID) (cf. Table 4.1).
5. A sequence β can have a sub-sequence α, thus β a super-sequence of α.
6. A sequential rule r, denoted $X \rightarrow Y$, is a relationship between two unordered itemsets $X, Y \subseteq I$ where $X \cap Y = \emptyset$. $X \rightarrow Y$ means that if items of X appear in a sequence, items of Y will also occur in the same sequence.
7. The support of a rule r in a sequence database SDB is defined as the number of sequences that contains $X \cup Y$ divided by the number of sequences in the database:
$$supSDB(r) = \frac{|\{s; s \in \text{SDB} \wedge r \wedge S\}|}{|\text{SDB}|} \tag{4.1}$$
8. The confidence of a rule r in a sequence database SDB is defined as the number of sequences that contains $X \cup Y$, divided by the number of sequences that contains X:

Table 4.1 A sequence database

SID	Sequence
1	$\langle\{a, b\}, \{c\}, \{f, g\}, \{g\}, \{e\}\rangle$
2	$\langle\{a, d\}, \{c\}, \{b\}, \{a, b, e, f\}\rangle$
3	$\langle\{a\}, \{b\}, \{f\}, \{e\}\rangle$
4	$\langle\{b\}, \{f, g\}\rangle$

$$conf\,SDB(r) = \frac{|\{s; s \in \text{SDB} \wedge r \vee S\}|}{|\text{SDB}|} \tag{4.2}$$

9. A rule r is a frequent sequential rule $iff\,supSDB(r) \geqslant minsup$, with $minsup \in [0, 1]$ is a threshold set by the user.
10. A rule r is a valid sequential rule iff it is frequent and $conf\,SDB(r) \geqslant minconf$, with $minconf \in [0, 1]$ is a threshold set by the user.
11. Apriori-based [47]: Many mining algorithms are based on this technique. The main idea is to create a list of the most frequent items with respect to $minsup$ and $minconf$. The list is increased progressively considering the support and the confidence.
12. Sequential rule mining is to find all frequent and valid sequential rules in a SDB [48].
13. Patten Growth [49] is a method for extracting frequent sequences by partitioning the search space and then saving the frequent item sets using a tree structure. Extraction is done by concatenating to the processed sequence (called prefix sequence) frequent items with respect to its prefix sequence. This method can be seen as depth-first traversal algorithm and eliminates the necessity to repetitively scan all the SDB.
14. Searching processes

    - Depth-First Search (DFS) is a searching process that traverses or searches tree or graph data structures. A node in the graph or tree is considered as the root where the search begins. In case of graph, some arbitrary nodes are selected as the root and explores as far as possible along each branch before backtracking.
    - Breadth-First Search (BFS) is a searching process for searching in trees or graph structures. It starts at the root like (DFS) and explores the neighbor nodes first, before starting exploring the next level neighbors.

    Let $\beta = \langle \beta_1 \ldots \beta_n \rangle$ and $\alpha = \langle \alpha_1 \ldots \alpha_m \rangle$ be two sequences where $m \leqslant n$.
15. Sequence α is called the prefix of $\beta iff \forall i \in [1 \ldots m], \alpha_i = \beta_i$
16. Sequence $\beta = \langle \beta_1 \ldots \beta_n \rangle$ is called the projection of some sequence S with regards to α, iff:

    - $\beta \preccurlyeq s$
    - α is a prefix of β
    - There exists no proper super-sequence β of β such that $\beta \preccurlyeq s$ and β also has a prefix

17. Sequence $\gamma = \langle \beta_{m+1} \ldots \beta_n \rangle$ is called the suffix of s with regards to α. β is then the concatenation of α and γ. Let SDB be a sequence database
18. Horizontal database: each entry in a horizontal database is a sequence as shown in Table 4.1.
19. Vertical database: each entry represents an item and indicates the list of sequences where the item appears and the position(s) where it appears [50] (cf. Table 4.2).

Table 4.2 A Vertical database for the sequence database of Table 4.1

a		b		c		d	
SID	Itemsets	SID	Itemsets	SID	Itemsets	SID	Itemsets
1	1	1	1	1	2	1	
2	1,4	2	3,4	2	2	2	1
3	1	3	2	3		3	
4		4	1	4		4	
e		f		g			
SID	Itemsets	SID	Itemsets	SID	Itemsets		
1	5	1	3	1	3,4		
2	4	2	4	2			
3	4	3	3	3			
4		4	2	4	2		

Table 4.3 Projected database with regards to prefix a

$\langle a \rangle$—projected database
$\langle \{\_, b\}, \{c\}, \{f, g\}, \{g\}, \{e\} \rangle$
$\langle \{\_, d\}, \{c\}, \{b\}, \{a, b, e, f\} \rangle$
$\langle \{b\}, \{f\}, \{e\} \rangle$
$\langle \rangle$

20. Projected database: the α-projected database, denoted by $\text{SDB}|_{a}$, is the collection of suffixes of sequences in SDB with regards to prefix α. Table 4.3 shows an example of the projected database considering "a" as prefix.

4.4.2 *Sequential Patterns Mining*

The only difference between Sequential pattern mining and Frequent pattern mining is that the first approach takes into consideration the notion of time, or the order of events. For example, in the diagnosis of a patient with heart disease, when measuring his vital signs such as weight and heart pressure for example, we care about which abnormal sign appears before the other and specifically in the prediction of a heart failure. Nowadays, data mining handles two types of data: sequences and time series. A time-series is an ordered list of numbers; a sequence is an ordered list of symbols [51]. Both of them take a sequence as input. Time series is used more to represent data, as it gives a graph representation for the data. Time series is very useful for temperature analysis representation, stock and items prices. Sequential pattern Mining is better in representing nominal values such an ordered list of items purchased by a customer, a sequence of words, a web page clicks, DNA research sequences

Sequential pattern mining was first introduced by Agrawal and Srikant [48], as a solution for the problem of discovering and mining interesting subsequences in a set of sequences. SPM can also be used with time series even if it was originally designed to support sequences. This is done by converting time-series to sequences using discretization techniques. This conversion can be done using transformation techniques such SAX and iSAX [52, 53] and others [54]. In this chapter we focus on Sequential Mining taking sequences as input. This chapter will be based on our classification that has been detailed in [46]. It classifies the SM into three categories: Sequential Pattern Mining, Sequential Rule Mining and Sequence Prediction. In [46], we mentioned some extensions of SM essential categories, in this chapter we will cover more extensions and put on the spot the techniques on which the algorithms are based.

All the SPM algorithms aim to discover the frequent subsequences which are important to the user. The traditional technique used in algorithms is to calculate the support of each sequence in a given Sequence database to find the one(s) who meet or overcome the minimum support threshold giving by the user. All the SPM algorithms have the same output for a same given database and a same minimum support. The difference between the algorithms is in the method of discovering patterns. A sequence containing n items in a sequence database for example, can have up to 2n – 1 distinct subsequences, then, we need to adopt methods that are efficient and realistic. These methods, including data structures and different strategies discussed further, decide which algorithm is more efficient, more performing and less memory consuming.

4.4.3 General and Specific Techniques Used by SPM Algorithms

In this section we will go from general to specific. As mentioned before, we classified the algorithms in three main categories. In this part, we will discuss the techniques used in these three categories: SPM, Sequential Rule Mining and Sequence Prediction.

Starting from the general, all SPM algorithms follow a lexicographical order, which is the order of processing items in the search spaces (defined in the important terms and definition section).

All the SPM algorithms explore the search space of sequential patterns by performing two basic strategies called s-extensions and i-extensions that are used to generate a $(k+1)$-sequence from a k-sequence.

A sequence $S_a = \{A_1, A_2, \ldots, A_{ni}\}$ is a prefix of a sequence $S_b = \{B_1, B_2, \ldots, B_{mi}\}$, if $n < m$, $A_1 = B_1$, $A_2 = B_2$, ..., $A_{n-1} = B_{m-1}$ and A_n is equal to the first $|A_n|$ items of B_n according to the $\neg$ order [49]. The following example will clarify the idea: the sequence (b) is a prefix of the sequence (a, b), (c), and the sequence $(a)(c)$ is a prefix of the sequence (a), (c, d). S_b is said to be an s-extension of a sequence $S_a = \{I_1, I_2, \ldots, I_n\}$ with an item x, if $S_b = \{I_1, I_2, \ldots, I_h, \{x\}\}$, i.e. S_a is a prefix of S_b and the item x appears in an item set occurring after all the item sets of S_a. For example, the sequences $\{(a), (a)\}$ and $\{(a), (b)\}$ and $\{(a), (c)\}$ are s-extensions of the sequence $\{(a)\}$. A sequence S_c is said to be an i-extension of S_a with an item x, if $S_c = Sa = \{I_1, I_2, ...I_h \cup \{x\}\}$, S_a is a prefix of S_c and the item x is appended to the last item set of S_a, and the item x is the last one in I_h according to the lexicographical order. For example, the sequences $\{(a, b)\}$ and $\{(a, c)\}$ are i-extensions of the sequence $\{(a)\}$.

Referring to [47], the basic SPM algorithms are GSP (1996), SPADE (2001), SPAM (2002), PrefixSpan (2004), LAPIN (2005), CM-SPADE and CM-SPAM (2014). Only the GSP algorithm [48] uses the breadth-first search strategy, while the others use depth-first search strategy. An efficient algorithm must be designed in a way to avoid scanning all the searching space, which is called Pruning. The first pruning mechanism was used by the extension of the well-known Apriori, Apriori-ALL (the first sequential Pattern mining algorithm) [50], and then in GSP.

After AprioriALL, the GSP, an Apriori-like studied SPs. The database in GSP is scanned multiple times. The first pass determines the support of each item, which is the number of data sequences that include the item. It counts the occurrences of singleton transactions (containing one element) in the given database (one scan of the whole database). After this process, non-frequent items are removed, and each transaction consists now of its original frequent items. This result will be the input of the GSP algorithm. Like Apriori, GSP algorithm makes multiple database scans. At the first pass, all single items of length 1 sequences (1-sequences) are counted. At the second pass, frequent 1-sequences are used to define the sets of candidate 2-sequences are mined, and another scan is made to calculate their support. Same process is used to discover the candidate 3-sequences but using frequent 2-sequences, and so on until no more frequent sequences are found. GSP algorithm is composed of two techniques:

1. Candidate Generation: Only candidates with minimum support or above are conserved until no new candidates are found. This technique generates an enormous number of candidate sequences and then tests each one with respect of the user-defined *minsup*. After the first scan of the database and obtaining frequent (k – 1)-frequent sequences F(k – 1), a joining procedure of F(k – 1) with itself is made and any infrequent sequence is pruned if at least one of its subsequences is not frequent.
2. Support Counting: a hash treebased search is used. Finally non-maximal frequent sequences are removed.

The GSP algorithm also allows frequent sequences discovery with time constraints. It can calculate the difference between the end-time of the element just found and the start-time of the previous element. This time is user defined and called maximum and minimum gap. Furthermore, it supports the concept of a sliding window (defines the interval of time between items in the same transaction).

SPADE [42] is an alternative of GSP based on the Eclat algorithm [55], an algorithm for mining frequent item sets. SPADE is based on a vertical id-list database format (refer to Table 4.2) in which each sequence is associated to a list of items in which it appears: each subsequence is originally associated to its occurrence list. The frequent sequences can be found by using the intersection on id-lists. The size of the id-lists is the number of sequences in which an item appears. Thus, the IDList of any pattern allows to directly calculating its support. The support of a pattern Sa is simply the number of distinct sequence identifiers in its IDList. SPADE reduces the search space by aggregating SPs into equivalent classes and thus reduces the execution time. Thereby, two k-length sequences are in the same equivalence class if they share the same $k - 1$ length prefix.

In his first step, SPADE computes the support of length 1 sequences, and this is done in a single database scan. In its second step, SPADE computes the support of 2-sequences and this is done by transforming the vertical representation into a horizontal representation in memory. This counting process is done with one scan of data and uses a bi-dimensional matrix. The idea consists of joining $(n - 1)$ sequences using their id-lists to obtain n-subsequences. If the size of id-list is greater than min-sup, then the sequence is frequent. The algorithm can use a breadth-first or a depth-first search method for finding new sequences. The algorithm stops when no more frequent sequences are found [46].

All the other algorithms, PrefixSpan, CM-SPADE, CM-SPAM, LAPIN adopts although the properties of the vertical database representation, that is what make them faster than GSP and Apriori-All. The ID-Lists approach shows to be very efficient, and better than the breadth-first search approach.

Sequential PAttern Mining, SPAM [56], improved the IDLists technique by introducing the bitmap representation (1 if the item exists and 0 if not) check the paper in [56] for more details. This popular optimization of the IDList structure used in the Spam and BitSpade [56, 57] algorithms is to encode IDLists as bit vectors. SPAM is a memory-based algorithm and uses vector of bytes (bitmap representation) to study the existence (1) or absence (0) of an item in a sequence after loading the database into the memory. Candidates are generated in a tree by s-extension that adds an item in another transaction, and by an I-extension that appends the item in the same transaction. The candidates are verified by counting the bytes with a value of one with the defined min-sup. Depth-first search is used to generate candidate sequences, and various I-step pruning and s-Step pruning are used to reduce the search space, this makes SPAM efficient for mining long sequential patterns. Vertical bitmap representation is used to store the transactional given data, which allows an efficient support counting as well as significant compression using bitmap. One new feature is introduced with SPAM is that it incrementally outputs new frequent itemSets in an online fashion.

New algorithms, CM-SPADE and CM-SPAM where designed to improve the SPAM and BitSpade algorithms: The CM-SPAM and CM-SPADE [58] are respectively the extensions of the two well know algorithms SPAM and SPADE. A new structure called Co-Occurrence MAP (C-MAP) is added. The latter is used to store co-occurrence information by dividing them into CMAPi and CMAPs sub-structures. The first stores the items that succeed each item by i-extension and the second stores the items that succeed each item by s-extension at least minsup times. Let S be the sequence $\{I_1, I_2, \ldots, I_n\}$. An item k is said to succeed by i-extension to an item j in S, $iffj$ and $k \in I_x$ for an integer x such that $1 \leqslant x \leqslant n$ and $k >_{lex} j$. An item k is said to succeed by s-extension to an item j in S, $iffj \in I_v$ and $k \in I_w$ for some integers v and w such that $1 \leqslant v < w \leqslant n$. The i-extension of pattern P with an item x is considered non-frequent if there exist an item i in the last itemset of P such that (i, x) is not in CMAPi. Same for the pruning of s-extension: The s-extension of a pattern P with an item x is infrequent if there exist an item i in P such that (i, x) is not in CMAPs.

PrefixSpan [59] based on FPGrowth algorithm [60] that mines frequent itemset uses a must-discussed approach namely the Pattern-growth. Algorithms that adopt this approach avoid recursively scanning the database to find larger patterns. Thus, they only consider the patterns actually appearing in the database. Performing database scans is costly. Pattern-growth algorithms are based on the concept of projected database that aims to the size of databases as larger patterns are considered by the depth-first search. The algorithm studies the prefix subsequences instead of exploring all the possible occurrences of frequent subsequences. Then, it performs a projection on their corresponding postfix subsequences. Frequent sequences will grow by mining only local frequent patterns showing the efficiency of this algorithm.

Till now, we have covered three main strategies in sequential pattern mining algorithms (most sequential pattern mining algorithms extends these three main strategies): breadth-first algorithms that perform candidate generation used in GSP and Aprioriall, depth-first search algorithms that perform candidate generation using the IDList structure and its variations used in SPADE, SPAM, BitSpade, CM-SPADE and CM-SPAM and finally the pattern-growth strategy used in FreeSpan and PrefixSpan.

The number of patterns in the search space influence directly on the time complexity of SPM and on the cost (memory consumption) of the operations used for generating and processing each itemset or subsequence. The pattern-growth technic used in SPM appears to be more efficient than other methods, because it only considers patterns actually appearing in database. But CM-SPADE outperforms the PrefixSpan algorithm. It may be, because the high cost of scanning the database and performing projections. In [58] the study shows that CM-SPADE is faster than PrefixSpan.

In Table 4.4, we present a comparison between some frequent sequential pattern algorithms based on the dataset existing in [61]. This comparison cover the size of the database, the value of the given minimum support, the execution time I seconds and the memory consumption in MB. This Study was done in our previous work in [46].

Table 4.4 Comparing some sequential pattern extraction algorithms based on [61] and extracted from [47]

Algorithm	Condition		Properties	
	Dataset size	MinSup size	Execution time (s)	Memory consumption (MB)
AprioriAll	AVG	Low	Very slow	Huge
		Medium	Slow	Huge
	Large	Low	Fail	Fail
		Medium	Very slow	Huge
GSP	AVG	Low	>3600	800
		Medium	2126	687
	Large	Low	Fail	Fail
		Medium	Fail	Fail
SPAM	AVG	Low	Fail	Fail
		Medium	136	574
	Large	Low	Fail	Fail
		Medium	674	1052
SPADE	AVG	2.5 times less efficient than SPAM		
	Large	More efficient than SPAM		
CM-SPADE		8 times more efficient than SPADE		
		Small memory overhead with large datasets		
PrefixSpan	AVG	Low	31	13
		Medium	5	10
	Large	Low	1958	525
		Medium	798	320
FreeSpan		Less efficient than PrefixSPan		
		Less memory consumption in case of "disk based projection"		
		Less efficiency in case of "memory based projection"		

4.4.4 Extensions of Sequential Pattern Mining Algorithms

The essential problem in SPM is generating a lot of unwanted patterns. In some applications, we are not interested to extract the whole frequent sequential pattern, or we would like to extract the longest patterns where a subsequence occurs for example. To overcome this problem, some extensions of SPM have been developed. Many of those extensions are discussed in [46] with an overview of the most important algorithms. In the following we will present a quick overview of those extensions and we will mention some new extensions to SPM and sequential rule categories.

Every time we think about new extensions in SM, we will be based on the limitations of the traditional sequential pattern mining algorithms. The basic problem in

SPM is the huge number of patterns that may be found by the algorithms, depending on a databases characteristics and the minsup threshold set by the users.

Outputting a huge number of patterns is a wide issue because users do not have time or the capacity to analyze a large amount of patterns. Hence, a large number of patterns will directly decrease the algorithms performance specifically memory and runtime conditions.

In this section we will introduce the extensions of SPM, giving a light overview of the existing extensions and the idea behind them. The main purpose of the following extensions is to minimize the search space and maximize the efficiency of algorithms according to the need of the applications.

4.4.4.1 Closed Sequential Pattern Mining

CloSpan [62], BIDE+ [63], ClaSP [64] and CM-ClaSP are closed SPM algorithms. To have a light idea about those algorithms, refer to our chapter in [46]. For more details it is recommended to check the references of the papers related to each one.

In general, a Closed Sequential Pattern (CSP) is not necessary included in another pattern having the same support. The set of CSPs is much smaller than the set of SPs making mining more efficient. There exists no super pattern S of pattern S having the same support of S. Then S is a closed sequential pattern, in another word, Closed Pattern Mining means that, for the same support, the mining process will mine the longest Pattern.

4.4.4.2 Generator Sequential Patterns

Generator sequential patterns are the set of sequential patterns that that do not subsequence having the same support:

GS = $\{s_a | s_a \in FS \wedge \nexists s_b \in FSsuchthat s_b \sqsubset s_a \wedge sup(s_a) = sup(s_b)\}$. The output of algorithms based on sequential generators technic is a subset of SP. This subset can be larger, smaller or equal as the set outputted with a closed pattern based algorithms [65]. Generators are the smallest subsequences that describe a group of sequences in a sequence database [66–68].

Thus, Generators are better than other representations according to the Minimum Description Length (MDL) principle [69]. Based on [70], the generators pattern can be also combined with close patterns to generate rules with a minimum antecedent and a maximum consequent. This approach allows obtaining a maximum amount of information based on a lower amount of information.

Many application such market basket analysis and classification have used generators and they showed to be more efficient comparing to the use of all patterns or only closed Patterns [67, 68, 71].

The generators algorithms that use the efficient Pattern-growth technic are Gen-Miner [67], FEAT [71] and FSGP [72]. The most recent algorithm, extended from CM-SPAM named VGEN [66] outperforms FEAT and FSGP.

4.4.4.3 Mining Maximal Sequential Patterns

Mining maximal Sequential patterns is a concept that was introduced as an attempt to solve the problem involving large number of patterns returned by sequential pattern mining. This can become a sever constraint in the eyes of the uses who has to analyze a wide amount of data. It also deprives the concept of data mining from its initial goal of achieving effective results which makes it useless in the eyes of the user. A Maximal SP represents a pattern that is not included in another pattern. Maximal Patterns mining algorithms are shown below: The MaxSP [73] is based on the PrefixSpan algorithm. It helps save time by reducing or eliminating redundancy. The BIDE-like mechanism that checks if a pattern is maximal is used in MaxSP. Although in Maximal Sequential pattern Mining category, we find VMSP [74] that is based on the SPAM search procedure that creates the pattern and detects candidate patterns having same prefix in a recursive or repetitive manner. VMSP adopts three main strategies: Efficient Filtering of Non-Maximal Patterns (EFN), Forward Maximal Extension Checking (FME) and Candidate Pruning by Co-Occurrence Map (CPC).

4.4.4.4 Compressing Sequential Patterns Mining

Compressing sequential patterns mining is a kind of algorithm used to minimize the size of mining results by aiming to reduce redundancies. GoKrimp and SeqKrimp [75] are two examples of compressing SPs mining algorithms. Based on the Krimp algorithm, they explore directly compressing patterns and help choosing the least costly branch by avoiding resource-consuming candidate generations.

SeqKrimp uses a frequent closed SPs mining algorithm to produce candidate patterns. It then absorbs this candidate and returns an adequate subset of compressing patterns. Then, through a greedy approach, calculates the benefits of adding/extending a given pattern from the candidates. This procedure goes on loop until useful patterns run out.

In the other hand, GoKrimp adopts a similar approach but comes as an improved extension of SeqKrimp. First to intervene is a procedure to search and find a set of sequential patterns that compresses the data most. Second to intervene is the Minimum description length principle which indicates that the best model is the one that compresses the data the most. GoKrimp does not rely on any parameters, which makes it stand out from the previously stated patterns mining procedures. Users are thus liberated of setting a basis for the support or taking difficult decisions which can be at the same time inaccurate and time consuming. In addition to that, a dependency test is executed in order to take into account only the related patterns to extend a given pattern. This technique aims to avoid the excessive tests of all possible extensions and makes the GoKrimp more optimal than SeqKrimp.

4.4.4.5 Top-k Sequential Patterns Mining

In SP mining algorithm, this allows the minsup parameter to be tuned to get enough plausible patterns. It can however turn into pain as far as the user-friendliness time factors are concerned. To alleviate this issue, Top-k Sequential Patterns mining algorithms were implemented to return k SPs. TSP (Top-k Closed Sequential Patterns) [76] performs a multi-pass mining to find and grow efficient patterns by relying on the concept of pattern growth and projection based SP mining of PrefixSpan algorithm. It then proceeds to apply a verification phase by checking the minimum length constraint verification, which minimizes the search space.

For its database representation, TKS [77] adopts a vertical bitmap environment. It adapts the SPAM search procedure to explore the search space of patterns to transform it to a top-k algorithm. Then, TSK prolongs the most interesting patterns, meaning that it finds patterns with high support in an early stage and discards infrequent items. It finishes by using PMAP (Precedence MAP) data structure to eliminate any unnecessary events in the search space.

4.4.4.6 High Utility Sequential Pattern Mining

The most popular extension in SPM is High Utility sequential Pattern Mining. This set of frequency-based techniques creates an output consisting of several patterns. Most of these though are not informative enough for any decision making process in general, be it business, medical, data. In recent years, high utility pattern mining has been combined with many algorithms that show potential in the universe of decision making, among which we find: EFIM [4], FHM [5] and FHM+ [74] (for mining high-utility item sets with length constraints). Many other algorithms are gradually emerging and proving to be taken into consideration for their ability to determine patterns which answer common business concerns such as dollar value. For instance, lets talk about the USPAN algorithm which entered the field of SPM: it incorporates utility into sequential pattern mining, and is designed for the purpose of mining high utility sequential patterns. It can extract a full set of high utility sequences by calculating the utility of a node in a lexicographic quantitative sequence tree. Thanks to that, it will be able to provide a highly effective pruning strategy for the node and its children. Tests were made on synthetic and real datasets that USpan efficiently identifies high utility sequences from large-scale data with very low minimum utility.

Adapting the behavior of sequential pattern which was recently established mining into sequential pattern mining aims to solve the issue mentioned in the last sentence. In total, only three papers were found in the literature. UMSP [78] for example was established in this perspective and allows high utility mobile sequential patterns. In this case, each item set in a sequence is associated with a location identifier. This means that the utility in this case is contained in a single value. UMSP searches for patterns within a structure called MTS-Tree, which is practical. Still, this algorithm is slowed down because of the specific constraint on the sequences which permit only the handling of certain sequences with simple structures (single item in each sequence element, and a single utility per item). In [79], the algorithm specifically targets utility web log sequences. The utility of a pattern can be coupled with several values, and the users can decide of the utility through the optimal values, which makes up for a representation with two tree structures. This is the case for UWAS-tree and IUWAS-tree. However, for sequence elements containing multiple items such as [(c, 2) (b, 1)] and cannot be supported, a simple scenario is established and consists of limiting the algorithms applicability for complex sequences. UI and US [80] are based on traditional and usual sequential pattern mining. The calculation of pattern utility can go in two ways: The distinct occurrences of utilities of sequences are grouped together. The rest of the duplicate occurrences are filtered to get the ones with highest values, which will be used to calculate the utilities. However, the problem in [80] is that no generic framework is proposed to make a transition from sequential pattern mining to high utility sequence analysis.

Sequential Pattern Mining algorithms, can only mine frequent sequences or sub-sequences from a given sequence database or dataset. The importance or the usage of these algorithms in medical field can be in term of mining frequent sequences, to use them as training sequences for others SM mining algorithms that are important for prediction and decision-making. Make sure you use the appropriate algorithm for your study and your dataset. For more details check the original papers of the algorithms.

In the following, we will discuss Sequence Prediction and the sequential Rule Mining that are also very important in Sequential mining. In fact, even though SPM is important, but it can only generate frequent sequence depending on the category of the used algorithms. Thus, other extension of SM was introduced to be used in prediction and decision-making

4.4.4.7 Sequential Rules Mining

As stated before, mining frequent sequential patterns does not always prove to be useful for decision-making. Sequential Rule mining for sequence prediction was established as an alternative of SPM to answer this problem. A sequential rule states that in return to every item that occurs in a sequence, there are other items that might occur afterward under a certain probability. Among these we find: CMDeo [81], created to explore rules in a single sequence by engaging in a breadth-first search to determine the space and extract all valid rules of size 1 * 1 respecting minimum support and confidence. Similar to Apriori, CMDeo generates a huge amount of valid rules by applying a left and a right expansion. There is also RuleGrowth [82] which explores sequential rules for more than just one sequence. It is based on the pattern-growth approach in finding the sequential rules that explores rules between two items and expands their left and right. We also find CMRules [83], a rational alternative to CMDeo, since it searches for the association rules to minimize the search environment, and then excludes any rules that do not respect minimum support and confidence. Finally, theres the ERMiner (Equivalence class based sequential Rule Miner) [84] algorithm that offers a database representation based on a vertical representation. It mines the search space through equivalent classes to generate rules with the same antecedent or consequent.

4.4.4.8 Top-k Sequential Rules Mining

Specifying the number of sequential rules to be found can help overcoming the difficulties faced when fine-tuning sequential rules parameters like minsup and minconf.

TopSeqRule [85] was the initiator in dealing with the top-k sequential rules mining. Thanks to its RuleGrowth search strategy integrated with the general process for mining top-k patterns, it manages to generate rules for several sequences. To maximize results, it isolates most interesting rules and narrows the search space by increasing minsup. TNS [86] is on the other hand able to discover the top-k non-redundant sequential rules. It does that by relying on and adaptation of the TopSeqRule to mine the top-k rules.

4.4.4.9 Sequential Rules with Window Size Constraints

Those algorithms are used to mine patterns that are limited within a time interval. TRuleGrowth [49] is based on RuleGrowth algorithm as its name suggests but with a sliding window constraint.

4.4.4.10 Mining High-Utility Sequential Rules Mining

The problem of mining high-utility sequential rule mining is similar to high-utility sequential pattern mining. However, a key advantage of high-utility sequential rule mining is that discovered rules provide information about the probability of a consequent knowing the antecedent. High-utility sequential patterns do not consider the confidence that a pattern will be followed. HUSRM [87] is the latest algorithm, published in 2015 to mine sequential rule. The input of HUSRM is a sequence database with utility information, a minimum utility threshold called min_utility that is a positive integer, and similar to the other sequential rule mining algorithms, minimum confidence threshold (a double value in the [0,1] interval, a maximum antecedent size (a positive integer) and a maximum consequent size (a positive integer).

4.4.4.11 Sequence Prediction

A huge importance is accorded to determining the next element in a sequence. Given a set of sequences, the main purpose of sequence Prediction Algorithms is to predict the next element in a sequence S. this extension of SM can be and showed to be very efficient in medical predictions and classifications The main algorithms in that regard are CPT and CPT+.

CPT (Compact Prediction Tree) [88] is prediction model that guarantees no losses by using all information in the sequence for its prediction. It consists of two phases: training and prediction. The first has a role of compressing the sequences into a prediction tree. A given sequence S is found by finding all sequences that contains the last x items from S regardless of the order and the position. CPT is more efficient than other existent algorithms PPM (Prediction by Partial Matching) [89], DG (Dependency Graph) [90] and All-K-th-Order Markov [91]. CPT+ [92] is a more elaborated version of CPT where Frequent Subsequence Compression (FSC), Simple Branch compression (SBC) and Prediction with improved Noise Reduction (PNR) strategies were added to improve prediction time and accuracy.

4.5 Discussion

Many classification and clustering algorithms are used in medical domain as shown in (Data mining and healthcare section). We also introduced some applications where Sequential Mining has been used and showed a great accuracy in prediction and classification. Medical datasets are based on attributes and values, and any patient diagnosis is influenced by the notion of time, and the order of the happening events. For further work we can think about merging the classification principal with the Sequential Mining Principal, so we respect the notion of time and still benefit from IF-Else-based strategy of Classification algorithms, which take the values of many attributes in order to classify the data.

On other hand, and after introducing the high utility Sequential Rule mining, that has a sequential dataset and a minimum utility value as input, and which helps mining the informative patterns, why not thinking about giving multiple minimum utility values where we have multiple attributes and mining patterns according to those specific minimum utility for each attribute specified by the user. This idea may help in analyzing sequential medical data and also in the prediction of any abnormal state for a patient.

4.6 Conclusion

Data mining is a very wide and important field, discovering and analyzing patterns; Medical data mining appeared recently to help in medical diagnosis and decision-making. In this chapter, we focused in data mining techniques used in the diagnosis of heart diseases and the prediction of Heart Failure one of the most dangerous disease causing the higher percentage of death worldwide. Few studies in medical field use sequential mining principal, due to its importance in prediction and decision making, we chose to present a study about SM and its algorithms in order to implement them in medical purposes and use them in the prediction of heart diseases and heart failure.

In the discussion paragraph, we gave some ideas about the usage of SM in medical field. As future works, we will try to develop a SM algorithm that handles the values of attributes of medical data [42].

References

1. Ponikowski, P., Voors, A.A., Anker, S.D., Bueno, H., Cleland, J.G., Coats, A.J., Falk, V., González-Juanatey, J.R., Harjola, V.P., Jankowska, E.A., et al.: 2016 ESC guidelines for the diagnosis and treatment of acute and chronic heart failure: the task force for the diagnosis and treatment of acute and chronic heart failure of the European society of cardiology (ESC) developed with the special contribution of the heart failure association (HFA) of the ESC. Eur. Heart J. **37**(27), 2129–2200 (2016)
2. Aljaaf, A., Al-Jumeily, D., Hussain, A., Dawson, T., Fergus, P., Al-Jumaily, M.: Predicting the likelihood of heart failure with a multi level risk assessment using decision tree. In: 2015 Third International Conference on Technological Advances in Electrical, Electronics and Computer Engineering (TAEECE), pp. 101–106. IEEE (2015)
3. Cowie, M.: The Heart Failure Epidemic. Medicographia (2012)
4. Son, C.S., Kim, Y.N., Kim, H.S., Park, H.S., Kim, M.S.: Decision-making model for early diagnosis of congestive heart failure using rough set and decision tree approaches. J. Biomed. Inform. **45**(5), 999–1008 (2012)
5. Roger, V.L.: The heart failure epidemic. Int. J. Environ. Res. Public Health **7**(4), 1807–1830 (2010)
6. Hartmann, C., Varshney, P., Mehrotra, K., Gerberich, C.: Application of information theory to the construction of efficient decision trees. IEEE Trans. Inf. Theory **28**(4), 565–577 (1982)
7. Quinlan, J.R.: C4.5: Programs for Machine Learning. Elsevier (2014)
8. Quinlan, J.R.: Induction of decision trees. Mach. Learn. **1**(1), 81–106 (1986)
9. Cortes, C., Vapnik, V.: Support-vector networks. Mach. Learn. **20**(3), 273–297 (1995)
10. Murty, M.N., Devi, V.S.: Bayes Classifier, pp. 86–102. Springer, London (2011)
11. Haykin, S.: Neural Networks: a comprehensive foundation, 2nd edn. Prentice Hall PTR, Upper Saddle River, NJ, USA (1998)
12. Breiman, L., Friedman, J., Stone, C.J., Olshen, R.A.: Classification and Regression Trees. CRC Press (1984)
13. Hall, M., Frank, E., Holmes, G., Pfahringer, B., Reutemann, P., Witten, I.H.: The weka data mining software: an update. SIGKDD Explor. Newsl. **11**(1), 10–18 (2009)
14. Zadeh, L.A.: Fuzzy sets. In: Lotfi A.Z. (ed.) Fuzzy Sets, Fuzzy Logic, and Fuzzy Systems, pp. 394–432. World Scientific (1996)
15. Hartigan, J.A., Wong, M.A.: Algorithm as 136: a K-means clustering algorithm. J. R. Stat. Soc. Ser. C (Appl. Stat.) **28**(1), 100–108 (1979)
16. Ho, T.K.: Random decision forests. In: Proceedings of the Third International Conference on Document Analysis and Recognition, vol. 1, pp. 278–282. IEEE (1995)
17. Ho, T.K.: The random subspace method for constructing decision forests. IEEE Trans. Pattern Anal. Mach. Intell. **20**(8), 832–844 (1998)
18. Suganya, S., Selvy, P.T.: A proficient heart diseases prediction method using fuzzy-cart algorithm. Int. J. Sci. Eng. Appl. Sci. **2**(1) (2016)
19. Masetic, Z., Subasi, A.: Congestive heart failure detection using random forest classifier. Comput. Methods Progr. Biomed. **130**, 54–64 (2016)
20. Goldberger, A.L., Amaral, L.A., Glass, L., Hausdorff, J.M., Ivanov, P.C., Mark, R.G., Mietus, J.E., Moody, G.B., Peng, C.K., Stanley, H.E.: Physiobank, physiotoolkit, and physionet. Circulation **101**(23), e215–e220 (2000)
21. Pandey, A.K., Pandey, P., Jaiswal, K.: A heart disease prediction model using decision tree. IUP J. Comput. Sci. **7**(3), 43 (2013)

22. UCI-Repository: Heart disease dataset, Center for Machine Learning and Intelligent Systems. http://archive.ics.uci.edu/ml/datasets/heart+disease
23. Bashir, S., Qamar, U., Javed, M.Y.: An ensemble based decision support framework for intelligent heart disease diagnosis. In: 2014 International Conference on Information Society (i-Society), pp. 259–264. IEEE (2014)
24. Chaurasia, V., Pal, S.: Early prediction of heart diseases using data mining techniques. Caribb. J. Sci. Technol. **1**, 208–217 (2013)
25. Gharehchopogh, F.S., Khalifelu, Z.A.: Neural network application in diagnosis of patient: a case study. In: 2011 International Conference on Computer Networks and Information Technology (ICCNIT), pp. 245–249. IEEE (2011)
26. Uppin, S., Anusuya, M.: Expert system design to predict heart and diabetes diseases. Int. J. Sci. Eng. Technol. **3**(8), 1054–1059 (2014)
27. Shouman, M., Turner, T., Stocker, R.: Integrating decision tree and k-means clustering with different initial centroid selection methods in the diagnosis of heart disease patients. In: Proceedings of the International Conference on Data Mining (DMIN), The Steering Committee of The World Congress in Computer Science, Computer Engineering and Applied Computing (WorldComp), pp. 1. (2012)
28. Bohacik, J., Kambhampati, C., Davis, D., Cleland, J.: Alternating decision tree applied to risk assessment of heart failure patients. J. Inf. Technol. **6**(2), 25–33 (2013)
29. Melillo, P., De Luca, N., Bracale, M., Pecchia, L.: Classification tree for risk assessment in patients suffering from congestive heart failure via long-term heart rate variability. IEEE J. Biomed. Health Inform. **17**(3), 727–733 (2013)
30. Sathish, M., Sridhar, D.: Prediction of heart diseases in data mining techniques. Int. J. Comput. Trends Technol. **24** (2015)
31. Isler, Y.: Discrimination of systolic and diastolic dysfunctions using multi-layer perceptron in heart rate variability analysis. Comput. Biol. Med. **76**, 113–119 (2016)
32. Shah, S.J., Katz, D.H., Selvaraj, S., Burke, M.A., Yancy, C.W., Gheorghiade, M., Bonow, R.O., Huang, C.C., Deo, R.C.: Phenomapping for novel classification of heart failure with preserved ejection fraction. Circulation **114** (2014)
33. Srinivas, K., Rao, G.R., Govardhan, A.: Analysis of attribute association in heart disease using data mining techniques. Int. J. Eng. Res. Appl. 1680–1683 (2012)
34. Nahar, J., Imam, T., Tickle, K.S., Chen, Y.P.P.: Association rule mining to detect factors which contribute to heart disease in males and females. Expert Syst. Appl. **40**(4), 1086–1093 (2013)
35. Methaila, A., Kansal, P., Arya, H., Kumar, P.: Early heart disease prediction using data mining techniques. Comput. Sci. Inf. Technol. J. 53–59 (2014)
36. Sudhakar, K., Manimekalai, D.M.: Study of heart disease prediction using data mining. Int. J. Adv. Res. Comput. Sci. Softw. Eng. **4**(1) (2014)
37. Ilayaraja, M., Meyyappan, T.: Efficient data mining method to predict the risk of heart diseases through frequent itemsets. Procedia Comput. Sci. **70**, 586–592 (2015)
38. Subramanian, S., Mohanapriya, S., Nagasandhiyalakshmi, B., Shanmugapriya, N.: Prediction of outbreak heart diseases using text mining. Discovery 1070–1077 (2016)
39. Yang, G., Ren, Y., Pan, Q., Ning, G., Gong, S., Cai, G., Zhang, Z., Li, L., Yan, J.: A heart failure diagnosis model based on support vector machine. In: 2010 3rd International Conference on Biomedical Engineering and Informatics (BMEI), vol. 3, pp. 1105–1108. IEEE (2010)
40. Bou Rjeily, C., Badr, G., El Hassani, A.H., Andres, E.: Sequence prediction algorithm for heart failure prediction. In: International Conference e-Health, pp. 109–116 (2017)
41. Bou Rjeily, C., Badr, G., El Hassani, A.H., Andres, E.: Predicting heart failure class using a sequence prediction algorithm. In: 2017 International Conference on Advances in Biomedical Engineering, IEEE (2017)
42. Zaki, M.J.: Spade: an efficient algorithm for mining frequent sequences. Mach. Learn. **42**(1), 31–60 (2001)
43. Reps, J., Garibaldi, J.M., Aickelin, U., Soria, D., Gibson, J.E., Hubbard, R.B.: Discovering sequential patterns in a UK general practice database. In: 2012 IEEE-EMBS International Conference on Biomedical and Health Informatics (BHI), pp. 960–963. IEEE (2012)

44. Batal, I., Valizadegan, H., Cooper, G.F., Hauskrecht, M.: A pattern mining approach for classifying multivariate temporal data. In: 2011 IEEE International Conference on Bioinformatics and Biomedicine (BIBM), pp. 358–365. IEEE (2011)
45. Wright, A.P., Wright, A.T., McCoy, A.B., Sittig, D.F.: The use of sequential pattern mining to predict next prescribed medications. J. Biomed. Inf. **53**, 73–80 (2015)
46. Bou Rjeily, C., Badr, G., El Hassani, A.H., Andres, E.: Overview on Sequential Mining Algorithms and Their Extensions. Springer (2017)
47. Bou Rjeily, C., Badr, G., El Hassani, A.H., Andres, E.: Sequential mining classification. In: IEEE International Conference on Computer and Applications (ICCA), pp. 190–194. IEEE (2017)
48. Srikant, R., Agrawal, R.: Mining sequential patterns: generalizations and performance improvements. In: Advances in Database TechnologyEDBT'96, pp. 1–17 (1996)
49. Fournier-Viger, P., Lin, J.C.W., Kiran, R.U., Koh, Y.S., Thomas, R.: A survey of sequential pattern mining. Data Sci. Pattern Recognit. **1**(1), 54–77 (2017)
50. Agrawal, R., Srikant, R.: Mining sequential patterns. In: Proceedings of the Eleventh International Conference on Data Engineering, pp. 3–14. IEEE (1995)
51. Han, J., Pei, J., Kamber, M.: Data Mining: concepts and techniques. Elsevier (2011)
52. Lin, J., Keogh, E., Wei, L., Lonardi, S.: Experiencing sax: a novel symbolic representation of time series. Data Min. Knowl. Discov. **15**(2), 107–144 (2007)
53. Camerra, A., Palpanas, T., Shieh, J., Keogh, E.: isax 2.0: Indexing and mining one billion time series. In: 2010 IEEE 10th International Conference on Data Mining (ICDM), pp. 58–67. IEEE (2010)
54. Fu, T.c.: A review on time series data mining. Eng. Appl. Artif. Intell. **24**(1), 164–181 (2011)
55. Zaki, M.J.: Scalable algorithms for association mining. IEEE Trans. Knowl. Data Eng. **12**(3), 372–390 (2000)
56. Ayres, J., Flannick, J., Gehrke, J., Yiu, T.: Sequential pattern mining using a bitmap representation. In: Proceedings of the Eighth ACM SIGKDD International Conference on Knowledge Discovery and Data Mining, pp. 429–435. ACM (2002)
57. Aseervatham, S., Osmani, A., Viennet, E.: bitspade: A lattice-based sequential pattern mining algorithm using bitmap representation. In: Sixth International Conference on Data Mining, ICDM'06, pp. 792–797. IEEE (2006)
58. Fournier-Viger, P., Gomariz, A., Campos, M., Thomas, R.: Fast vertical mining of sequential patterns using co-occurrence information. In: Pacific-Asia Conference on Knowledge Discovery and Data Mining, pp. 40–52. Springer (2014)
59. Han, J., Pei, J., Mortazavi-Asl, B., Pinto, H., Chen, Q., Dayal, U., Hsu, M.: Prefixspan: mining sequential patterns efficiently by prefix-projected pattern growth. In: Proceedings of the 17th International Conference on Data Engineering, pp. 215–224 (2001)
60. Han, J., Pei, J., Yin, Y., Mao, R.: Mining frequent patterns without candidate generation: a frequent-pattern tree approach. Data Min. Knowl. Discov. **8**(1), 53–87 (2004)
61. Mabroukeh, N.R., Ezeife, C.I.: A taxonomy of sequential pattern mining algorithms. ACM Comput. Surv. **43**(1), 3 (2010)
62. Yan, X., Han, J., Afshar, R.: Clospan: mining: closed sequential patterns in large datasets. In: Proceedings of the 2003 SIAM International Conference on Data Mining, SIAM, pp. 166–177 (2003)
63. Wang, J., Han, J.: Bide: efficient mining of frequent closed sequences. In: Proceedings of the 20th International Conference on Data Engineering, pp. 79–90. IEEE (2004)
64. Gomariz, A., Campos, M., Marin, R., Goethals, B.: Clasp: an efficient algorithm for mining frequent closed sequences. In: Pacific-Asia Conference on Knowledge Discovery and Data Mining, pp. 50–61. Springer (2013)
65. Lee, Y.S., Yen, S.J.: Incremental and interactive mining of web traversal patterns. Inf. Sci. **178**(2), 287–306 (2008)
66. Fournier-Viger, P., Gomariz, A., Šebek, M., Hlosta, M.: Vgen: fast vertical mining of sequential generator patterns. In: International Conference on Data Warehousing and Knowledge Discovery, pp. 476–488. Springer (2014)

67. Lo, D., Khoo, S.C., Li, J.: Mining and ranking generators of sequential patterns. In: Proceedings of the 2008 SIAM International Conference on Data Mining, SIAM, pp. 553–564 (2008)
68. Pham, T.T., Luo, J., Hong, T.P., Vo, B.: Msgps: a novel algorithm for mining sequential generator patterns. In: International Conference on Computational Collective Intelligence, pp. 393–401. Springer (2012)
69. Barron, A., Rissanen, J., Yu, B.: The minimum description length principle in coding and modeling. IEEE Trans. Inf. Theory **44**(6), 2743–2760 (1998)
70. Fournier-Viger, P., Gomariz, A., Gueniche, T., Soltani, A., Wu, C.W., Tseng, V.S.: SPMF: a java open-source pattern mining library. J. Mach. Learn. Res. **15**(1), 3389–3393 (2014)
71. Gao, C., Wang, J., He, Y., Zhou, L.: Efficient mining of frequent sequence generators. In: Proceedings of the 17th International Conference on World Wide Web, pp. 1051–1052. ACM (2008)
72. Yi, S., Zhao, T., Zhang, Y., Ma, S., Che, Z.: An effective algorithm for mining sequential generators. Procedia Eng. **15**, 3653–3657 (2011)
73. Fournier-Viger, P., Wu, C.W., Tseng, V.S.: Mining maximal sequential patterns without candidate maintenance. In: International Conference on Advanced Data Mining and Applications, pp. 169–180. Springer (2013)
74. Fournier-Viger, P., Wu, C.W., Gomariz, A., Tseng, V.S.: VMSP: Efficient vertical mining of maximal sequential patterns. In: Canadian Conference on Artificial Intelligence, pp. 83–94. Springer (2014)
75. Lam, H.T., Mörchen, F., Fradkin, D., Calders, T.: Mining compressing sequential patterns. Stat. Anal. Data Min. ASA Data Sci. J. **7**(1), 34–52 (2014)
76. Tzvetkov, P., Yan, X., Han, J.: Tsp: Mining top-k closed sequential patterns. Knowl. Inf. Syst. **7**(4), 438–457 (2005)
77. Fournier-Viger, P., Zida, S., Lin, J.C.W., Wu, C.W., Tseng, V.S.: Efim-closed: Fast and memory efficient discovery of closed high-utility itemsets. In: Machine Learning and Data Mining in Pattern Recognition, pp. 199–213. Springer (2016)
78. Shie, B.E., Hsiao, H.F., Tseng, V.S., Philip, S.Y.: Mining high utility mobile sequential patterns in mobile commerce environments. In: International Conference on Database Systems for Advanced Applications, pp. 224–238. Springer (2011)
79. Ahmed, C.F., Tanbeer, S.K., Jeong, B.S.: Mining high utility web access sequences in dynamic web log data. In: 2010 11th ACIS International Conference on Software Engineering Artificial Intelligence Networking and Parallel/Distributed Computing (SNPD), pp. 76–81. IEEE (2010)
80. Ahmed, C.F., Tanbeer, S.K., Jeong, B.S.: A novel approach for mining high-utility sequential patterns in sequence databases. ETRI J. **32**(5), 676–686 (2010)
81. Deogun, J., Jiang, L.: Prediction mining–an approach to mining association rules for prediction. In: Rough Sets, Fuzzy Sets, Data Mining, and Granular Computing, pp. 98–108 (2005)
82. Fournier-Viger, P., Nkambou, R., Tseng, V.S.M.: Rulegrowth: mining sequential rules common to several sequences by pattern-growth. In: Proceedings of the 2011 ACM Symposium on Applied Computing, pp. 956–961. ACM (2011)
83. Fournier-Viger, P., Faghihi, U., Nkambou, R., Nguifo, E.M.: CMRules: mining sequential rules common to several sequences. Knowl. Based Syst. **25**(1), 63–76 (2012)
84. Fournier-Viger, P., Gueniche, T., Zida, S., Tseng, V.S.: Erminer: sequential rule mining using equivalence classes. In: International Symposium on Intelligent Data Analysis, pp. 108–119. Springer (2014)
85. Fournier-Viger, P., Tseng, V.S.: Mining top-k sequential rules. In: International Conference on Advanced Data Mining and Applications, pp. 180–194. Springer (2011)
86. Fournier-Viger, P., Tseng, V.S.: Tns: mining top-k non-redundant sequential rules. In: Proceedings of the 28th Annual ACM Symposium on Applied Computing, pp. 164–166. ACM (2013)
87. Zida, S., Fournier-Viger, P., Wu, C.W., Lin, J.C.W., Tseng, V.S.: Efficient mining of high-utility sequential rules. In: International Workshop on Machine Learning and Data Mining in Pattern Recognition, pp. 157–171. Springer (2015)

88. Gueniche, T., Fournier-Viger, P., Tseng, V.S.: Compact prediction tree: a lossless model for accurate sequence prediction. In: ADMA, vol. 2, pp. 177–188 (2013)
89. Cleary, J., Witten, I.: Data compression using adaptive coding and partial string matching. IEEE Trans. Commun. **32**(4), 396–402 (1984)
90. Padmanabhan, V.N., Mogul, J.C.: Using predictive prefetching to improve world wide web latency. ACM SIGCOMM Comput. Commun. Rev. **26**(3), 22–36 (1996)
91. Pitkow, J., Pirolli, P.: Mining longest repeatin g subsequences to predict World Wide Web surfing. In: Proceedings of USENIX Symposium on Internet Technologies and Systems, pp. 1 (1999)
92. Gueniche, T., Fournier-Viger, P., Raman, R., Tseng, V.S.: CPT+: Decreasing the time/space complexity of the compact prediction tree. In: Pacific-Asia Conference on Knowledge Discovery and Data Mining, pp. 625–636. Springer (2015)

Chapter 5
Machine Learning Methods for the Protein Fold Recognition Problem

Katarzyna Stapor, Irena Roterman-Konieczna and Piotr Fabian

Abstract The protein fold recognition problem is crucial in bioinformatics. It is usually solved using sequence comparison methods but when proteins similar in structure share little in the way of sequence homology they fail and machine learning methods are used to predict the structure of the protein. The imbalance of the data sets, the number of outliers and the high number of classes make the task very complex. We try to explain the methodology for building classifiers for protein fold recognition and to cover all the major results in this field.

Keywords Supervised learning algorithm · Classifier · Features · Protein fold recognition

5.1 Introduction

The information coding system in living systems uses nucleotide sequences in DNA. Each nucleotide carries two bits of information. Thus, the human body operates basing on approximately one gigabyte of information in each cell. The information contained in the DNA strand is transcribed into the language of amino acids, of which there are 20. The DNA codes may be treated as symbols storing information. The coding of information in the world of amino acids is a record of properties. Each amino acid carries a different characteristic expressed by its physico-chemical characteristics. Proteins are biochemical compounds consisting of one or more polypeptides which are single linear polymer chain of amino acids bound together by peptide

---

K. Stapor (✉) · P. Fabian
Silesian University of Technology, Gliwice, Poland
e-mail: katarzyna.stapor@polsl.pl

P. Fabian
e-mail: piotr.fabian@polsl.pl

I. Roterman-Konieczna
Jagiellonian University, Kraków, Poland
e-mail: iroterman@cm-uj.krakow.pl

G. A. Tsihrintzis et al. (eds.), *Machine Learning Paradigms*, Intelligent Systems Reference Library 149, https://doi.org/10.1007/978-3-319-94030-4_5

bonds between the carboxyl and amino groups of adjacent amino acid residues [45]. The sequence of amino acids in a protein is known as primary structure and is defined by the sequence of genes (which is encoded in the genetic code). One of the most distinguishing features of polypeptides is their ability to fold to their native structures, i.e. the 3D (tertiary) structures, spontaneously [3] and thus it follows a decrease of the internal energy of the system [13].

The information contained in amino acids is used for the forming of an appropriate 3D structure of a protein. The distribution of appropriate features in space is needed to determine the function that each protein guarantees for the activity of such a complex system like the human body. Protein's function is strongly influenced by its structure [5, 13, 43, 45].

However, the activity of proteins limited only to the cell is a system which is so complex that it has not been fully reproduced in a computer program. Throughout this multistep process of coding information leading to the construction of the right tool, which is every protein, the transition from the amino acid sequence to the 3D form is still a mystery despite the activities of world scientific centers (including the CASP project, in particular—http://predictioncenter.org/).

Currently, sequencing projects rapidly produce protein sequences, but the number of 3D protein structures increases slowly due to the expensive and time-consuming conventional laboratory methods, namely X-ray crystallography and nuclear magnetic resonance (NMR). Moreover, not all proteins are amenable to experimental structure determination. The protein sequence data banks such as Universal Protein Resource (UniProtKB/TrEMBL) [4] contain now more than 16 000 000 protein sequence entries, while the number of stored protein structures in Protein Data Bank (PDB) [7] is over 126 000 (current statistics is available at: http://www.rcsb.org/pdb/statistics/holdings.do).

This leads to the necessary alternative to experimental determination of 3D protein structures, the computational methods like **ab initio** or homology modeling ones.

Methods of predicting structures based on **homology** (also known as **comparative**) **modeling** solve the problem by referring to structures of proteins of known structure by transferring structural elements for a given sequence occurring in the predicted protein [63]. The search for a set of proteins with similar sequences may be based on the principle of amino acid distribution identification. Another technique is looking for similar sequences, but with restriction to homologous proteins. The assumption for such a strategy lies in the fact that if the evolution kept the given sequence, it was probably also about the behavior of the structure. If the "target" protein (the protein for which the structure is predicted) can be additionally located as part of the phylogenetic tree, then prediction of the structure of such a protein is much easier due to heuristic knowledge of the evolutionary processes. Structures of close evolutionary neighbours are treated as patterns for structure construction of the "target" protein. Determination of the structure of proteins using homologous sequences is carried out based on the genetic algorithm technique [57, 71]. A set of proteins with a similar sequence (including of course the sequence of homologous proteins) is treated as a set of "parents". Fragments of sequences that have correspondingly high similarity are copied to the "child", i.e. to the protein whose structure is determined. If there are many "parents", the right fragment, and therefore the appropriate part of the spatial structure, is inherited from each of them. The structure of the remaining "non-inherited" fragments (with a locally different sequence) is determined basing on other techniques, e.g. for predicting loop structures. To a structure determined in that way, conformational variability is introduced in so-called "mutation" positions, i.e. in positions with lower reproducibility among the compared sequences or in fragments without the equivalent in other sequences. For the "mutation" type position, a random selection of conformation is used (random selection of Phi and Psi angles). Of course, it is possible to combine several methods [59]. But the hurdle exists in homology modeling when the query protein does not have any structure-known homologous protein in the existing databases!

Another task is faced by specialists who try to predict the structure in the **ab initio** system, without referring to other structures and trying to develop a model that would be universal and which would result from the mechanism of the process itself reproducing it in the in silico version. Here the importance of proposing a structure for detailed analysis is critical. Because of the Levinthal paradox [30] searching the conformational space is not feasible. In addition, the duration of the folding process observed experimentally is much shorter [3] than that would be required in systematic search all over the conformational space. There are two possible solutions of the problem of hypervisible character of conformational space. One of them is a technique of simplifying the structure itself, replacing the accordingly grouped atoms with one so-called pseudo-atom [46]. The reduction of structure elements significantly reduces the calculation time. Quickly searched space with a significantly reduced number of variables eliminates unlikely conformations. The danger of such an approach lies in the necessity of using a pseudo-force field equipped with pseudo-parameters—variables and parameters not measurable experimentally. Another technique to reduce the complexity of the system is to reduce the dimensions of the conformational space by introducing a large step for conformational changes, which also significantly reduces the calculation time by limiting space. Simplified structures obtained by means of the aforementioned techniques, also referred to as coarse-grained [44, 48]. After this reduced search in coarse-grained representation the all atom model is applied for selected reasonable set of structures. The model treating the process of folding as a two-stage introduces the concept of an early and late intermediate, determining the conditions and specificity of each of them was proposed in [5, 43, 61]. The specificity of each stage was determined based on experimental research and interpretation of phenomena at the cell and body level [45].

The review of the techniques given above is a justification for looking for a technique that allows **predicting the structure of a polypeptide chain** preferred by a given set of amino acids. Facing this problem, predicting 3D structure of a protein very often is converted to a problem of **protein fold recognition**, mainly using **machine learning based (ml-based) methods**. Fold can be defined as a three-dimensional pattern characterized by a set of major protein secondary structure conformations with certain arrangement and their topological connections (the secondary structure is the characterization of a protein with respect to certain local structural conformations like α-helices, β-sheets and other such as loops, turns and coils). Protein fold recognition methods have taken central stage as fold information could facilitate the identification of a protein tertiary structure and function. Machine learning based methods for protein fold recognition assume that the number of protein folds in the universe is limited, according to [20] about 1000 and therefore, the protein fold recognition can be viewed as a **fold classification problem**: the construction of a **classifier** using the set of protein derived features whose fold (class) is known.

Many different fold recognition classifiers have been proposed with still increasing accuracy and the main aim of this article is to cover all the major results in this exciting field.

The organization of the rest of this paper is as follows. The next, second section explains the main ideas and approaches to supervised learning of classifiers, the single and the ensemble-based ones. Third section is the introduction to the use of deep learning methods for generating new, "deep" features that can be effectively used in pattern recognition systems of new generation. Fourth section deals with the protein feature generation methods dedicated for the purpose of fold classification in the described later systems/methods of fold classification (fifth section). In the last section, current problems as well as the directions for future research in the field of machine learning-based protein fold classification are identified.

5.2 Supervised Learning

Machine learning is a branch of artificial intelligence which is concerned with the development of learning algorithms that allow computers to evolve their behavior based on the empirical data—the examples [1]. Based on the examples a learning algorithm captures characteristics of interest, for example the underlying probability distribution to automatically learn to recognize complex patterns and make intelligent decisions based on data. **Supervised learning** is the machine learning task of inferring a function from labeled set of examples, a training set. In machine learning and statistics, **classification** is the problem of identifying to which of a set of categories a new observation belongs, on the basis of a training set of data containing observations whose category membership is known. In the terminology of machine learning, classification is considered an instance of supervised learning [1].

The **classification problem** [8] can be formally stated as follows. Suppose we have a **training dataset**:

$$U_n = \{(x_1, y_1), \ldots, (x_n, y_n)\}$$

where each $x_i = (x_{i1}, \ldots, x_{id})$ represents an observation, $y_i \in \{1, \ldots, c\}$ is a categorical variable, a class label. We seek for a function $d(x)$ such that the value of $d(x)$ can be evaluated for any new observation x and such that label $\hat{y} = d(x)$ predicted for that new observation x is as close as possible to the true class label y of x. The function $d(x)$ known as **classifier** is an element of some space of possible functions, usually called the hypothesis space. **A classification learning algorithm** is a general methodology that can be used, given a specific training dataset, to learn a specific classifier (we don't make here a distinction between a classier and classification learning algorithm).

The procedure of building a classifier typically comprises the following steps [70]:

(1) data collection (on appropriate features),
(2) data preprocessing (for example normalization, outlier detection),
(3) feature selection/extraction (to avoid curse of dimensionality [8]),
(4) classifier training and validation using classification learning algorithm,
(5) classifier testing to estimate its performance (most frequently by the resampling technique, the k-fold cross-validation [8], or if available, on a separate testing set).

The built classifier can then be used for making predictions on new, unknown observations.

Classifiers come in a great diversity of techniques and algorithms. Below, we only sketch the most representative approaches, for a more complete description see the literature (for example [1, 8, 37, 70]).

Single Classifiers

In the Bayesian decision framework, in order to measure how well a function fits the training data, a **loss function** $L(y, d(x))$ for penalizing errors in prediction is defined. By far, the most common is 0/1 loss function, where all misclassifications are charged a single unit. This leads to a criterion for choosing $d(x)$ as the expected prediction error $L(y, d(x))$ (Bayes classifier):

$$d(x) = \arg\max_i d_i(x) = \arg\max_i f(x|i) \cdot P(i)$$

where $f(x|i)$ is a class-conditional density, $P(i)$ is a priori probability of class i.

Many classification methods are based on the Bayes classifier including parametric and nonparametric ones according to the estimation method used. One of the most popular parametric classifiers are Gaussian, especially, if covariance matrices in all classes are identical, this leads to the popular **Linear Discriminant Analysis** (**LDA**) classifier:

$$d_i(x) = x^T \Sigma^{-1} \mu_i - \frac{1}{2}\mu_i^T \Sigma^{-1} \mu_i + \ln P(i) d_i(x)$$

Among nonparametric Bayes classifiers, the most popular is ***k* nearest neighbor** (**knn**) classifier, in which the predicted class is the most frequently represented among the k nearest neighbors from a training set.

Linear **support vector machine** (**SVM**) binary classifier [8, 70] is defined by the optimal separating hyperplane (OSH), i.e., the one which maximizes the separation margin which is the distance between the hyperplane and the closest training observations (called support vectors). In the case when the data are not linearly separable, a non linear transformation is used to map indirectly the input data vectors into a higher dimensional Hilbert space using a **kernel function** K which leads to a classifier:

$$d(x) = \mathrm{sgn}\left(\sum_{i=1}^{n} \alpha_i y_i K(x_i, x) + b\right)$$

where $0 \leq \alpha_i \leq C$ $(i = 1, \ldots, n)$ are Lagrange multipliers, C is a regularization parameter, b is a constant, all obtained through a numerical optimization during learning. One of the most widely used "standard" kernel functions is the Gaussian kernel:

$$K(x_i, x_j) = \exp\left(-\frac{1}{2\sigma^2}\left\|x_i - x_j\right\|^2\right)$$

where parameter σ means the width of the kernel. The originally defined SVM is a binary classifier and one way for using it in a multi-class classification problem is to adopt standard techniques for combining the results of binary classifiers. The most popular are one versus all and one versus one [8].

The **multi-layer perceptron (MLP)** [8] also termed **feed-forward neural network** is a generalization of the single-layer perceptron. In fact, just three layers (including the input layer) are enough to approximate any continuous function which leads to a classifier of the following form:

$$d_i(x) = f_2\left(\sum_{j=0}^{M} w_{ij}^2 f_1\left(\sum_{r=0}^{d} w_{jr}^1 x_r^0\right)\right) \quad i = 1, \ldots, c$$

where x_r^0 are inputs, w_{ij}^1, w_{jr}^2 are components of two layers of network weights, d is the dimensionality of the input pattern, the univariate functions f_1 and f_2 are typically each set to:

$$f(x) = \frac{1}{1 + e^{-x}}$$

The parameters of the network (i.e. weights) are modified during learning to optimize the match between outputs and targets, typically by minimizing the total square error using a variant of gradient descent which is conveniently organized as **a backpropagation of errors** [8].

A **tree classifier**, also known as a **classification tree** or a **decision tree**, is a formal representation of a sequence of conditions leading to the classification of a test sample. Classification begins in the root node. A condition regarding the feature vector is associated with each inner node of the tree. After determining the value of the condition, the analysis moves along the appropriate edge down the tree to one of the leaves, representing the classification result (one of the classes). In graphical form, the tree is most often represented with rectangle shaped decision nodes and with triangle shaped leaves containing class labels. If each condition divides the set of objects into two separate subsets, we talk about binary classification trees. If each condition concerns only one component of the feature vector, an attribute, then we talk about a **one-dimensional tree**. The condition has a form of comparison of the attribute with a constant value. For **multidimensional** trees, the condition is a combination of many features. The decision tree is constructed in several steps basing on the training set. In the first phase a tree is created by recursively dividing the training set. In the next phase, called pruning, some branches are truncated to improve the accuracy of the classification. The **tree-building/inducing algorithm** has to solve three problems: the choice of attributes in successive steps (this affects the division into subsets), the stop condition for divisions (this affects the size of the tree), the allocation of elements in the leaves to classes. The first problem can be solved by determining the measure of heterogeneity of elements belonging to a given node, e.g. using the Gini index like in the CART method (Classification and Regression Trees) [11]. The second problem (stop condition for divisions) can be solved by assuming a minimum number of elements in a node. Another way to keep the tree small is pruning—removing earlier created branches. The third problem (assigning leaf elements to appropriate classes) is usually done by selecting the most represented class in the leaf.

Ensemble-Based Classifiers

Ensemble methods are considered as one of the most efficient and relevant general techniques to solve a classification problem. In a classifier ensemble, individual (base) classifiers are integrated in some way in order to produce a final classifier that outperforms every one of them.

There is no unifying theoretical framework for ensemble methods. For most of them, there is actually no clear understanding of their exact mechanisms. Basically, the process of creating an ensemble classifier consists of two main stages: **ensemble building** (i.e. providing with a set of base classifiers) and **output combination** (providing a combination method) [29].

Two basic methodologies for building ensemble classifiers can be distinguished.

In the first building methodology, called the **dependent**, the output of a classifier is used to build the next classifier. Two approaches can be used here. In the incremental batch learning procedure, classification of the former classifier is used by the learning algorithm to build the next classifier. In model-guided instance selection procedure, classifier from the previous iteration is used to manipulate the training dataset for the next iteration. The representative example is **boosting**. There are many boosting algorithms. The original ones, proposed by Robert Schapire [66] works by repeatedly running learning algorithm, for example a decision tree inducer on various distributed training datasets and then combining the resulted classifiers. The improved adaptive boosting method, AdaBoost [34] gives more focus to the misclassified examples by assigning them the increasing weight.

In the second building methodology known as **independent**, the base classifiers are first trained on the disjoint or overlapping datasets created by the division of the original training dataset and next their decisions are combined to output the final one. The first representative ensemble method here is **bagging** [9] in which each classifier is trained on a sample of examples taken with a replacement from a training dataset. The final decision of the ensemble is the class that has been predicted the most often (voting approach). The second well known classifier ensemble is **Random Forest** [10] which uses a large number of individual, unpruned decision trees. The individual decision tree is constructed by any top-down decision tree induction algorithm, as for example that described in the previous section, but with the following modification: the decision tree is not pruned and at each node, instead of choosing the best split among all the attributes, the induction algorithm randomly samples subset of the attributes. The majority voting is then used for final decision. There are of course many other methods for randomization of decision trees.

It is generally accepted that each method for creating the ensemble classifier consists of the following four building blocks [60]:

(1) training dataset,
(2) learning algorithm(s) responsible for building base classifiers on a training dataset,
(3) diversity generator responsible for generating diverse classifiers,
(4) combiner for combining the decisions from base classifiers.

The **diversity** of base classifiers (i.e. having different error patterns) is even more important than their high accuracy because extending an ensemble with new classifiers whose performances are identical provides no additional information to the ensemble of classifiers [29].

According to Brown et al. [12], the following approaches to diversity creation in ensembles can be distinguished: (1) manipulating the training sample (each base classifier is trained from a different training set), (2) manipulating the learning algorithm differently for each base classifier, (3) changing the target attribute representation, (4) partitioning the search space, (5) using various learning algorithms or ensemble strategies. The mentioned bagging and boosting are examples of homogenous classifiers which are learned on different training datasets.

Basically, there are two group of methods or combining outputs of the base classifiers in the ensemble: **weighting** and **meta-learning methods**. Weighting methods can be divided into [60] (a) voting or support-based, (b) trainable or untrainable, (c) static or dynamic. Among the most representative weighting methods are majority voting, performance weighting, distribution summation, Bayesian combination, Dempster-Shafer, Naive Bayes [60]. Meta-learning relies on the classifiers built by learning algorithms and classifications of these classifiers on training dataset. Stacking, arbiter trees and grading are the best examples [60].

An important aspect of ensemble methods is to determine how many and which base classifiers should be included in an ensemble. Several algorithms have been proposed (for an overview see [60]).

5.3 Deep Learning Methods in Pattern Recognition

The efficiency of machine learning depends to a large extent on the way how the data (attributes) is represented and passed as input to the classification algorithms. Therefore, the initial stage of processing the raw data usually involves transformations like reduction of the dimensionality of the feature vector, filtering etc. The selection of features and the way how they are transformed is usually not done automatically. A desirable feature of a recognition system would be automatic **"understanding" of low-level data**. This would allow much more efficient construction of classifiers. The deep learning concept assumes transforming features in the subsequent stages of processing so that the results obtained at a given stage are passed as inputs for the next stage. We can talk here about a **hierarchical** representation of data.

Neural networks with many hidden layers are the most commonly used architecture for deep learning. The signal propagation in such networks looks similar to that used in the multi-layer perceptron (MLP) described before. However, we have **many hidden layers** here (more than one). Layers lying close to the input neurons deal with low-level signal processing (e.g. pixel values in the image recognition task). The higher hidden layers are responsible for transformations of features into spaces more and more similar to the assumed result of classification.

For example, in the image recognition task, the initial layer of neurons may deal with low-level filtering of pixel values, the next layer may discover simple shapes—fragments of lines of different directions, the next may combine fragments of lines with similar directions into longer sections, the next may identify geometric figures constructed from lines, next—may classify the obtained shapes.

The basic difference between traditionally understood machine learning and deep learning is the way how features are processed. Traditional ML methods extract features at the initial stage and possibly transform them with algorithms developed by humans. These algorithms are not the result of automatic learning. At the stage of learning, transformed features are passed as input data to the learning algorithm and finally we get a trained classifier. This is called "shallow architecture".

In the case of deep learning methods, the system contains **many layers responsible for transforming features**. Low-level features are subsequently transformed into higher-level features, walking up the hierarchy of "abstraction levels". Only the features obtained at the highest level are passed to the training algorithm or later—recognition algorithm. The method of transforming features may be partially controlled by the network designer, e.g. by pre-assigning weights to connections in individual layers of the neural network. This architecture, containing many hidden layers responsible for different layers of data abstraction, is called "deep" and resembles the "deep architecture" of the human brain. The multilayer architecture of neural networks was known for a long time, as it is an obvious extension of a perceptron. But before the year 2006, the results obtained with more than two hidden layers were disappointing. Training deeper networks gave worse results. Gradient-based algorithms got stuck in local minima and adding layers even worsened the accuracy of classification. Connections between hidden layers away from the output are only slightly modified and therefore are not converting the input feature space well. In [38] a method of training deep networks was presented. The algorithm trains one layer at a time using unsupervised learning.

The multilayer neural network may be treated as a **sequence of simpler networks**: Restricted Boltzmann Machines (RBM) or autoencoders. An autoencoder is a kind of MLP with a hidden layer containing fewer neurons than the input.

Practical tests have shown that deep layer methods implemented in the form of a multilayer neural network cope very well with classification tasks like handwriting recognition, speech recognition and image recognition. During the training phase, connection weights between neurons are iteratively modified to minimize the error. However, the meaning of these weights is usually unknown to the operator. We get a system consisting of a known (designed) architecture of interconnected neurons and knowledge encoded in the form of connection weights and additional parameters. We can treat this system as a "black box" giving correct answers, but it is not easy to tear down the rules that the network has learned.

5.4 Features of the Amino Acid Sequence

For machine learning-based protein fold classification it is necessary to represent the underlying protein as a feature vector, i.e. a vector composed of values of the features representing a protein. To realize that, one of the keys is to find an effective model to represent a sample of a protein, because the performance of a fold classifier critically depends on the features used. Several methods for the extraction of features of amino acid sequences for protein fold classification have been developed.

The most straightforward sequential model relies on representing a query protein as a sequence of consecutive symbols of amino acids in a specific order, but it would fail when the query protein did not have significant homology with proteins of known characteristics. The simplest non-sequential or *discrete model* of a protein $P = R_1 \ldots R_L$ with L amino acid residues R_i can be expressed by its amino acid composition (AAC):

$$P = [f_1, \ldots, f_{20}]$$

where $f_j (j = 1, \ldots, 20)$ are the normalized occurrence frequencies of the 20 native amino acids in P.

Dubchak et al. [31, 33] first proposed a way to extract global physical and chemical propensities of the amino acid sequence as fold discriminatory features. Together with AAC a protein sequence is represented by a set of following 126 parameters divided into six groups: (1) AAC plus the sequence length (21 features collectively denoted by a letter C), (2) the predicted secondary structure (21 features denoted by S), (3) hydrophobicity (21 features denoted as H), (4) normalized van der Waals volume (21 features denoted as V), (5) polarity (21 features denoted by P) and (6) polarisability (21 features denoted by Z).

Table 5.1 Amino acid attributes and corresponding groups

Attribute	Group 1	Group 2	Group 3
Secondary structure	Helix	Strand	Coil
Hydrophobicity	Polar R, K, E, D, Q, N	Neutral G, A, S, T, P, H, Y	Hydrophobic C, V, L, I, M, F, W
Polarizability	(0–2.78) G, A, S, C, T, P, D	(2.95–4.0) N, V, E, Q, I, L	(4.43–8.08) M, H, K, F, R, Y, W
Polarity	(4.9–6.2) L, I, F, W, C, M, V, Y	(8.0–9.2) P, A, T, G, S	(10.4–13.0) H, Q, R, K, N, E, D
Van der Waals volume	(0–0.108) G, A, S, D, T	(0.128–0.186) C, P, N, V, E, Q, I, L	(0.219–0.409) K, M, H, F, R, Y, W

Secondary structural information based on the three-state model: helix, strand and coil could be accomplished using one of the existing methods for secondary structure prediction, for example PSI-PRED [42].

Apart from AAC characteristics (C set of features), all other features were extracted based on the classification of all amino acids into three classes (for example polar, neutral, and hydrophobic for hydrophobicity attribute, see Table 5.1) in the following way. The descriptors a-composition, transition and distribution were calculated for each attribute to describe the global percent composition of each of the three groups in a protein, the percent frequencies with which the attribute changes its index along the entire length of the protein, and the distribution pattern of the attribute along the sequence, respectively. In the case of hydrophobicity, for example, the a-composition descriptor AC consists of three numbers—the global percent compositions of polar, neutral and hydrophobic residues in the protein (because regarding to hydrophobicity attribute, all amino acids are divided into three groups: polar, neutral and hydrophobic). The transition descriptor T consists of the following three numbers—the percent frequency with which: a polar residue is followed by a neutral one or a neutral by a polar residue and similarly with the other two types of residues. The distribution descriptor D consists of five numbers for each of the three groups: the fractions of the entire sequence, where the first residue of a given group is located, and where 25, 50, 75, and 100% of those are contained. The complete parameter vector contains $3aC + 3T + 5 \cdot 3D = 21$ components. Therefore the full feature vector (C, S, H, V, P, Z) counts $6 \cdot 21 = 126$ features.

Pseudo amino acid composition (PseAA) was originally proposed [21] to avoid completely losing the sequence-order information as in AAC-discrete model. In PseAA model the first 20 factors represent the components of AAC while the additional ones incorporate some of its sequence-order information via various modes (i.e., as a series of rank-different correlation factors along a protein chain). The PseAA C-discrete model can be formulated as $20 + \lambda$ components [22]:

$$p_u = \begin{cases} \frac{f_u}{\sum_{i=1}^{20} f_i + w \sum_{k=1}^{\lambda} \tau_k} & 1 \leq u \leq 20 \\ \frac{w \tau_{u-20}}{\sum_{i=1}^{20} f_i + w \sum_{k=1}^{\lambda} \tau_k} & 20 + 1 \leq u \leq 20 + \lambda \end{cases}$$

where w is the weight factor and τ_k the k-th tier correlation factor that reflects the sequence order correlation between all the k-th most contiguous residues:

$$\tau_k = \frac{1}{L-k} \sum_{i=1}^{L-k} J_{i,i+k} \quad (k < L)$$

with

$$J_{i,i+k} = \frac{1}{3} \{ [H_1(R_{i+k}) - H_1(R_i)]^2 + [H_2(R_{i+k}) - H_2(R_i)]^2 + [M(R_{i+k}) - M(R_i)]^2 \}$$

where $H_1(R_i)$, $H_2(R_i)$ and $M(R_i)$ are respectively the hydrophobicity, hydrophilicity and side chain mass values for the amino acid R_i, λ is a parameter (before substituting these values special normalization is used, for details see [22]).

The n-th order amino acid pair composition proposed by Shamim et al. [65] is calculated using the following formula:

$$f\left(D^{i,i+n}\right)_j = \frac{N\left(D^{i,i+n}\right)_j}{L-n}$$

where $N\left(D^{i,i+n}\right)_j$ is the number of the n-th order amino acid pair j ($j = 1, \ldots, 400$) in protein sequence of length L. These features encapsulate the interaction between the i-th and $(i+n)$-th amino acid residues and give the local order information in a protein. A special case of these features are bigram and spaced-bigram features proposed by Huang et al. [39], both derived from the N-gram concept.

Besides the features extracted directly from amino acid sequences, some features are constructed by exploiting information such as predicted secondary structure, predicted solvent accessibility, functional domain and sequence evolution.

Secondary structure-based features are generated based on the (predicted) secondary structure profile, for example generated by PSIPRED method [42]. Such profile comprises a state sequence, i.e. a sequence of the three possible symbols representing states: helix (H), strand (E) and coil (C) and the three probability sequences, each for one state, being the probability values with which the states occur along the query amino acid sequence.

The first examples of these features are those used by Dubchak et al. [31, 33] as described at the beginning of this section. Chen and Kurgan [15] proposed two new features: (1) the number of different secondary structure segments (DSSS), being the numbers of occurrences of distinct helix, strand and coil structures which length is above a certain threshold, (2) the arrangement of DSSS: there are $3^3 = 27$ possible segment arrangements, i.e. class-class-class where *class*= '*H*', '*E*' and '*C*'. Similarly to *n*-th order amino acid pair composition features, Shamim et al. [65] defined the secondary structural state frequencies of amino acids pairs which are calculated as:

$$f\left(D_k^{i,i+n}\right)_j = \frac{N\left(D_k^{i,i+n}\right)_j}{L-n}$$

where $N\left(D_k^{i,i+n}\right)_j$ is the number of the *n*-th order amino acid pair $j\,(j = 1, \ldots, 400)$ found in the state $k{=}(H,\ E,\ C)$. Treating amino acid sequences as a time series Yang and Chen [74] proposed the following procedure for the extraction of new features from the PSIPRED profile. For each of the three state sequences of secondary structural elements they first applied chaos game representation, analyzed them by a nonlinear technique, the recurrence quantification analysis (for details see [74]) and then applied autocovariance (AC) transformation which is the covariance of the sequence against a time-shifted version of itself:

$$AC_{l,t} = \sum_{i=1}^{L-l} (t_i - \bar{t})(t_{i+l} - \bar{t})/(L-l) \quad l = 1, \ldots, l_{\max}$$

where $t = (t_1, \ldots, t_L)$ is the input sequence, $\bar{t}$ is the average of all t_i, l is the distance between two positions along the sequence, $l_{\max}$ is the maximum of l being the value of the shift.

Shamim et al. [65] proposed the solvent accessibility state frequencies of amino acids calculated as follows:

$$f_i^k = \frac{N_i^k}{L}$$

where $k = (B,\ E)$ are solvent accessibility states: *B*—buried, *E*—exposed, N_i^k is the number of amino acid *i* in solvent accessibility state *k*. For calculating the frequencies they used solvent accessibility states predicted by the method [16]. Similarly to *n*-th order of amino acid pair composition features, they defined the solvent accessibility state frequencies of amino acids pairs:

$$f\left(D_k^{i,i+n}\right)_j = \frac{N\left(D_k^{i,i+n}\right)_j}{L-n}$$

where $N\left(D_k^{i,i+n}\right)_j$ is the number of the *n*-th order amino acid pair $j\,(j = 1, \ldots, 400)$ found in the accessibility state $k{=}(B,\ E,\ I)$ (*I*—partially buried state) or secondary structural state $k{=}(H,\ E,\ C)$.

Proteins often contain several modules (domains), each with a distinct evolutionary origin and function. Several databases were developed to capture this kind of information, for example CDD database (version 2.11) [54] which covers 17 402 common protein domains and families. In [69] the functional domain (FunD) composition vector for representing a given protein sample was proposed. It is extracted through the following procedure: (1) use RPS-BLAST (reverse PSI-BLAST [64]) to compare the protein sequence with each of the 17 402 domain sequences in CDD database, (2) if the significance threshold value is less than 0.001 for the i-th profile that means the hit is found and the i-th component of the protein in 17 402-dimensional space is assigned 1, otherwise 0.

Evolutionary-based features mainly are extracted from position-specific scoring matrix (PSSM) profile generated by the PSI-BLAST program [2]. PSI-BLAST aligns a given query amino acid sequence to the NCBI's non-redundant database. Using multiple sequence alignment PSI-BLAST counts a frequency of each amino acid at each position for the query sequence and generates a 20-dimensional vector of amino acid frequencies for each position in the query sequence, thus the element S_{ij} of PSSM matrix reflects the probability of amino acid i occurring at the position j. More often than the absolute frequencies, the relative frequencies are tabulated in a profile (i.e. relative to a probability of a sequence in a random functional site). The generated profile considers evolutionary information, i.e. it can be used to identify key positions of conserved amino acids and positions that undergo mutations.

Chen and Kurgan [15] extracted a profile-based composition vector (PCV) from 20-dimensional PSSM profile in a way by which the negative elements of PSSM profile are first replaced by zeroes, and then each column is averaged. However, in such representation valuable evolutionary information would be definitely lost. To avoid this, Shen and Chou [69] proposed pseudo position-specific scoring matrix (PsePSSM) by adding to the profile-based composition vector the correlation factors defined as:

$$\Phi_j^{\xi} = \frac{1}{L-\xi}\sum_{i=1}^{L-\xi}\left[S_{ij} - S_{(i+\xi)j}\right]^2 \quad j = 1, 2, \ldots, 20; \ \xi < L$$

Φ_j^1 is the correlation factor by coupling the most contiguous PSSM scores along the protein chain for the amino acid type j, Φ_j^2—the same as previous but for the second-most contiguous PSSM scores, and so forth. Another approach is proposed in [74]. Global features are extracted from PSSM matrix by first using a special normalization followed by the consensus sequence (CS) transformation:

$$\mu(i) = \arg\ \max\left\{f_{ij} : 1 \leq j \leq 20\right\} \quad 1 \leq i \leq L$$

where f_{ij} denotes the normalized value of the element S_{ij} of PSSM, and then computing:

$$AACCS(j) = \frac{n(j)}{L} \quad 1 \leq j \leq 20$$

where $n(j)$ is the number of the amino acid j occurring in the CS. Additional two global features represent the entropy of the feature set:

$$ECS = -\sum_{j=1}^{20} AACCS(j) \ln AACCS(j)$$

$$EFM = -\frac{1}{L}\sum_{i=1}^{L}\sum_{j=1}^{20} f_{ij} \ln f_{ij}$$

the last computed on the raw, normalized PSSM. To extract local features, they first divide the raw, normalized PSSM into non-overlapping fragments of equal length. Then, for each fragment s, the 20 features are computed as the average occurrence frequency of the amino acid j in the fragment s during the evolution process (for details see [74]). Each residue in the amino acid sequence has many physical and chemical properties, so a sequence may be viewed as a time sequence of the corresponding properties. In Dong et al. [32] proposed features extracted from PSSM using AC transformation. The result measures the correlation of the same property between two residues separated by a distance l along the sequence:

$$AC(i, l) = \sum_{i=1}^{L-l} \left(S_{ij} - \overline{S}_i\right)\left(S_{ij+l} - \overline{S}_i\right)/(L-l)$$

where i is one of the residues, L is the protein sequence length, S_{ij} is the PSSM score of amino acid i at position j, $\overline{S}_i$ is the average score of amino acid i along the whole sequence. They also proposed the AC transformation for two different properties between two residues separated by l along the sequence:

$$CC(i1, i2, l) = \sum_{i=1}^{L-l} \left(S_{i1j} - \overline{S}_{i1}\right)\left(S_{i2j+l} - \overline{S}_{i2}\right)/(L-l)$$

where $i1$, $i2$ are two different amino acids.

Slightly different from the methods described above are the feature extraction methods based on kernels. A core component of each kernel methods (for example the described SVM) is the kernel function, which measures the similarity between any pair of examples. Different kernels correspond to different notions of similarity and can lead to discriminative functions with different performance. One of the early approaches for deriving a kernel function for protein classification was the SVM-pairwise scheme [50] which presents each sequence as a vector of pairwise similarities to all sequences in the training set. A relatively simpler feature space that contains all possible short subsequences ranging from 3 to 8 amino acids (k-mers) is explored in [47]. A sequence x is represented here as a vector in which a particular dimension u (k-mer) is present in x vector (has non-zero weight) if x contains a substring that differs with u in at most a predefined number of positions (mismatches). An alternative to measuring pairwise similarity through a dot-product of vector representations is to calculate an explicit protein similarity measure. The method [62] measures the similarity between a pair of protein sequences by taking into account all the optimal local alignment scores with gaps between all of their possible subsequences. In the work described in [58] they developed new kernel functions that are derived directly from explicit similarity measures and utilize sequence profiles constructed automatically via PSI-BLAST [2] and employed a profile-to-profile scoring scheme developed by extending profile alignment method [64]. The first kernel function, window-based, determines the similarity between the pair of sequences by using different schemes to combine ungapped alignment scores of certain fixed-length subsequences. The second, local alignment-based, determines the similarity between the pair of sequences using Smith-Waterman alignments and a position independent affine gap model, optimized for the characteristics of the scoring system. Experiments with fold classification problem show that these kernels together with SVM [58] are capable of producing excellent results, the overall performance measured on DD dataset is 67.8%.

Sharma et al. proposed in [67] a bi-gram based feature extraction method. The method is based on counting bi-gram frequencies of occurrences from PSSM derived from PSI-BLAST. Probabilities of transitions between elements of 400 possible pairs of amino acids are computed. The feature vector was then passed to an SVM classifier. The highest accuracy achieved by the bi-gram technique on the DD dataset in this case was 69.5%.

In [14], authors propose a method called proFold with a feature vector composed of four feature groups: DSSP features, the amino acid composition and physicochemical properties (AAsCPP), the PSSM feature group and the functional domain composition (FunD) feature group. DSSP features are derived from secondary structure states, reduced from eight to four groups. There are 40 DSSP-based features: state composition, group composition, number of continuous states, continuous groups etc. The second group contains 188 features combining amino acid composition and physiochemical properties. Next 20 features come from the PSSM matrix, calculated as the average value of each column. The last group contains 17 402 binary features computed as in [54]. These features used with an ensemble classifier gave 76.2% accuracy on the DD dataset and even more on the EDD-dataset and TG-dataset: 93.2% and 94.3% respectively.

5.5 Protein Fold Machine Learning-Based Classification Methods

5.5.1 *Datasets Used in the Described Experiments*

Most implementations of the machine learning-based protein fold classification methods have adopted the **SCOP** (*Structural Classification of Proteins*) architecture [53], with which a query protein is classified into one of the known folds. Most of these methods use the **DD dataset** for the construction of a protein fold classifier. This dataset (training and testing one) was developed by Ding and Dubchak [31, 33]. The DD dataset contains 311 and 383 proteins for training and testing, respectively. This dataset has been formed such that, in the training set, no two proteins have more than 35% sequence identity to each other and each fold has seven or more proteins. In the test dataset, proteins have no more than 40% sequence identity to each other and have no more than 35% identity to proteins of the training set.

The proteins from training and testing datasets belong to 27 different folds (according to SCOP [31]), representing all major structural classes $\alpha, \beta, \alpha+\beta$ and α/β. These are the following 27 fold types: (1) globin-like, (2) cytochrome c, (3) DNA-binding 3-helical bundle, (4) 4-helical up-and-down bundle, (5) 4-helical cytokines, (6) EF-hand, (7) immunoglobulin-like-sandwich, (8) cupredoxins, (9) viral coat and capsid proteins, (10) ConA-like lectins/glucanases, (11) SH-3 like barrel, (12) OB-fold, (13) beta-trefoil, (14) trypsin-like serine proteases, (15) lipocalins, (16) (TIM)-barell, (17) FAD (also NAD)-binding motif, (18) flavodoxin like, (19) NAD(P)-binding Rossmann fold, (20) P-loop, (21) thioredoxin-like, (22) ribonuclease H-like motif, (23) hydrolases, (24) periplasmic binding protein-like, (25) β—grasp, (26) ferredoxin-like, (27) small inhibitors, toxins, lectins. Of the above 27 fold types, types 1–6 belong to all α structural class, type 7–15 to all β class, type 16–24 to α/β class and types 25–27 to $\alpha + \beta$ class.

Later, researchers (see for example [74]) found some duplicate pairs between the training and testing sequences in the DD dataset. After excluding such sequences, a new dataset called *revised DD dataset* (**RDD**) was created. Another extended DD dataset (called **EDD**) was constructed by populating additional protein samples. It is based on the Astral SCOP (http://astral.berkeley.edu/) in which any two sequences have less than 40% identity. To cover more folds, they constructed other datasets, comprising 86, 95, 194 and 199 folds respectively (**F86**, **F95**, **F194**, **F199**); for the detailed description see [32, 74].

5.5.2 *Methods*

Supervised machine learning-based methods for protein fold prediction have gained great interest since the work described in Craven et al. [25]. Craven et al. obtained several sequence-derived features, i.e., average residue volume, charge and polarity composition, predicted secondary structure composition, isoelectric point, Fourier transform of hydrophobicity function, from a set of 211 proteins belonging to 16 folds and used the sequence attributes to train and test the following popular classifiers: decision trees, *k* nearest neighbor and neural network classifiers in the 16-class fold assignment problem.

Ding and Dubchak [31, 33] first experimented with unique one-versus-others and one-versus-all methods using neural networks or SVMs as classifiers in multiple binary classification tasks on a DD dataset of proteins using global description of amino acid sequence described in the previous section. They were able to recognize the correct fold with the accuracy of approximately 56%.

Here, the **accuracy** refers to the percentage of proteins whose fold has been correctly identified on the test set.

Other researchers have tried to improve prediction performance by either incorporating new features (as described in the previous section) or developing novel algorithms for multi-class classification (for example fusion of the different classifiers).

A modified nearest neighbor algorithm called K-local hyperplane (HKNN) was used by Okun [56], with the overall accuracy 57.4% on the DD dataset).

Classifying the same dataset as in Dubchak et al. [33] and input features, employing a Bayesian Network-based approach [35], Chinnasamy et al. [17] improves to 60% on the average fold recognition results reported by Dubchak et al. [33].

Nanni [55] proposed a specialized ensemble called *SE* of K-local hyperplane based on random subspace and feature selection and achieved 61.1% total accuracy on the DD dataset. Classifiers in this ensemble can be built on different subsets of features, either disjoint or overlapping. Feature subsets for a given classifier with a "favourite" class are found as those that best discriminate this class form others (i.e. in the context of the defined distance measure).

For the prediction of protein folding patterns Shen and Chou proposed in [69] the ensemble classifier, known as PFP-Pred, constructed from the nine individual ET-KNN [27] (evidence-theoretic k-nearest neighbors) classifiers, each operating on only one of the inputs (in order not to reduce the cluster-tolerant capacity) and obtained the accuracy 62%. The ET-KNN rule is a pattern classification method based on the Dempster-Shafer theory of belief functions. Near-optimal parameters of each such component classifier were obtained using optimization procedure from [77] resulting in the OET-KNN optimized classifier. As a protein representation they used features from Dubchak et al. [31, 33] (except the composition) as well as the different dimensions of pseudo-amino acid composition, i.e. with four different values of parameter λ (see the description in the previous section), together nine groups of features. Rather than using a combined correlation function they proposed the alternate correlation function between hydrophobicity and hydrophilicity of the amino acids components to reflect sequence-order effects (for details see [69]). The outcomes of the individual classifiers were combined through a weighted voting to give a final determination for classifying a query protein.

Chmielnicki and Stapor [18, 19] proposed a hybrid classifier of protein folds composed of regularized Gaussian and SVM. Using feature selection algorithm to select the most informative features from those designed by Dubchak et al. [31] they obtained accuracy 62.6% on the DD dataset.

In [36] Guo and Gao presented a hierarchical ensemble classifier named GAOEC (Genetic-Algorithm Optimized Ensemble Classifier) for protein fold recognition. As the component classifier they proposed a novel optimized GAET-KNN classifier which uses GA to generate the optimal parameters in ET-KNN to maximize the classification accuracy. Two-layer GAET-KNNs are used to classify query proteins in the 27 folds. As in Dubchak et al. [31] six kinds of features are extracted from every protein in the DD dataset. Six component GAET-KNN classifiers in the first layer are used to get a potential class index for every query protein. According to the result of the first layer, every component classifier of the second layer generates a 27-dimensional vector whose elements represent the confidence degrees of 27 folds. The genetic algorithm is used for generating weights for the outputs of the second layer to get the final classification result. The overall accuracy of GAOEC is 64.7%.

In [32] the protein fold recognition approach using SVM and features containing evolutionary information extracted from PSSM by using the described in the previous section AC transformation is presented. Two kinds of the AC transformation were proposed resulting in two kinds of features: (1) measuring the correlation between the same and (2) the two different properties. Two versions of a classifier were examined: with features (1) and the combination of (1) and (2) resulting in the performance 68.6% and 70.1% respectively (on the DD dataset using 2-fold cross-validation). Using the EDD, F86, F199 datasets the performances computed by 5-fold cross-validation for the combination of (1) and (2) sets of features gain 87.6%, 80.9% and 77.2%, respectively.

Kernel matrices comprise the similarity between data objects within a given input space. Using kernel-based learning methods, the problem of heterogeneous data sources integration can be transformed into the problem of learning the most appropriate combination of their kernel matrices. The approach proposed in [26] "utilizes" four of the state-of-the-art string kernels built for proteins and combines them into an overall composite kernel where the multinomial probit kernel machine operates. The approach is based on the ability to embed each object description via the kernel trick [68] into a kernel space. This produces a similarity measure between proteins in every feature space and then, having a common measure, they combined informatively these similarities into a composite kernel space on which a single multi-class kernel machine can operate effectively. The performance obtained using this method on the DD dataset is 68.1%.

Chen and Kurgan [15] proposed the fold recognition method PFRES using features generated from PSI-BLAST and PSI-PRED profiles (as described in the previous section) and the voting-based ensemble of the three different classifiers: SVM, Random Forests [10] and K-star [23]. Using the entropy-based feature selection algorithm resulting in the compact representation (36 features) [76] they obtained 68.4% accuracy on the DD dataset.

In [51] two-level classification strategy called hierarchical learning architecture (HLA) using neural networks for protein fold recognition was proposed. It relies on two indirect coding features (based on bigram and spaced-bigram as described in the previous section) as well as combinatorial fusion technique to facilitate feature selection and combination. The resulting accuracy is 69.6% for 27 folding classes from the DD dataset. One of the novelties is the notion of a diversity score function between a pair of features. This parameter is used to select appropriate and diverse features for the combination. It may be possible to achieve better results with a combination of more than two features.

Shamim et al. [65] developed a method for protein fold recognition based on an SVM classifier (with three different multi-class methods: one versus all, one versus one and Crammer/Singer method [24]) that uses secondary structural state and the solvent accessibility state frequencies of amino acids and amino acid pairs as feature vectors (as described in the previous section). Among the feature combinations, a combination of the secondary structural state and solvent accessibility state frequencies of amino acids and first-order amino acid pairs gave the highest accuracy 70.5% (measured using 5-fold cross-validation) on the EDD dataset.

Shen and Chou developed new method PFP-FunD for protein fold pattern recognition using functional domain (FunD) composition vector and features extracted from PsePSSM matrix (as described in the previous section) with the previously designed OET-KNN ensemble classifier obtaining accuracy 70.5%.

Yang and Chen [74] developed fold recognition method TAXFOLD that extensively exploits the sequence evolution information form PSI-BLAST profiles and the secondary structure information form PSI-PRED profiles. A comprehensive set of 137 features is constructed as described in the previous section which allows for depiction of both global and local characteristics. They tested different combinations of the extracted features. It follows that PSI-BLAST and PSI-PRED features make complementary contributions to each other, and it is important to use both kinds of features for enhanced protein fold recognition. The consensus sequences contain much more evolution information than the amino acid sequences, thereby leading to more accurate protein fold recognition. The best accuracy of TAXFOLD is 71.5% on the DD dataset (83.2%, 86.9%, 76.5% and 72.6% on RDD, EDD, F95 and F194 datasets, respectively).

In the kernel-based learning method [75] they proposed a novel information-theoretic approach to learn a linear combination of kernel matrices through the use of a Kullback-Leibler divergence between the output kernel matrix and the input one. Based on the position of the input and output kernel matrices, there are two formulations of the resulted optimization problem: by a difference of convex programming method and by a projected gradient descent algorithm. The method improves the fold discrimination accuracy to 73.3% (on the DD dataset).

Another, but different from those described in section two group of methods used for protein fold recognition are those based on Hidden Markov Model (HMM) [35]. The most promising result here was achieved by Deschavanne and Tuffery [28] who employed the hidden Markov structure alphabet as an additional feature and got accuracy 78% (on the EDD database).

Li et al. [49] created the feature vector from sequence-based features: composition vectors, predicted structure descriptors, the secondary structure information and features based on BLAST. Additionally, a method for calculating features based on sequence motifs was proposed. A total number of 60 features was used with a classifier based on random forests. The best result achieved in experiments was 73.7%.

Wei et al. proposed in [72] a novel taxonomic method combining a new feature set with an ensemble classifier. The set of 473 features is created from the sequential evolution information from the profiles of PSI-BLAST, local and global secondary structure information from the profiles of PSI-PRED. The method, called PFPA, achieves an overall accuracy of 73.6%.

The use of Deep Learning Networks brings surprisingly good results, exceeding even 91.2% as shown in [41]. The result was obtained on the LINDAHL dataset [52].

Chen et al. proposed in [14] using a composition of features: amino acids composition, hydrophobicity, Van der Waals volume, polarity, polarizability, charge, surface tension, secondary structure, solvent accessibility (21 features in most groups). The feature vector was passed to an ensemble classifier involving random forests, rotation trees and functional trees. The accuracy reaches 76.2% on the DD dataset or even 93.2 and 94.3% on specific datasets.

Authors of [40] proposed a three-stage framework PCA-DELM-LDA to extract feature vectors from the amino acid sequences. The second stage of their methodology, called DELM (*Deep Extreme Learning Machine*) is a deep model of Extreme Learning Machine (ELM), being a single layer forward network in which the hidden neurons are randomly generated and the output weights are tuned based on the least-squares solutions. DELM is a stack of six layers of ELM (the detailed pseudocode of its training is given in [40]). They applied the six original descriptors from [31] to extract six new feature vectors from protein sequences: autocovariance, Moran, Geary, Moreau Broto autocorrelations, conjoin trias and a local descriptor. Then PCA has been used for dimensionality reduction. The extracted, new feature vectors have been used with the original features by DELM to generate four, new useful „deep features" used by Linear Discriminant Analysis (LDA) to classify an input protein sequence into one of 27 classes. The specificity and sensitivity on the DD dataset is 98% and 52% respectively.

Table 5.2 shows a comparison of results achieved by machine learning methods [73].

5.6 Discussion, Conclusions and Future Work

This paper is not a comparison among the existing protein fold classification methods, it is only a review. To make such a comparison between performance of different classifiers one should implement the described methods and use the special, dedicated statistical tests to make sure that the differences are statistically significant. The presented machine learning-based methods for fold recognition can solve, to a certain degree, the intrinsic limitations of experimental methods (being time-consuming and expensive), but several challenges remain to be addressed.

Table 5.2 Results achieved by ml-based methods, according to [73]

Method	Year	Accuracy (%)
Nanni et al., Ensemble	2006	61.1
PFP-Pred, Ensemble	2006	62.1
Shamim et al., Single (SVM)	2007	60.5
PFRES, Ensemble	2007	68.4
Damoulas et al., Single (SVM)	2008	68.1
ALHK, Ensemble	2008	61.8
GAOEC, Ensemble	2008	64.7
PFP-FunDSeqE, Ensemble	2009	70.5
ACCFold_AC, Single (SVM)	2009	65.3
ACCFold_ACC, Single (SVM)	2009	66.6
Ghanty et al., Ensemble	2009	68.6
TAXFOLD, Single (SVM)	2011	71.5
Alok Sharma et al., Single (SVM)	2012	69.5
Marfold, Ensemble	2012	71.7
Kavousi et al., Ensemble	2012	73.1
PFP-RFSM, Single (RF)	2013	73.7
Feng and Hu, Ensemble	2014	70.2
PFPA, Ensemble	2015	73.6
Deep Learning Networks (T. Jo et al.)	2015	98/52[a]
Feng et al., Ensemble	2016	70.8
ProFold, Ensemble	2016	76.2

[a]Denotes specificity/sensitivity values in this one case

First, it should be noted that the reported accuracies of classifiers were estimated by using different methods: by applying the independent test dataset or 2 (5) cross-validation on the training dataset and sometimes using different datasets. The accuracies obtained using different estimation methods have different bias and variance. Moreover, these are only the values of the point estimators, the confidence intervals for example would give valuable information on the standard deviation of the prediction error (**problem 1**).

Second, the benchmark dataset (e.g. DD dataset) used to evaluate the performance of classifiers is highly imbalanced (for example the largest fold class contains 30 samples while the smallest—only 6 examples). Classifiers learned on such imbalanced and small dataset are easily overfitting (**problem 2**).

Considering the nature of the protein fold prediction problem, where the fold type of a protein can depend on a large number of protein characteristics and also noting that the number of fold types approaches 1000, it is straightforward to see the need for a methodological framework that can cope with a large number of classes and can incorporate as many feature spaces as they are available. As mentioned above, the existing protein fold classification methods produced several fold discriminatory data sources (i.e. groups of attributes such as amino acid composition, predicted secondary structure, and selected structural and physicochemical properties of the constituent amino acids). One of the problems is how to integrate many fold discriminatory data sources systematically and efficiently, without resorting to ad hoc ensemble learning. One solution are the kernel methods that have been successfully used for data fusion in many biological applications. But the problem of integrating heterogeneous data sources is still a challenging problem (**problem 3**).

It is known that performance of many classifiers (like for example widely used SVM) depends on the size of the dataset used—a look at the DD dataset reveals that many folds are sparsely represented. Training in such a case becomes skewed towards populous folds labeled as positive rather than less populated folds labeled as negative (**problem 4**). An alternative class structure should be developed, for example by re-evaluation of the current class structure to determine classes which should be aggregated or discarded, or by incorporating larger sets of folding classes.

Another source of errors in the described methods are inappropriate (i.e. with small discriminatory power) features of the protein sequence. Moreover, the incorrectly predicted features like the secondary structure or solvent accessibility ones, could decrease classification performance. Extracting a set of highly discriminative set of features from amino acid sequences remains a challenging problem (**problem 5**), though the newest work [74] shows, that even using a single classifier but with the carefully designed features (most discriminative), it is possible to obtain a very good classification performance, even greater than 80%. The obtained result is very well acceptable accuracy for the 27-class classification problem! A random classifier would have a 3.7% ($1/27 \times 100$) only. The main source of the achieved improvement is attributed to the application of features extracted from PSI-BLAST profile which considers evolutionary information, with suitable transformations.

New ensemble-based classifiers have demonstrated their classification power in protein fold recognition, achieving accuracy up to 76.2%. The use of deep learning algorithms for generating new "deep" features for classification tasks have been successfully applied in protein fold recognition giving even over 90% correct recognitions for specific data sets. Deep learning networks combined with ensemble-based classifiers will, probably, be a good alternative to improve the efficiency of protein fold recognition. This combination is a challenge for the authors of protein fold recognition methods (**problem 6**).

The probability of correctness obtained with the use of the discussed tools is not satisfactory. This fact may, however, stem from the very essence of the objects to which the method has been applied. Among chemical molecules, proteins demonstrate their desirability in the form of the encoded function they perform. This function has a very high specificity. This means that you probably should not look for a general model based on which very diverse structures could be designed. The division into forms called the secondary structure is a big simplification. Evolution does not create structural forms. She creates tools. The same secondary structural form can be used for many tasks that differentiate it. The treatment of common secondary forms is therefore a fundamental simplification. The creation of secondary forms is aimed at obtaining a corresponding effect, which is not a structure as such. The high similarity of structures with a common name "sandwich" (according to CATH nomenclature 2.60.40.10) turns out to be highly diversified from the point of view of local stability, which significantly affects the flexibility of the structure and the associated function of these domains [6].

The given sequences take the right structure not because of purely energy reasons. Since the function is encoded in the structure, then the status of a given element of the structure (secondary) must be varied despite the topological similarity. With such a highly diversified and sophisticated functionality of proteins, one common model for generating their structures cannot be binding.

References

1. Alpaydin, E.: Introduction to Machine Learning. MIT Press (2009)
2. Altschul, S.F., et al.: Gapped BLAST and PSI-BLAST: a new generation of protein database search programs. Nucleic Acids Res. 3389–3402 (1997)
3. Anfinsen, B.C.: Principles that govern the folding of protein chains. Science, 223–230 (1973)
4. Apweiler, R., Bairoch, A., Wu, C.H., et al.: UniProt: the universal protein knowledgebase. Nucleic Acids Res. D115–D119 (2004)
5. Banach, M., Konieczny, L., Roterman, I.: The late-stage intermediate. In: Protein Folding in Silico, pp. 21–38
6. Banach, M., Konieczny, L., Roterman, I.: The fuzzy oil drop model, based on hydrophobicity density distribution, generalizes the influence of water environment on protein structure and function. J. Theor Biol. 6–17 (2014)
7. Berman, H.M., et al. The protein databank. Nucleic Acids Res. 235–242 (2000)
8. Bishop, MCh.: Pattern Recognition and Machine Learning. Springer, New York (2006)
9. Breiman, L.: Bagging predictors. Mach. Learn. 123–140 (1996)
10. Breiman, L.: Random Forests. Mach. Learn. 5–32 (2001)
11. Breiman, L., Friedman, J., Olshen, R., Stone, C.: Classification and Regression Trees (1984)
12. Brown, G., et al.: Diversity creation methods: a survey and categorization. Inf. Fusion, 5–20 (2005)
13. Chan, H.S., Dill, K.: The protein folding problem. Phys. Today, 24–32 (1993)
14. Chen, D., Tian, X., Zhou, B., Gao, J.: ProFold: protein fold classification with additional structural features and a novel ensemble classifier. BioMed. Res. Int. (2016)
15. Chen, K., Kurgan, L.: PFRES: protein fold classification by using evolutionary information and predicted secondary structure. Bioinformatics, 2843–2850 (2007)
16. Cheng, J.: SCRATCH: a protein structure and structural feature prediction server. Nucleid Acid Res. 72–76 (2005)
17. Chinnasamy, A., Sung, W.K., Mittal, A.: Protein structure and fold prediction using tree-augmented naïve Bayesian classifier. In: Proceedings of PSB, Stanford CA (2004)
18. Chmielnicki, W., Stapor, K.: Protein fold recognition with combined RDA-SVM classifier. Lecture Notes on Artificial Intelligence, pp. 162–169 (2010)
19. Chmielnicki, W., Stapor, K.: A hybrid discriminative/generative approach to protein fold recognition. Neurocomputing, 194–198 (2012)
20. Chothia, C.: One thousand families for the molecular biologist. Nature, 543–544 (1992)
21. Chou, K.C.: Prediction of protein cellular attributes using pseudo-amino acid composition. Proteins, 246–255 (2001)
22. Chou, K.C.: Pseudo amino acid composition and its applications in bioinformatics, proteomics and system biology. Curr. Proteomics, 262–274

23. Clearly, J.G., Trigg, I.E.: K*: an instance-based learner using an entropic distance measure. Proc. Int. Conf. Mach. Learn. 108–114 (1995)
24. Crammer, K., Singer, Y.: On the learnability and design of output codes for multiclass problems. In: 13th Computational Learning Theory Conference, pp. 35–46 (2000)
25. Craven, M.W., Mural, R.J., Hauser, L.J., Uberbacher, E.C.: Predicting protein folding classes without overly relying on homology. In: Proceedings of Intelligent Systems in Molecular Biology (ISMB), pp. 98–106 (1995)
26. Damoulas, T., Girolami, M.: Probabilistic multi-class multi-kernel learning: on protein fold recognition and remote homology detection. Bioinformatics, 1264–1270 (2008)
27. Denoeux, T.: A k-nearest neighbor classification rule based on Dempster-Shafer theory. IEEE Trans. Syst. Man Cybern. 804–813 (1995)
28. Deschavanne, P., Tuffery, P.: Enhanced protein fold recognition using a structural alphabet. Proteins, 129–137 (2009)
29. Dietterich, T.G.: Ensemble methods in machine learning. In: 1st International Workshop on Multiple Classifier Systems, pp. 1–15 (2000)
30. Dill, K.A., Chan, H.S.: From Levinthal to pathways to funnels. Nat. Struct. Biol. 10–19 (1997)
31. Ding, C.H., Dubchak, I.: Multi-class protein fold recognition using support vector machines and neural networks. Bioinformatics, pp. 349–358 (2001)
32. Dong, Q., Zhou, S., Guan, J.: A new taxonomy-based protein fold recognition approach based on autocross-covariance transformation. Bioinformatics, 2655–2662 (2009)
33. Dubchak, I., Muchnik, I. Holbrook, S.R., Kim, S.H.: Prediction of protein folding class using global description of amino acid sequence. Proc. Natl. Acad. Sci. USA, 8700–8704 (1995)
34. Freund, Y., Shapire, R.: A decision-theoretic generalization of online learning and an application to boosting. J. Comput. Sys. Sci. 119–139 (1997)
35. Ghahramani, Z.: An introduction to Hidden Markov Models and Bayesian networks. Int. J. Pattern Recognit. Artif. Intell. 9–42
36. Guo, X., Gao, X.: A novel hierarchical ensemble classifier for protein fold recognition. Protein Eng. Des. Sel. 659–664 (2008)
37. Hastie, T., Tibshirani, R., Friedman, J.: The Elements of Statistical Learning: Data Mining, Inference, and Prediction. Springer (2009)
38. Hinton, G.E., Osindero S., Teh, Y.: A fast learning algorithm for deep belief nets. Neural Comput. 1527–1554 (2006)
39. Huang, C.D., Lin, C.T., Pal, N.R.: Hierarchical learning architecture with automatic feature selection for multiclass protein fold classification. IEEE Trans. Nanobiosci. 221–232 (2003)
40. Ibrahim, W., Abadeh, M.S.: Extracting features from protein sequences to improve deep extreme learning machine for protein fold recognition. J. Theor. Biol. 1–15 (2017)
41. Jo, T., Hou, J., Eickholt, J., Cheng, J.: Improving protein fold recognition by deep learning networks. Sci. Rep. (2015)
42. Jones, D.T.: Protein secondary structure prediction based on position-specific scoring matrices. J. Mol. Biol. 195–202 (1999)
43. Jurkowski, W., Baster, Z., Dulak, D., Roterman, I.: The early-stage intermediate. In: Protein Folding in Silico, pp. 1–20 (2012)
44. Kmiecik, S., Gront, D., Kolinski, M., Wieteska, L., Dawid, A.E., Kolinski, A.: Coarse-grained protein models and their applications. Chem. Rev. 7898–7936 (2016)
45. Konieczny, L., Roterman-Konieczna, I., Spólnik, P.: The structure and function of living organisms. Syst. Biol. 1–32 (2013)
46. Krupa, P., Sieradzan, A.K., Rackovsky, S., Baranowski, M., Olldziej, S., Scheraga, H.A., Liwo, A., Czaplewski, C.: Improvement of the treatment of loop structures in the UNRES force field by inclusion of coupling between backbone- and side-chain-local conformational states. J. Chem. Theory Comput. (2013)
47. Leslie, C.S., et al.: Mismatch string kernels for discriminative protein classification. Bioinformatics, 467–476 (2004)
48. Levitt, M.: Accurate modeling of protein conformation by automatic segment matching. J. Mol. Biol. 507–533 (1992)

49. Li, J., Wu, J., Chen, K.: PFP-RFSM: protein fold prediction by using random forests and sequence motifs. J. Biomed. Sci. Eng. 1161–1170 (2013)
50. Liao, L., Noble, W.S.: Combining pairwise sequence similarity and support vector machines for detecting remote protein evolutionary and structural relationships. J. Comput. Biol. 857–868 (2003)
51. Lin, K.L., Lin, C.Y., Huang, C.D., Chang, H.M., Yang, C.Y., Lin, C.T., Hsu, D.F.: Feature selection and combination criteria for improving accuracy in protein structure prediction. IEEE Trans. NanoBiosci. 186–196 (2007)
52. Lindahl, E., Elofsson, A.: Identification of related proteins on family, superfamily and fold level. J. Mol. Biol. 613–625 (2000)
53. Lo Conte, L., Ailey, B., Hubbard, T.J.P., Brenner, S.E., Murzin, A.G., Chothia, C.: SCOP: a structural classification of protein database. Nucleic Acids Res. 257–259 (2000)
54. Marchler-Bauer, A., et al.: CDD: a conserved domain database for interactive domain family analysis. Nucleid Acid Res. D237–D240 (2007)
55. Nanni, L.: A novel ensemble of classifiers for protein fold recognition. Neurocomputing, 2434–2437 (2006)
56. Okun, O.: Protein fold recognition with k-local hyperplane distance nearest neighbor algorithm. In: Proceedings of the Second European Workshop on Data Mining and Text Mining in Bioinformatics, pp. 51–57 (2004)
57. Pedersen, J.T., Moult, J.: Genetic algorithms for protein structure prediction. Curr. Opin. Struct. Biol. 227–231 (1996)
58. Rangwala, H., Karypis, G.: Profile-based direct kernels for remote homology detection and fold recognition. Bioinformatics, 4239–4247 (2005)
59. Rashid, M.A., Newton, M.A.H., Hoque, M.T., Sattar, A.: Mixing energy models in genetic algorithms for on-lattice protein structure prediction. BioMed. Res. Int. (2013)
60. Rokach, L.: Ensemble-based classifiers. Artif. Intell. Rev. 1–39 (2010)
61. Roterman, I., Bryliński, M., Konieczny, L., Jurkowski, W.: Early-stage protein folding—in silico model. Recent Adv. Struct. Biol. (2007)
62. Saigo, H., et al.: Protein homology detection using string alignment kernels. Bioinformatics, 1682–1689 (2004)
63. Sali, A., Blundell, T.L.: Comparative protein modelling by satisfaction of spatial restraints. J. Mol. Biol. 779–815 (1993)
64. Schaffer, A., et al.: Improving the accuracy of PSI-BLAST protein database searches with composition-based statistics and other refinements. Nucleid Acids Res. 2994–3005 (2001)
65. Shamim, M., et al.: Support vector machine-based classification of protein folds using the structural properties of amino acid residues and amino acid residue pairs. Bioinformatics, 3320–3327 (2007)
66. Shapire, R.: The strength of weak learnability. Mach. Learn. 197–227 (1995)
67. Sharma, A., Lyons, J., Dehzangi, A., Paliwal, K.: A feature extraction technique using bi-gram probabilities of position specific scoring matrix for protein fold recognition. J. Theor. Biol. 41–46 (2013)
68. Shawe-Taylor, J., Cristiannini, N.: Kernel Methods for Pattern Analysis. Cambridge University Press (2004)
69. Shen, H.B., Chou, K.C.: Predicting protein fold pattern with functional domain and sequential evolution information. J. Theor. Biol. 441–446 (2009)
70. Stapor, K.: Classification methods in computer vision (in Polish). Scientific Publishing House PWN, Warsaw (2011)
71. Unger, R., Moult, J.: Genetic algorithms for protein folding simulations. J. Mol. Biol. 75–81 (1993)
72. Wei, L., Liao, M., Gao, X., Zou, Q.: Enhanced protein fold prediction method through a novel feature extraction technique. IEEE Trans. Nanobiosci. 649–659
73. Wei, L., Zou, Q.: Recent progress in machine learning-based methods for protein fold recognition. Int. J. Mol. Sci. (2016)

74. Yang, J.-Y., Chen, X.: Improving taxonomy-based protein fold recognition by using global and local features. Proteins, 2053–2064 (2011)
75. Ying, Y., Huang, K., Campbell, C.: 2009. Enhanced protein fold recognition through a novel data integration approach. BMC Bioinformat. 267–287
76. Yu, L., Liu, H.: Feature selection for high-dimensional data: a fast correlation-based filter solution. In: Proceedings of 10th International Conference Machine Learning, pp. 856–863
77. Zouhal, L.M., Denoeux, T.: An evidence-theoretic kNN rule with parameter optimization. IEEE Trans. Syst. Man Cybern. 263–271 (1998)

Chapter 6
Speech Analytics Based on Machine Learning

Grazina Korvel, Adam Kurowski, Bozena Kostek and Andrzej Czyzewski

Abstract In this chapter, the process of speech data preparation for machine learning is discussed in detail. Examples of speech analytics methods applied to phonemes and allophones are shown. Further, an approach to automatic phoneme recognition involving optimized parametrization and a classifier belonging to machine learning algorithms is discussed. Feature vectors are built on the basis of descriptors coming from the music information retrieval (MIR) domain. Then, phoneme classification beyond the typically used techniques is extended towards exploring Deep Neural Networks (DNNs). This is done by combining Convolutional Neural Networks (CNNs) with audio data converted to the time-frequency space domain (i.e. spectrograms) and then exported as images. In this way a two-dimensional representation of speech feature space is employed. When preparing the phoneme dataset for CNNs, zero padding and interpolation techniques are used. The obtained results show an improvement in classification accuracy in the case of allophones of the phoneme /l/, when CNNs coupled with spectrogram representation are employed. Contrarily, in the case of vowel classification, the results are better for the approach based on pre-selected features and a conventional machine learning algorithm.

G. Korvel · A. Kurowski · A. Czyzewski
Faculty of Electronics, Telecommunications and Informatics, Multimedia Systems Department, Gdańsk University of Technology, G. Narutowicza 11/12, 80-233 Gdańsk, Poland

G. Korvel
Institute of Data Science and Digital Technologies, Vilnius University, Akademijos str. 4, LT-04812 Vilnius, Lithuania

A. Kurowski
Faculty of Electronics, Telecommunications and Informatics, Gdańsk University of Technology, G. Narutowicza 11/12, 80-233 Gdańsk, Poland

B. Kostek (✉)
Faculty of Electronics, Telecommunications and Informatics, Audio Acoustics Laboratory, Gdańsk University of Technology, G. Narutowicza 11/12, 80-233 Gdańsk, Poland
e-mail: bozenka@ssound.eti.pg.gda.pl; bokostek@audioacoustics.org

G. A. Tsihrintzis et al. (eds.), *Machine Learning Paradigms*, Intelligent Systems Reference Library 149, https://doi.org/10.1007/978-3-319-94030-4_6

6.1 Introduction

Speech signal, the most natural way of human communication, is characterized by phonemic variation, temporal structure, prosody, voice timbre and quality. It encompasses also aspects of the speaker's profile such as emotions or sentiments [1–4]. All these elements are interdependent and gathered in one-dimensional signal. The most important task of the speech-related research is automatization and detection of the above-mentioned elements of speech. In the speech domain, vectors of parameters are extracted and machine learning methods are used for classifying each problem, separately. Well-known methods such as Hidden Markov Models, Nearest Neighbors, Support Vector Machines, Artificial Neural Networks, etc. and combinations of the above are employed for that purpose [2, 5, 6]. It should also be mentioned that a renewed interest in phonemic-level-based analyses appeared recently [7, 8], that are applied to various areas, such as for example biometry [9].

Before speech classification takes place, a thorough analysis of speech elements is first performed. To that end speech analytics, defined as a solution that can automatically discover and analyze words, phrases, categories and themes spoken during calls to reveal hidden insights to help surface valuable intelligence, is employed. In this chapter we are looking at speech analytics from a new perspective. The primary goal of the experiments carried out is an analysis, that checks whether a typical approach to automatic speech recognition based on machine learning retains information described above after feature extraction and compares it to the deep learning approach, in which feature extraction process is discarded. For that purpose, standard speech acoustic parameters along with descriptors from the music information retrieval (MIR) domain are used. We also explore Deep Neural Networks (DNNs) approach without performing feature extraction. The DNNs let us combine the two last above mentioned phases: feature extraction and application of soft computing methods into one step. We convert the audio data to different spaces and then export them as images. These two approaches are applied to phoneme and allophone classification.

6.2 Speech Phoneme Signal Analysis

In this section speech phoneme analysis is first performed. We consider speech signal as an acoustic signal. We assume that each phoneme can be represented as a vector of acoustic features.

First of all, before we move onto the extraction of characteristics of the speech signal, we should show to what extent phoneme groups differ from each other. In a general case, the character of vowel and semivowel phonemes is periodic. An example of these phoneme signals is given in Fig. 6.1.

From Fig. 6.1 we see, that the periods of vowel and semivowel phoneme signals are not exactly the same, but they are very similar. Meanwhile, some of the consonant phonemes (e.g. /p/, /k/, /t/) can be considered as quasi-periodic signals in noise, and others (e.g. /s/, /ʃ/, /f/) are aperiodic signals (see Fig. 6.2).

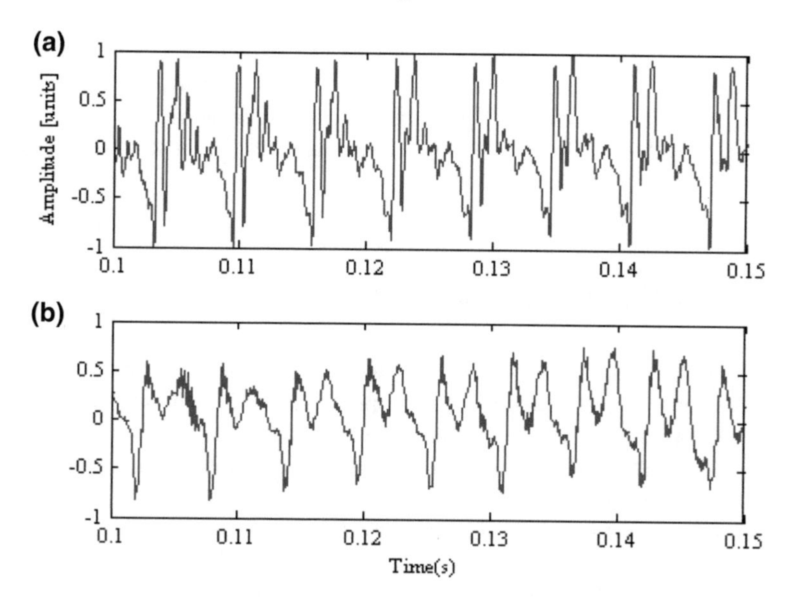

Fig. 6.1 The periodic character of phonemes: **a** the plot of the vowel /a/, **b** the plot of the semivowel /l/

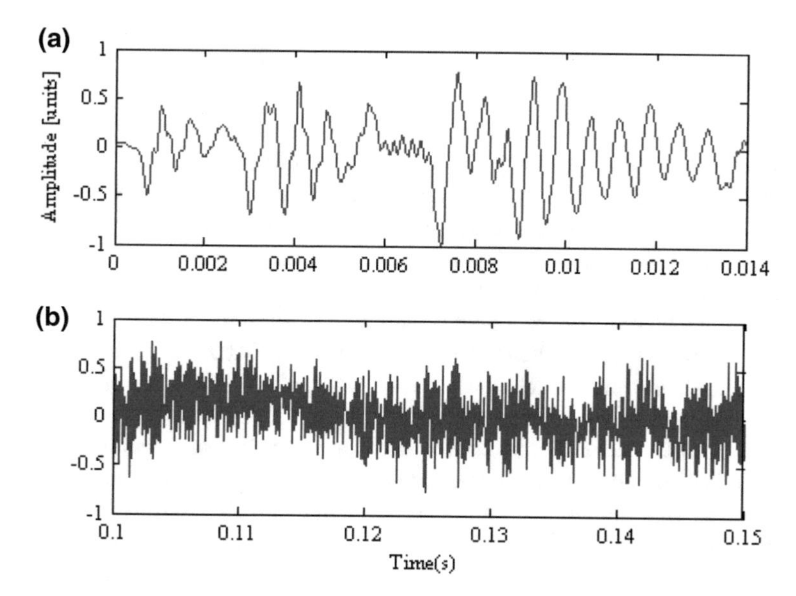

Fig. 6.2 The consonant phonemes: **a** the plot of the consonant /k/, **b** the plot of the consonant /s/

We see from examples shown in Figs. 6.1 and 6.2 that the character of phonemes differs. Moreover, it is worth mentioning that phonemes have different variants depending on the environment in which they occur. One should recall that phonetically distinct variants of a phoneme are allophones, the occurrence of which is usually determined by its position in the word. Therefore, in the context of speech analytics, we should determine all properties of phonemes such as voiced, unvoiced, noise like, periodic, etc. In addition, the same phonemes may also differ depending on the type of a sentence.

6.3 Speech Signal Pre-processing

The state-of-the-art methods applied to speech analysis involve several steps. First of all, the signal is converted to the appropriate space domain and preprocessing is carrying out. In the second step, an extraction of features is performed. Finally, Artificial Intelligence Algorithms (AIA, i.e. machine learning) are used to obtain new knowledge from these features.

In this chapter we consider utterances recorded with the following parameters: PCM 48 kHz; 16 bit; stereo, stored as a wav file. The sampling frequency is 48 kHz, which means that 48,000 samples are captured per second in order to represent the waveform. Since both channel data are identical, we examined a single channel data.

Let:

$$x = [x(1),\ x(2),\ \ldots\ , x(N)]^T \tag{6.1}$$

be a sequence of samples of the analyzed phoneme, where N is the number of samples.

Our goal is to extract important speech acoustic information from the data gathered. Extracting the useful information from speech signal increases the performance of the application. A broad scope of speech signal features is used for classification tasks [10], automatic speech recognition [11], emotion recognition [2], speech modeling [12], and other applications [13].

Since in the literature there is a great variety of features that have been introduced during the last years, the focus is to be on features that have successfully been used in Music Information Retrieval (MIR) systems [14]. Our performed research show that using standard speech parameters along with descriptors from the music area, the phoneme recognition accuracy is better, regardless of singular and specific features of voice exhibited by a speaker [11]. These speech signal features can be divided into two main groups: time- and frequency-domain features. Time domain representation shows the time-varying behavior of signal, while the frequency domain representation shows how the energy of the signal is contained within the frequency range. Features, extracted in one domain provide domain-related information, thus they are not universal.

Feature extraction is to be performed on a signal divided into short-time segments. We use this approach, which is typical for audio analysis and signal analytics, in order to get more accurate information. The segment-based features yield a short time description of the signal at any given time instance. It is to be noted that the choice of the time segment length affects two aspects: quality of spectral coefficients

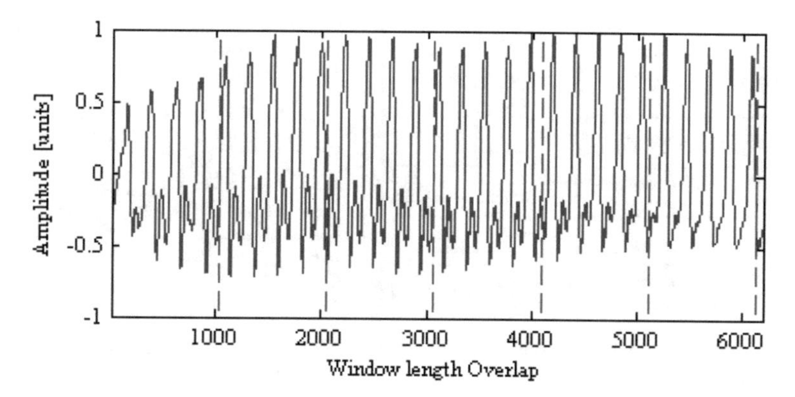

Fig. 6.3 Dividing speech phoneme into short-time segments

and temporal resolution. Longer segments give a higher frequency resolution of the Fourier spectrum. Therefore, derived coefficients are more accurate. As a result of shorter segments, we obtain smoothing coefficients. This provides an entirely different perspective. In the case of longer segments greater time intervals between each of them are formed causing worse temporal resolution. Shorter segments give us a finer resolution in time. Choosing the optimum segment length depends on a practical application. If time domain information is to be more accurate, then shorter segment should be applied and the other way around. Dividing speech signal into segments is shown in Fig. 6.3.

Denoted by M is the number of the samples of the segment and by L is the number of intervals. Therefore, the sequence of samples of the analyzed phoneme short-time segment is as follows:

$$x_l = [x_l(1), x_l(2), \ldots, x_l(M)]^T \tag{6.2}$$

where the index $l = 1 \ldots L$.

The overview of processing steps for speech segments is given in the next section.

6.4 Speech Information Retrieval Scheme

One of the main research issues is the extraction of significant information from the speech signal. In this chapter two feature extraction ways are considered. The first one is based on the speech analysis employing acoustic feature extracting, followed by selected soft computing methods, as shown in Fig. 6.4. The speech signal is divided into L segments, and a feature vector (a vector of K acoustic coefficients) is extracted from each of these segments.

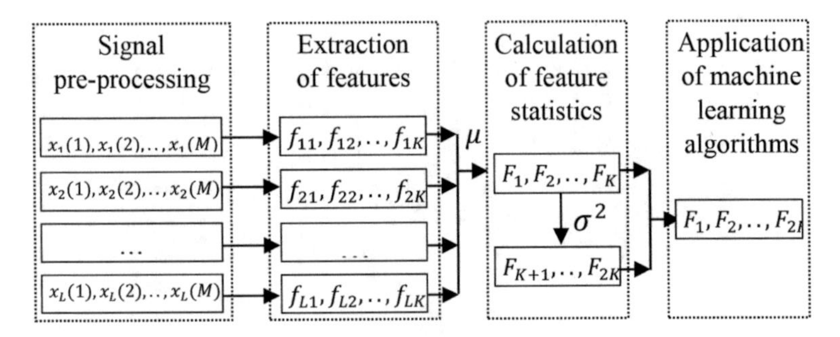

Fig. 6.4 Speech signal information retrieval based on recognition based on feature extraction and machine learning approach

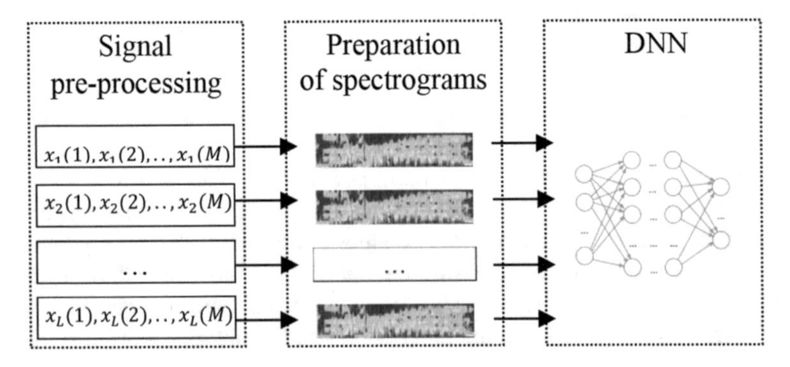

Fig. 6.5 Speech signal information retrieval based on a deep learning approach

Features $f_{lk}(1 \leq l \leq L, 1 \leq k \leq K)$ presented in Fig. 6.4 are derived from the MIR domain and are given in more detail in Sect. 6.5. It is shown in the literature that additional features are often found by computing the first and second order derivatives and by applying linear regression over 3–7 consecutive frames [15]. In Fig. 6.4, it is proposed to calculate mean and variance values of the features extracted.

In the further part of this chapter deep learning is considered without actually speech feature extracting. Deep learning has shown good results in various speech analysis applications such as speech recognition, emotion recognition, and others [1, 3, 4]. Inspired by the success of this approach, we propose a method of speech signal preparation for deep learning. An illustration of the application of a deep learning approach for phoneme signal is given in Fig. 6.5.

As seen in Fig. 6.5, Deep Neural Network models learn features directly from the spectrogram. The preparation of spectrogram for deep learning approach is described in more detail in Sect. 6.6.

6.5 Feature Extraction

As already mentioned, an essential step of speech phoneme analysis is feature extraction. The focus in this paper is on the most frequently used acoustic features borrowed from the music domain. In this section, we outline the application of these features for the speech signal.

6.5.1 Time Domain Features

Time domain features reflect the variation of signal amplitude with time. The most known time domain features are Temporal Centroid (TC), Zero Crossing Rate (ZCR), Root Mean Square (RMS). Since these parameters are commonly used in speech and music signal analysis and are widely presented in the rich literature [6, 13], therefore they are only mentioned here. Contrarily, in this section several dedicated parameters proposed by Kostek and her coworkers [14] are to be further discussed.

The entire set of dedicated parameters consists of 8 parameters. The first three parameters correspond to the number of samples exceeding levels: $r1, r2, r3$—equal to RMS, $2 \times$ RMS, $3 \times$ RMS and are defined by formula:

$$p_n = \frac{count(samples\_exceeding\_r_n)}{length(x(k))} \tag{6.3}$$

where $n = 1, 2, 3$ and $x(k)$ represent the analyzed signal fragment.

In the second step, the parameter called the 'peak to RMS' ratio is determined. This parameter is calculated as the mean value of the ratio calculated in 10 sub-frames.

The last group of dedicated parameters is related to the observation of the threshold crossing rate (TCR). The calculation procedure consists in computing number of signal crossings in relation to zero, $r1$, $r2$, and $r3$ values. Parameters contained in this group are the values resulting from the entire frame analysis.

An example of the given parameters for phonemes /a/ and /l/ is shown in Table 6.1.

Table 6.1 Time domain features for phonemes /a/ and /l/

	Fraction of samples p_n	Peak to RMS ratio	TCR values for the entire frame
/a/	0.28740 0.07060 0.00000	2.81140	0.03320 0.01680 0.00890 0.00000
/l/	0.32570 0.05050 0.00100	3.22070	0.01380 0.01180 0.00590 0.00069

6.5.2 Frequency Domain Features

In the frequency domain, features are most often derived from the Fourier spectrum. We should note that the edges of data become a step change what causes an appearance of noise after the segmentation procedure. In order to reduce noise, a window function is used. By multiplying the signal by the window function, we obtain a smooth transition between repeated intervals because this function returns zero at both edges. In addition, the window function makes the windowed Fourier transform data appearing more periodic than they really are. There are many window functions with different shapes proposed. The window functions that are commonly employed in the digital signal processing are described by Prabhu [16].

As a result of the window procedure, a significant part of the signal data is lost. The situation can be improved by making the segments overlap. In the work by Heinzel et al. [17] it is mentioned, that the segment overlap depends on the window function and in addition according to requirements of the analysis.

In conclusion, we can state that in the process of applying the Fourier transform to short signals, it is very important to set correctly processing parameters such as the window type, window size, and the overlap ratio. In this section, examples are generated using the following parameters: the input signal is segmented with a frame of 1024 samples, and then for each frame the Hamming window is chosen. We use an overlap of 50%. It should be remembered that the spectrum resolution of a 1024 point frame at 48 kHz is 47 Hz.

According to the Nyquist-Shannon sampling theorem, the sampling rate of f_s kHz allows us for reconstructing frequencies up to $f_s/2$ kHz. To convert audio data from the time to the frequency domain the Discrete Fourier Transform is applied. The Discrete Fourier transform of the phoneme lth short-time segment $x_l(\mathrm{n})$ $(n = 1, \ldots, M)$ is computed according to the formula:

$$X_l(k) = \sum_{m=1}^{M} x_l(m)w(m)e^{(-2\pi j)(m-1)\frac{k-1}{M}} \tag{6.4}$$

where $X_l(k)(k = 1, \ldots, M_{FT})$ are Fourier transform coefficients, M_{FT}—the number of Fourier transform coefficients ($M_{FT} \geq M$, M_{FT} is an integer power of 2), $w(m)$—is the window function.

Resulting from Eq. (6.4) is a sequence of complex numbers. In order to obtain the magnitude spectrum, the following formula is used:

$$|X_l(k)| = \frac{1}{M_{FT}}\sqrt{(X_l(k))_{re}^2 + (X_l(k))_{im}^2}. \tag{6.5}$$

The phase in radians is given as:

$$\angle X_l(k) = \mathrm{atan}\left(\frac{((X_l(k))_{im}}{((X_l(k))_{re}}\right). \tag{6.6}$$

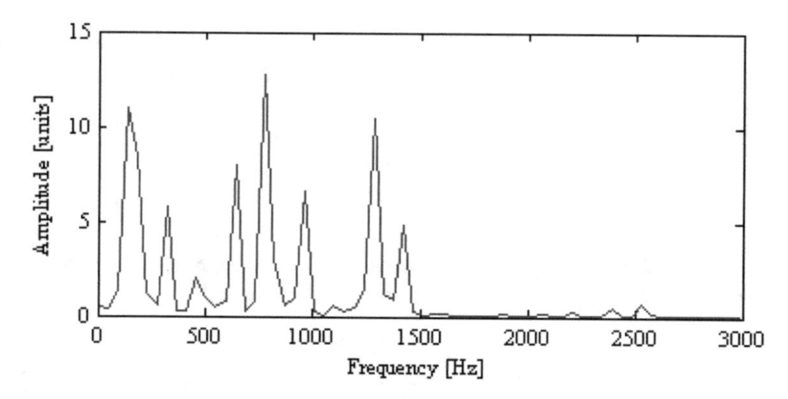

Fig. 6.6 The power spectrum of a short-time segment of the vowel /a/

The power spectrum $P_l(k)$ is defined as a square of the Fourier magnitude spectrum

$$PS_l(k) = |X_l(k)|^2. \tag{6.7}$$

An example of the power spectrum of the short-time segment (length of segment 1024 samples) is given in Fig. 6.6.

The short-time power spectrum is used in this section to describe the speech signal in term of flatness, envelope, and central moments such as centroid, spread, skewness and kurtosis.

6.5.2.1 Spectral Shape Parameters

Spectral shape parameters are derived from the central moments. Often they are computed from the magnitude spectrum [18, 19]. We are considering features that are based on the MPEG-7 audio content description standard [20, 21]. By this standard, the equations of spectral shape parameters are defined on log-frequency power spectrum.

Audio Spectral Centroid (ASC) provides a measure of the center of mass of the short-time power spectrum. It is calculated as the first order central moment. The formula is given below:

$$ASC = \frac{\sum_{i=1}^{M_{FT}/2} log_2\left(\frac{f(i)}{1000}\right) PS_l(i)}{\sum_{i=1}^{M_{FT}/2} PS_l(i)} \tag{6.8}$$

where $PS_l(i)$ is the power spectrum of lth short-time segment of the phoneme, $f(i)$—the frequency corresponding to bin i, M_{FT}—the number of Fourier transform coefficients.

As we see from Eq. (6.8), the measures are based on an octave frequency scale centered at 1 kHz. In order to simplify calculation of the remaining three moments, we need to define an auxiliary formula:

$$w(i) = \frac{PS_l(i)}{\sum_{i=1}^{M_{FT}/2} PS_l(i)}. \tag{6.9}$$

Audio Spectral Spread (ASSp) corresponds to the root square of the second order central moment of the spectrum and is defined by the formula:

$$ASSp = \sqrt{\sum_{i=1}^{M_{FT}/2} [log_2\left(\frac{f(i)}{1000}\right) - ASC]^2 w(i)} \tag{6.10}$$

with the Audio Spectral Centroid as defined by Eq. (6.8). As we see from Eq. (6.10), Audio Spectral Spread is the standard deviation of the power spectrum around the centroid. This means that a low ASSp value shows that the spectrum is concentrated around the centroid. In the case of a high value of ASSp, the power of spectrum is distributed across a wider range of frequencies.

Audio Spectral Skewness (ASSk) is the third order central moment divided by the third power of the standard deviation (i.e. Audio Spectral Spread) and is given by the formula:

$$ASSk = \frac{\sum_{i=1}^{M_{FT}/2} [log_2\left(\frac{f(i)}{1000}\right) - ASC]^3 w(i)}{ASS^3} \tag{6.11}$$

Equation (6.11) defines the spectral symmetry. In the case of skewness coefficient equals zero, a spectrum power distribution is symmetrical around its mean value.

Accordingly, the definition of Audio Spectral Kurtosis (ASK) is as follows:

$$ASK = \frac{\sum_{i=1}^{M_{FT}/2} [log_2\left(\frac{f(i)}{1000}\right) - ASC]^4 w(i)}{ASS^4} \tag{6.12}$$

The spectral kurtosis defines the flatness of spectrum distribution compared to the Gaussian distribution [6].

Examples of Spectral shape parameters and their variance values for the selected vowel, semivowel and consonant phonemes are given in Tables 6.2 and 6.3, respectively.

Form Tables 6.2 and 6.3 we see to what extent spectral shape parameters and their variance values differ for different types of phonemes.

Table 6.2 Spectral shape parameters

Phoneme	ASC	ASSp	ASSk	ASK
/a/	0.157	1.790	1.281	3.677
/u/	1.834	2.193	−0.778	3.755
/k/	2.911	1.753	−1.842	6.805
/s/	3.043	1.612	−2.003	7.847
/c/	3.114	1.453	−1.841	7.309
/t/	2.761	1.796	−1.541	5.582

Table 6.3 Variance values of spectral shape parameters

Phoneme	ASC	ASSp	ASSk	ASK
/a/	0.027	0.028	0.104	0.791
/u/	1.250	0.345	0.872	8.209
/k/	0.033	0.071	0.068	1.724
/s/	0.013	0.055	0.024	0.422
/c/	0.004	0.007	0.008	0.383
/t/	0.279	0.208	0.283	7.091

6.5.2.2 Spectral Flatness Measure

The Spectral Flatness lets us separate voiced and unvoiced speech. The spectral flatness values lie in the interval between 0 and 1. For the unvoiced signal such as white noise, this value is equal to 1. The more signal is voiced, the Spectral Flatness is closer to 0. For example, for the sinusoidal signal, this value is equal to 0. Often, the Spectral Flatness is applied to voice activity detection [22, 23].

The traditional definition of the Spectral Flatness is the ratio of the geometric mean and arithmetic mean of the magnitude spectrum. In this chapter, we consider a concept of the Spectral Flatness Measure (SFM), i.e., Spectral Flatness calculated on a sub-band level, as it is used in the MPEG7 standard [20, 21]. The Spectral Flatness Measure is defined as an averaged value for K frequency bands P_k, $(1 \leq k \leq K)$. The Spectral Flatness Measure in a single band k is calculated by the formula:

$$SFM\,(k) = \frac{\left[\prod_{i=P_k}^{P_{k+1}} PS_l(i)\right]^{1/N}}{\frac{1}{N_{sb}}\sum_{i=P_k}^{P_{k+1}} PS_l(i)} \tag{6.13}$$

where:

$$N_{sb} = P_{k+1} - P_k + 1 \tag{6.14}$$

$PS_l(i)$—the power spectrum of lth short-time segment of the phoneme. The calculation procedure of the frequency band edges is presented by the following pseudocode, shown as Algorithm 1.

Algorithm 1

INPUT:
p – the lower edge
j – multiplication factor of spectral resolution
fs – sampling frequency
M – the number of spectrum coefficients
K – the number of frequency bands
STEP 1. $r \leftarrow$ int (fs/M)
STEP 2. $k \leftarrow 0$, n $\leftarrow 0$
STEP 3. $w \leftarrow 2j$
STEP 4. WHILE $(n < M/2)$ and $(k < (K+2))$
    IF $(n*r > p)$
        p $\leftarrow p*w$
        $k \leftarrow k+1$
        $P(k) \leftarrow n$
    END IF
    $n \leftarrow n+1$
    END WHILE
OUTPUT:
$P_1, P_2, ..., P_K$ – the edges of the frequency bands

The comments on Algorithm 1:

- The value of the lower edge is equal to 250 Hz.
- The multiplication factor of spectral resolution must be of the interval [−4, 3], which means ranging from 1/16 of an octave to 8 octaves [20].

Examples of Spectral Flatness Measure values corresponding to the octave frequency bands for phonemes /a/ and /l/ are given in Fig. 6.7.

For the example presented in Fig. 6.7, the number of frequency bands $K = 18$ and a resolution of a quarter-octave were chosen.

As mentioned above, the Spectral Flatness Measure is defined as a value based on mean of values obtained in the frequency bands. Examples of Spectral Flatness Measure values for the chosen vowel, semivowel and consonant phonemes are given in Table 6.4.

Table 6.4 The SFM values for different phonemes

Phoneme	SFM	Phoneme	SFM	Phoneme	SFM
/e/	0.5703	/l/	0.7536	/p/	0.8048
/a/	0.6192	/j/	0.7739	/g/	0.8595
/r/	0.689	/k/	0.7741	/s/	0.8998
/i/	0.7151	/m/	0.7963	/f/	0.8998

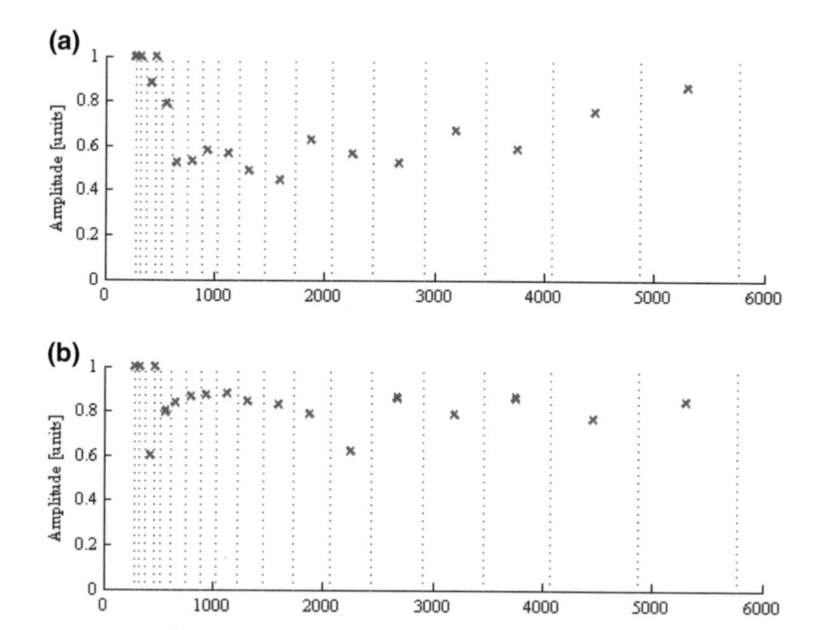

Fig. 6.7 The SFM values for phonemes /a/ and /l/

Again, from Table 6.4 we see differences in Spectral Flatness Measure values for different phonemes, which may be helpful in a discernibility analysis between phonemes.

6.5.2.3 Audio Spectrum Envelope

Audio Spectrum Envelope (ASE) gives us a compact representation of the power spectrum of the speech signal. Audio Spectrum Envelope as well as Spectral Flatness Measure is also considered on a sub-band level. The bands are logarithmically distributed corresponding to a specific octave frequency. The frequency band edges $P_k(1 \leq k \leq K)$ are calculated by Algorithm 1. The main difference is that the value of the lower edge, which is equal to 62.5 Hz.

Audio Spectrum Envelope is calculated as the sum of the power spectrum inside frequency bands. According to the MPEG-7 standard, ASE consists of a coefficient representing the power spectrum between 0 Hz and the low edge (i.e. between 0 and P_1), a number of coefficients representing power spectrum in bands between low and high edges (i.e. between P_1 and P_K), and a coefficient representing the power spectral density above the high edge (i.e. between P_K and $f_s/2$, where f_s is sampling frequency). With that in mind, ASE can be calculated by the following formula:

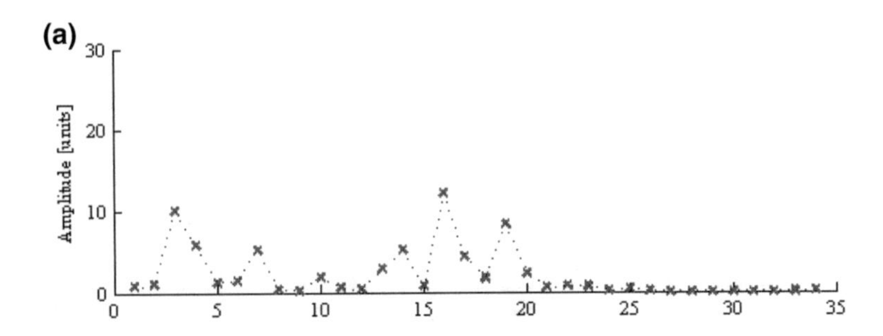

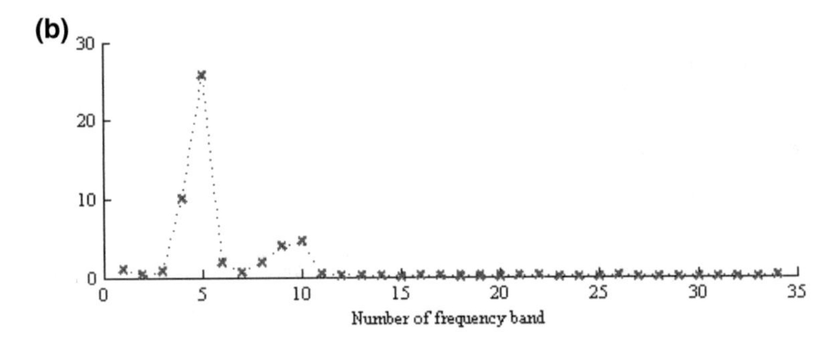

Fig. 6.8 The ASE comparison of two kinds of phonemes: **a** vowel /a/ and **b** the semivowel /l/

$$ASE(k) = \left\{ \begin{array}{c} \sum_{i=0}^{P_1} PS_l(i), k = 1 \\ \sum_{i=P_{k-1}}^{P_k} PS_l(i), 2 \leq k \leq K + 1 \\ \sum_{i=P_{k+1}}^{f_s/2} PS_l(i), k = K + 2 \end{array} \right\} \quad (6.15)$$

where: $PS_l(i)$ is the power spectral density of lth short-time segments of the phoneme, k—the frequency band number ($1 \leq k \leq K + 2$).

As we see from Fig. 6.8, Audio Spectrum Envelope highlights significant differences between phonemes, which are not so well visible by simply comparing power spectrum of these phonemes.

6.5.3 Mel-Frequency Cepstral Coefficients

Mel-Frequency Cepstral Coefficients (MFCCs) are the most commonly used acoustic features in speech processing. MFCCs were introduced by Mermelstein [24] as a tool for speech recognition, and these features are successfully employed in both speech and music areas [25].

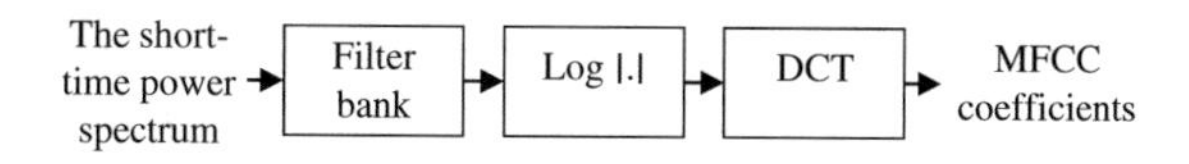

Fig. 6.9 The process of MFCCs extraction

The MFCCs represent the short-term power spectrum of sound and are often used for speech and speaker recognition [5, 26]. The MFCCs are also employed in speech emotion recognition process [2].

The process of MFCC coefficients extraction is shown in Fig. 6.9.

As we see from Fig. 6.9, the MFCC feature extraction process begins with filtering the short time power spectrum. The triangle bandpass filters that simulate characteristics of a human ear are used. The filters are spread over the frequency range from the lower to the upper frequency. Many studies show that the pitch is linear in low frequencies and logarithmic in high frequencies.

Denote the boundary points of the filter bank by $f[n]$. Let f_l be the lowest frequency, then the first boundary point $f[0] = f_l$. The boundary points in low frequencies are defined by the formula:

$$f[n] = f[n-1] + l_b \tag{6.16}$$

where l_b—the length of the linear band, $1 \le n \le H_l + 1$, H_l—the number of linear bands.

As we see from Eq. (6.16), the upper boundary of the frequency ranges, above which the scale becomes logarithmic, is defined by the number of linear bands and by their length. The boundary points in high frequencies are defined by the formula:

$$f[n] = Mel^{-1}\left(Mel(H_l \times (l_b + 1)_l) + n\frac{Mel(f_h) - Mel(H_l \times (l_b + 1))}{M + 1}\right) \tag{6.17}$$

where

$$Mel = 2595\log\left(1 + \frac{f}{700}\right), \tag{6.18}$$

$H_l + 1 < n \le H + 1$, $H = H_l + H_{nl}$, H_{nl}—the number of nonlinear bands, f_h—the highest frequency of the filter bank.

Log magnitude is calculated in order to obtain the real cepstrum. The energies from filters to cepstral coefficients are converted by Discrete Cosine Transform (DCT). DCT transforms the frequency domain into a time-like domain called quefrency domain.

For speech recognition task usually first 12–13 coefficients are used, for the speaker recognition—20 coefficients. Examples of the first 20 MFCCs and variance of these values are given in Figs. 6.10 and 6.11 for the phonemes /a/ and /l/ respectively.

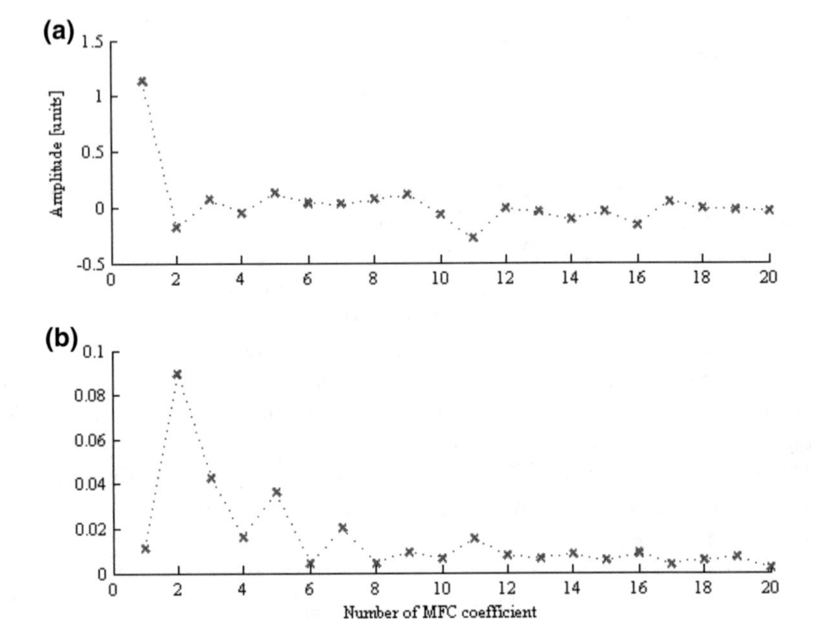

Fig. 6.10 MFCC values for the phonemes /a/: **a** the first 20 MFC coefficients and **b** variance of the MFC coefficients

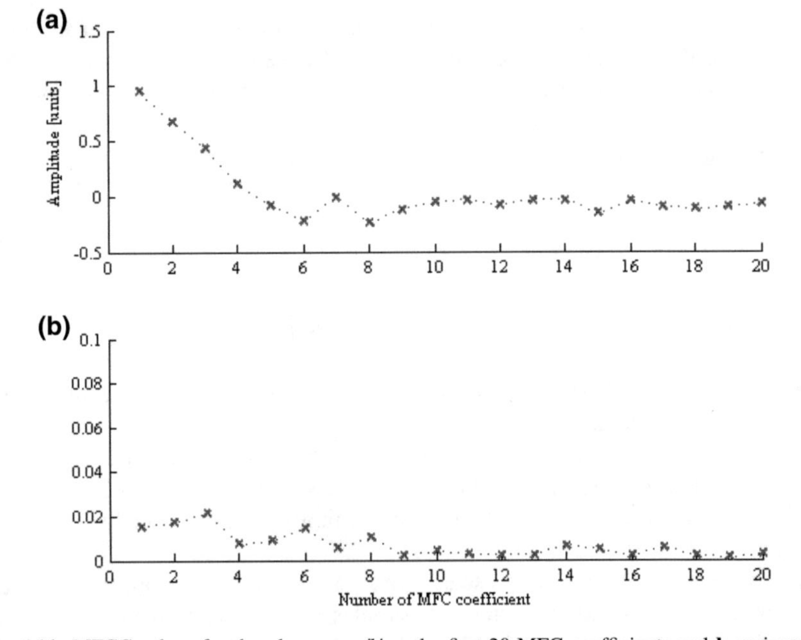

Fig. 6.11 MFCC values for the phonemes /l/: **a** the first 20 MFC coefficients and **b** variance of MFC coefficients

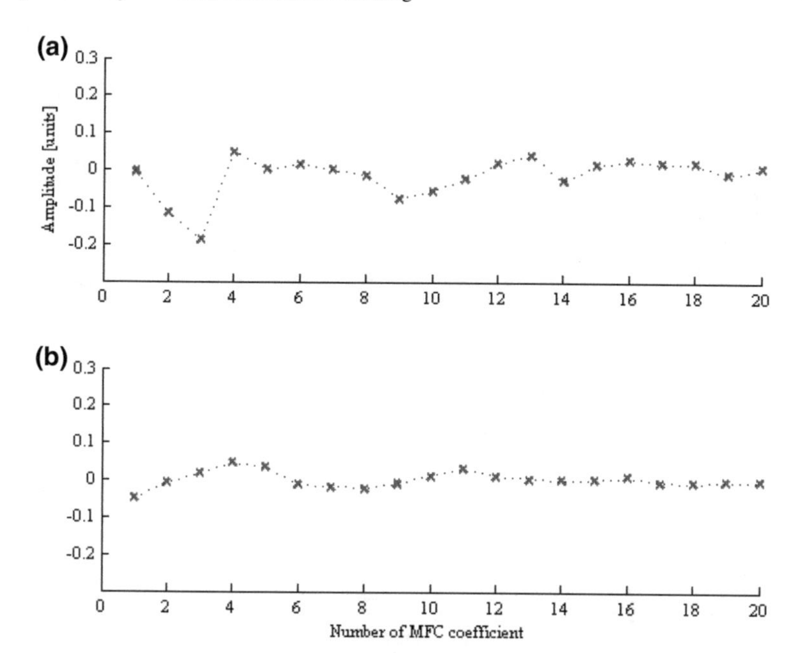

Fig. 6.12 Delta derivatives of the first 20 MFC coefficients for the phoneme /a/: **a** the first-order dynamic coefficients (DMFCC), **b** the second-order dynamic coefficients (DDMFCC)

In examples given in Figs. 6.10 and 6.11, the number of linear bands $H_l = 13$. The fixed length of the linear band $l_b = 66.67$ Hz was used. This value was chosen to set the threshold between the linear and logarithmic scale close to 1000 Hz as is often used in the literature [27].

Often the combination of MFCC coefficients are used together with its delta derivatives (DMFCC and DDMFCC) in various speech processing tasks [28]. The first-order dynamic coefficients (DMFCC) are calculated from the static MFCCs using the following regression formula:

$$d_t = \frac{\sum_{n=1}^{N} n(c_{t+n} - c_{t-n})}{2\sum_{n=1}^{N} n^2} \tag{6.19}$$

where c is nth cepstral coefficient calculated at time t. N is the regression window size. The second-order dynamic coefficients (DDMFCC) can be calculated by the same formula (6.19), where c is denoted as the first-order dynamic coefficients.

Examples of delta derivatives are given in Fig. 6.12 for the phoneme /a/ and in Fig. 6.13 for the phoneme /l/.

Data presented in Figs. 6.12 and 6.13 were obtained by employing Eq. (6.20), where the regression window size $N = 2$.

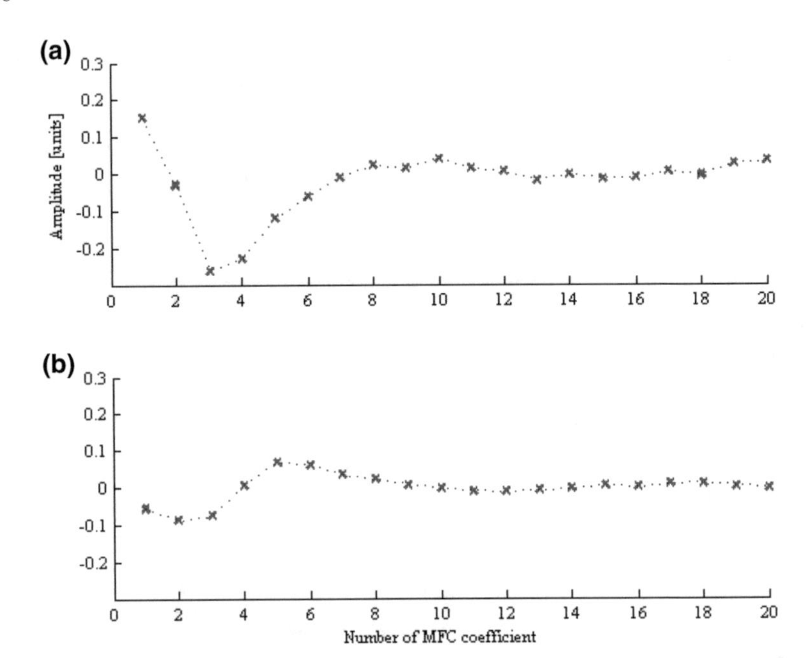

Fig. 6.13 Delta derivatives of the first 20 MFC coefficients for the phoneme /l/: **a** the first-order dynamic coefficients (DMFCC), **b** the second-order dynamic coefficients (DDMFCC)

6.6 Data Preparation for Deep Learning

As shown earlier, the speech signal features were extracted manually. In the deep learning approach, useful features are extracted automatically by the process—e.g. directly from the spectrum image. In this section we propose to use spectrogram images as the input dataset of Deep Neural Network (DNN).

A spectrogram is the visual representation of time and frequency domain features on one level, which has found a wide use in the speech processing [29, 30]. A spectrogram is constructed from a series of short-time Fourier transforms, which are computed along the time domain waveform of the analyzed phoneme. It should be emphasized that the phoneme signal is divided into short-time segments and the window function is applied as in the case of the calculation procedure of acoustic features described earlier. Depending on whether power or phase of each frequency within a given time frame is shown, two types of spectrogram are used: power spectrogram and phase spectrogram.

A review of the literature shows that the power spectrogram is often used as the phase spectrogram. Within each short-time Fourier segment log energy is calculated in the power spectrogram and then the obtained values of energy are converted into grade scale of color, which is positioned at time and frequency.

We can state that a spectrogram depends on the following properties of the Fourier transform:

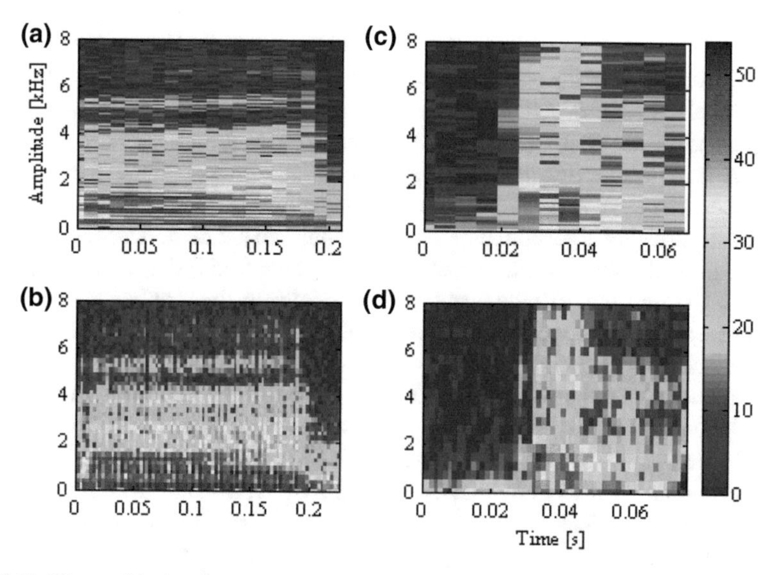

Fig. 6.14 The graphical representation of narrowband **a** phoneme /a/, **c** phoneme /l/ and wideband **b** phoneme /a/, **d** phoneme /l/ spectrograms

- The length of the segment;
- The overlap between contiguous segments;
- The shape of the window applied to each segment.

The properties mentioned above determine the form of the spectrogram. Spectrogram showing the broad spectral envelope is called wideband spectrogram, while the spectrogram using narrow filter bandwidth is identified as narrowband.

The bandwidth of the narrowband spectrogram filters is usually between 30 and 50 Hz, while the bandwidth of the wideband spectrogram filters is 300 and 500 Hz [31]. We must pay particular attention to the relationship between the bandwidth and the length of the segment. An example of the narrowband and wideband power spectrograms for phonemes /a/ and /l/ is shown in Fig. 6.14.

The top plots of Fig. 6.14 are the narrowband spectrogram with a window size of 512 samples, while the bottom plots are a wideband spectrogram with a window size of 128 samples. The Hamming window and 50% overlap are used.

As seen from Fig. 6.14, the spectrogram is a two-dimensional graph, with the third dimension represented by the degree of darkness. Time is shown along the x-axis, frequency along the y-axis, while the color indicates intensity of the speech signal against time and frequency. In grayscale the range of shades of gray refer to intensity, i.e. darker shade indicates larger value. Comparing the top and bottom plots of Fig. 6.14, one can see that the power spectrogram provides an analysis of the frequency components of the signal in terms of its harmonics or formants. The harmonic structure of spectrogram is obtained in the case of the narrow filter bandwidth, and the wideband spectrogram reflects formants of signals.

The power spectrogram uses the amplitude values of the Fourier transform but ignores information of the phase values. However, Leonard [30] shows that the phase values are also useful in signal analysis. This author presented an original approach for representing the phase spectrogram. In addition, he defines the term "frequency spectrogram", in which the time-frequency representation is obtained from the phase.

For the purpose of searching for the best phoneme classifier in terms of its accuracy, a deep learning algorithm based on convolutional neural networks (CNNs) and spectrograms with the frame length of 1024 samples and 75% overlap and Hamming windowing as the network input were applied. For the analysis only 128 of 512 components were used in order to restrict the analysis bandwidth to the most important frequency range for speech processing.

Calculations were performed for 936 samples originating from 9 speakers containing phonemes /l/ assigned to two groups: dark and clear. Therefore, an input bandwidth was up to 5512.5 Hz. This results in varying number of pixels for different spectrograms, but to train a CNN, the input dimensions of the neural network should be the same. That is why, unifying the size of each spectrogram had to be performed. Two methods were employed for the purpose of comparison: padding all shorter sequences with zeros to obtain spectrograms of the length (in the time domain) equal to the length of the longest example and applying linear interpolation to the same length. The resulting length of spectrograms was equal to 143.

Also, for the same reason, spectrograms based on MFCC coefficients (which we would refer to as MFCCgrams) were calculated and their dimensions were unified in the same manner as for spectrograms. The resulting length of MFCCgrams was equal to 73.

For each example 80 MFCC coefficients were calculated. Both types of spectrograms and MFCCgrams were normalized by subtracting the mean form the data and dividing by the standard deviation (z-normalization). A sample visualization of each stage of preprocessing is shown in Fig. 6.15.

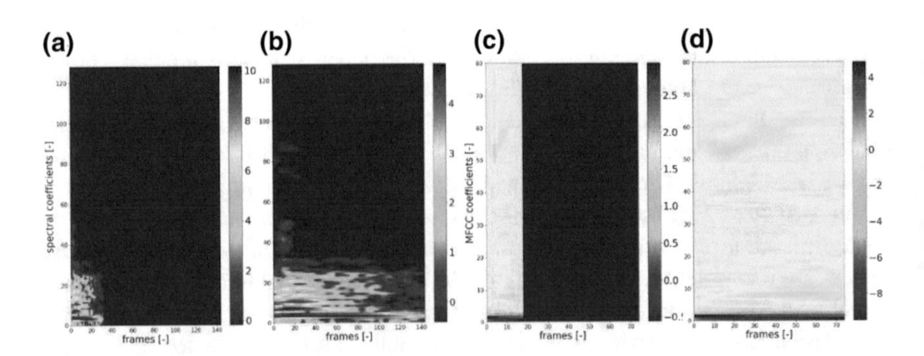

Fig. 6.15 Graphical representation of parameters calculated for an example of a dark /l/ phoneme: **a** and **c** spectral and MFCC coefficients padded with a minimum value present in original data are shown, **b** and **d** represent original data, which were interpolated in order to match their length to the length of the longest sample present in the database

6.7 Experiment Results

Examples of application of machine learning methods to speech phoneme classification are given in this section. For this purpose, words recorded as wav audio file with the following parameters: PCM 48 kHz, 16 bit are used. The recorded signals are segmented into phonemes or allophones, the latter ones divided into 'dark' and 'clear' /l/, and then the feature vector containing the above described parameters is extracted.

6.7.1 *Feature Vector Applied to Vowel Classification*

The first part of this research study concerns vowel (/a/, /e/, /i/, /o/, /u/) classification. A list of 209 words containing these phonemes was compiled. Each word was uttered by four speakers. An experiment was performed on vowel phonemes extracted from the recorded words. 70% of vowels sounds is used for training, 30% ones for testing. The starting point of the analyses was the parameterization of all considered phonemes. The parameters described in Sect. 6.5 are extracted. Then, the extracted parameters are normalized. It was decided to normalize the values to the range between 0 and 1. In order to reduce feature space dimensionality feature scoring algorithm based on MI (Mutual Information) criteria [32] is used. In generally, MI between two random variables X and Y can be defined by the formula:

$$MI = \sum_{x \in X} \sum_{y \in Y} p(y, z) \log \left(\frac{p(y, z)}{p(y)p(z)} \right), \tag{6.20}$$

where $p(\cdot)$ denotes the probability of the state.

In this experiment we have two type of variables: features and labels of these. The labels are integer, meanwhile features are not. In order to make Eq. (6.20) applicable, we have to quantize features. In this research the quantization approach proposed by Pohjalainen et al. [33] is used. A feature scoring algorithm gives us a score value for each feature to reflect its usefulness. 15 features with the highest scores were selected for further processing. These parameters are the variance of Audio Spectral Skewness and Audio Spectral Kurtosis, 3th and 5th variances of MFCC coefficients, 1st, 2nd, 4th, 5th and 10th delta derivatives of MFCC, 2nd, 3rd, 4th, 6th, 7th and 8th delta delta derivatives of MFCC.

A support vector machine (SVM) algorithm [34, 35] is applied as a classifier for this part of the experiment. The overall classification accuracy for vowel phonemes is 0.86. The accuracy of different vowel phonemes is shown in Fig. 6.16, where Multi-Class Confusion Matrix is depicted. The obtained results show that it is possible to discern between vowels by means of a machine learning algorithm with a sufficient accuracy.

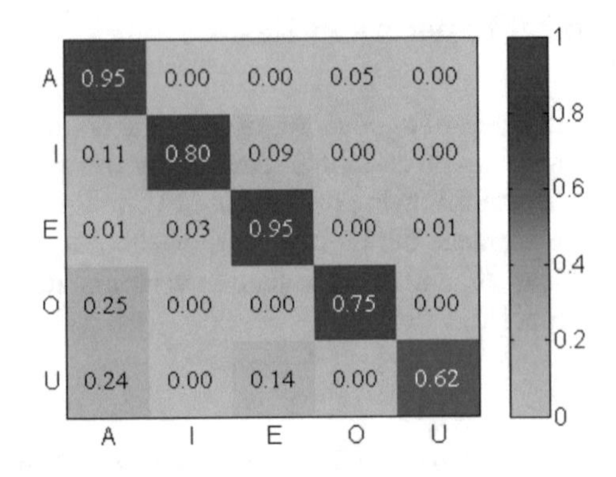

Fig. 6.16 The vowel phonemes classification accuracy

Table 6.5 *k*NN accuracy results for various sets of features

TDF + MFCC + SFM (%)	MFCC + SFM + ASE2 (%)	MFCC + SFM (%)	TDF + MFCC + SFM + DDMFCC (%)	MFCC + SFM + ASEV (%)	MFCC + SFM + SSP (%)	MFCC + SFM + DMFCC + DDMFCC (%)
92.27	92.23	92.19	92.18	92.12	92.04	92.02

6.7.2 *Feature Vector Applied to Allophone Classification*

In order to show that it is also applicable to employ machine learning methods to automatic classification of allophones, in the second part of the experiment, the study is performed in the context of the allophones of the phoneme /l/. For this purpose, a list of 'dark' and 'clear' /l/ allophones was created [36, 37]. To recall: 'dark' /l/ is articulated with the back of the tongue raised towards the soft palate and occurs word-finally or before another consonant, contrarily the 'clear' /l/ onset is voiceless and voicing starts at the end of the /l/ articulation. Sounds of allophones extracted from 1030 words uttered by nine speakers were considered in this experiment. The *k*-Nearest Neighbor (*k*NN) method [34, 35] was applied to the allophone classification.

Various sets of features were tested to achieve the best performance of the method used. In order to select the best parameter sets, the Decision Tree algorithm [38] was used. The results for sets of features, which resulted in the highest accuracies are presented in Table 6.5. The following abbreviations are used in Table 6.5: TDF—time domain features, SSP—spectral shape parameters, SFM—spectral flatness measure, ASE—mean of audio spectrum envelope, MFCC—Mel-Frequency Cepstral Coefficients, DMFCC—delta derivatives of MFCC, and DDMFCC—delta delta derivatives of MFCC.

As seen from Table 6.5, the optimized feature vectors enable to return high accuracies of the allophone classification.

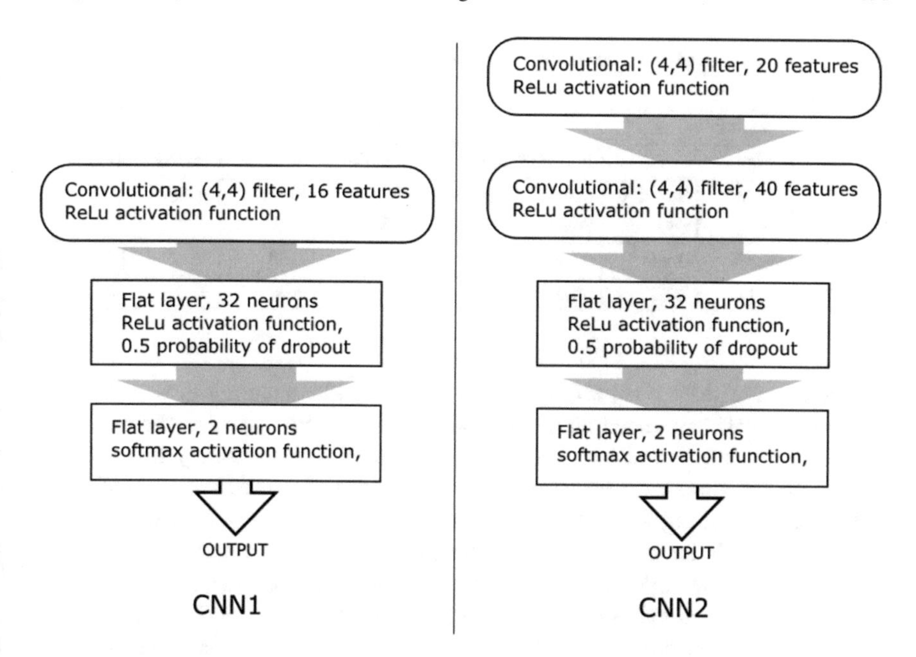

Fig. 6.17 Structure of the first network (CNN1) and the second one (CNN2)

6.7.3 Convolutional Neural Networks Applied to Allophone Classification

As already mentioned, another approach to the recognition of the type of phonemes consisted of using convolutional neural networks based on spectrograms as an input data. Such an approach is widely used for the purpose of computer vision classification with very good results [7]. Two architectures of CNNs were used for the purpose of classification. Experiments were performed using DGX Station. The network structure is depicted in Fig. 6.17.

The first network (CNN1) had only one convolutional layer and generated 16 features, which were then processed by a simple feed-forward neural network consisting of two layers: 32 neurons with rectified linear unit (ReLu) activation function and 2 output neurons with softmax activation function. The output was encoded in one-hot manner, therefore each output was associated with one of the two analyzed classes of objects. The final decision on classification was made by choosing a class associated with an output of a greater value. As a loss function a binary cross-entropy was chosen. For the learning rate optimizer the Adam algorithm, introduced by Kingma and Ba, was used [39]. The learning rate was set in both cases to 0.001, beta_1 and beta_2 parameters were set to 0.9 and 0.999, respectively. Calculations were performed with the use of Keras Python deep learning library with TensorFlow library as a back-end [40, 41].

Table 6.6 CNN classification accuracy results for all speakers

Network	Parameters							
	Spectral				MFCC			
	Padding		Interpolation		Padding		Interpolation	
	Mean (%)	Std. dev. (%)	Mean (%)	Std. dev. (%)	Mean (%)	Std. dev. (%)	Mean (%)	Std. dev. (%)
CNN1	96.37	0.44	97.24	0.58	97.11	2.00	96.69	0.36
CNN2	96.12	2.20	96.36	2.41	97.68	0.99	97.19	0.36

The training phase was performed with the use of 400 randomly chosen examples of the input data associated with both considered phoneme classes. The remaining 536 examples were used as a test set at the end of each efficiency test iteration. Both CNN networks were tested with the use of each of four datasets, namely consisting of either spectral or MFCC-based data (i.e. spectrograms/ MFCCgrams) and padding/interpolation preprocessing of these data. 1000 trials of training and tests of each pair of the network and a set of parameters were performed (Table 6.6).

Accuracies obtained for both network architectures are higher than for the approach based on pre-selected features. This occurs even in the case of a simpler network CNN1. For the more complicated structure of the network a slight improvement of the performance was achieved for MFCCgrams, however the standard deviation of the accuracy in such a case is nearly twice larger than in the case of the best case scenario for CNN1. In the case of CNN1 interpolation of spectral coefficients allowed for improvement of the achieved, results, which was not the case for CNN2, where the best performance was found for zero-padded MFCCgrams.

6.7.4 Convolutional Neural Networks Applied to Vowel Classification

Analogous experiment was performed for both architectures of CNN in context of identification of vowels. Only modifications in this case was modification of the last layer of each neural network to support five one-hot outputs associated with five possible vowels and the number of training epochs which in this case was set to 20. Results of training is depicted in the form of confusion matrices in Figs. 6.18 and 6.19.

In the case of CNN1 the most problematic vowels were ‘O’ and ‘U’, which were classes that contained the least number of examples available in the dataset. However, CNN2 was able to overcome this problem and achieve 68% classification accuracy. It is also worth noting that the training set in the case of both CNNs consisted of only 40% of the overall number of available examples and the rest of them was used as a test set. In the case of both networks preprocessing employing the interpolation

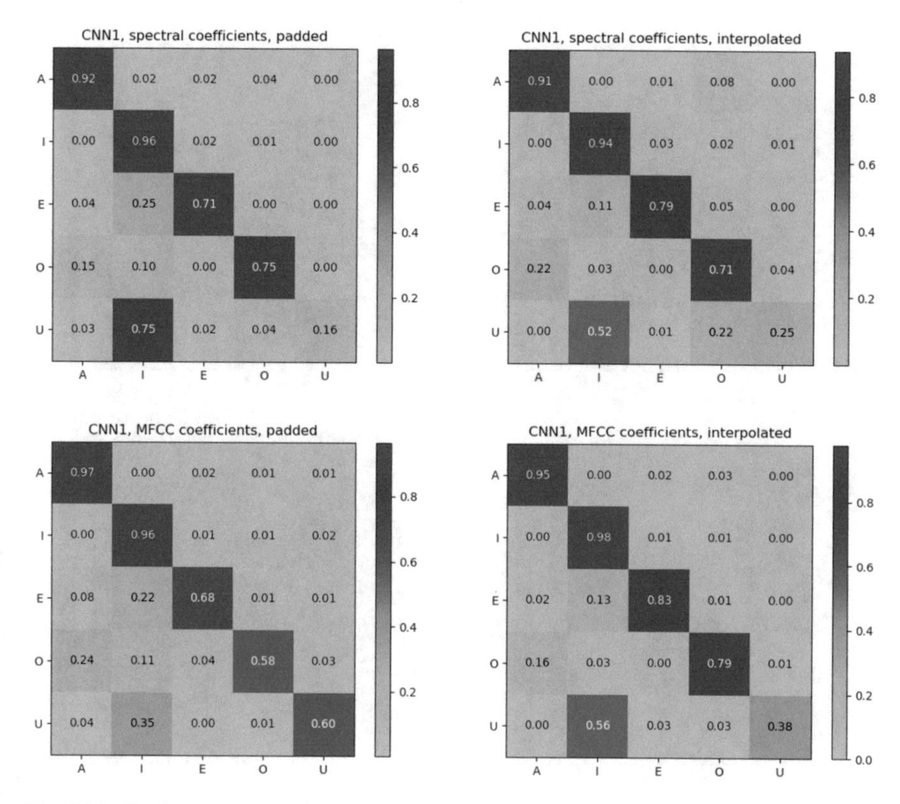

Fig. 6.18 Confusion matrices for tests performed for CNN1 with four types of the input data preprocessing

technique of the input parameters works better than padding the input with zeros. Also the use of MFCC as the input type of features permits to obtain higher accuracies.

6.8 Conclusions

In this chapter phoneme/allophone signal is considered as an acoustic signal, which can be described as a set of features. To obtain useful features, we apply two ways of data preparing. In the first approach, feature vectors are extracted, and then machine learning method is used, while in the second approach, the speech signal is converted into images and DNN is employed. To deal with the phoneme/allophone classification, features, which are widely used in Music Information Retrieval are adapted to the short speech signal elements, such as phonemes and allophones. The formulas and calculation algorithms of these features were formulated for that purpose.

In order to test whether features coming from MIR are useful for speech analytics, two experiments were performed. First, the created feature vector and SVM algo-

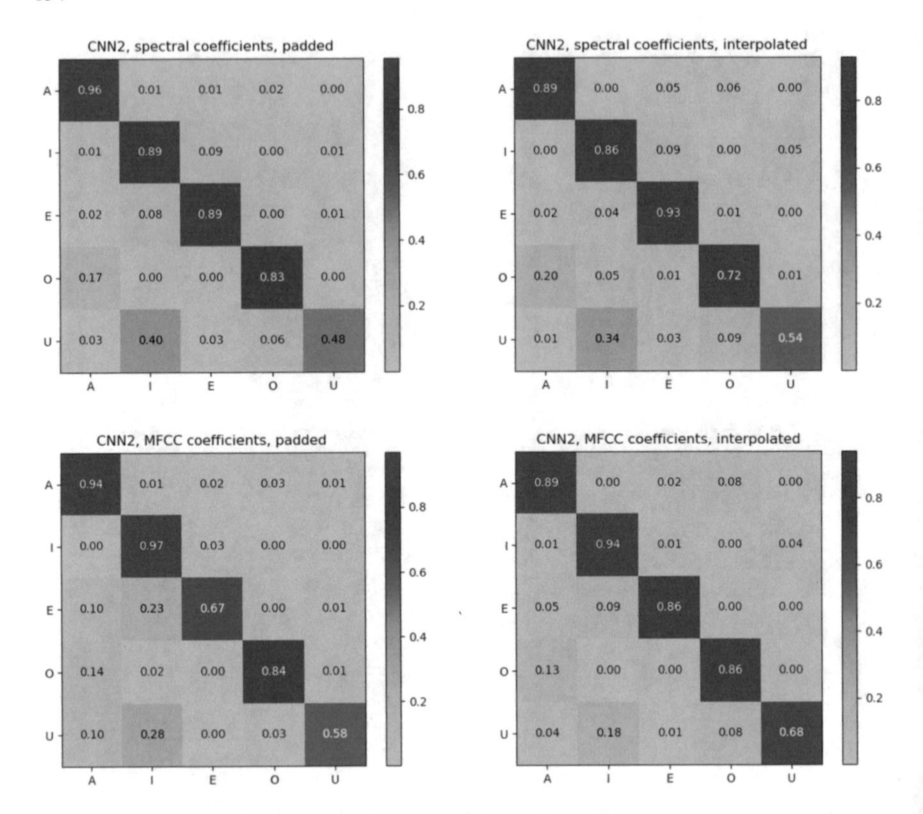

Fig. 6.19 Confusion matrices for tests performed for CNN2 with four types of input data preprocessing

rithm along with the feature scoring algorithm based on MI criteria were applied to the vowel classification. The overall classification accuracy for five different vowel phonemes (/a/, /e/, /i/, /o/, /u/) was equal to 86%.

In the second experiment the feature vector was applied to classification of the phoneme /l/ allophones. The *k*NN method was applied as a classifier. In order to select the best parameter sets, the Decision Tree algorithm was used. The highest accuracy (92.27%) has been achieved for a set of the following features: TDF, MFCC and SFM.

For the purpose of searching for the best phoneme/allophone classifier in terms of its accuracy, deep learning, based on convolutional neural networks (CNNs), was applied. Spectrograms and MFCCgrams were used as the network input. Also, two architectures of CNNs were used for the purpose of classification. The first network had only one convolutional layer and generated 16 features, while the second one had two convolutional layers, which generated 20 and 40 features respectively.

The CNN applied to the allophone /l/ classification outperformed the classical machine learning approach. The best result (97.68%) was achieved in the case of the network with two convolutional layers for zero-padded MFCCgrams.

Similarly, an experiment with CNNs was performed for the vowel classification. The best result (84.6%) was achieved in the case of the network with two convolutional layers for the interpolated MFCCgrams, which was nearly as good as the result of the SVM and MI approach. However, one should take into account that the amount of data available for this experiment was not sufficiently large for CNNs. Therefore, in the future experiments the dataset is to be expanded, moreover additional corpora are planned to be utilized (e.g. [42]).

Acknowledgements Research partially sponsored by the Polish National Science Centre, Dec. No. 2015/17/B/ST6/01874. This work has also been partially supported by Statutory Funds of Electronics, Telecommunications and Informatics Faculty, Gdansk University of Technology.

References

1. Badshah. A.M., Ahmad, J., Rahim, N., Baik, S.W.: Speech Emotion Recognition from Spectrograms with Deep Convolutional Neural Network International Conference on Platform Technology and Service (PlatCon), pp. 1–5 (2017)
2. Noroozi, F., Kaminska, D., Sapinski, T., Anbarjafari, G.: Supervised vocal-based emotion recognition using multiclass support vector machine, random forests, and adaboost. J. Audio Eng. Soc. **65**(7/8), 562–572 (2017). https://doi.org/10.17743/jaes.2017.0022
3. Sainath, T.N., Mohamed, A.-R., Kingsbury, B., Ramabhadran, B.: Deep convolutional neural networks for LVCSR. In: 2013 IEEE International Conference on Acoustics, Speech and Signal Processing, pp. 8614–8618 (2013)
4. Xu, Y., Du, J., Dai, L.R., Lee, C.H.: A regression approach to speech enhancement based on deep neural networks. IEEE/ACM Trans. Audio, Speech, Lang. Process. **23**(1), 7–19 (2015)
5. Alam, M.J., Kenny, P., O'Shaughnessy, D.: Low-variance multitaper mel-frequency cepstral coefficient features for speech and speaker recognition systems. Cognit. Comput. **5**(4), 533–544 (2013)
6. Lerch, A.: An Introduction To Audio Content Analysis: Applications in Signal Processing and Music Informatics, p. 248. Wiley, Hoboken, N.J (2012)
7. Biswas, A., Sahu, P., Chandra, M.: Multiple camera in car audio–visual speech recognition using phonetic and visemic information. Comput. Electr. Eng. **47**, 35–50 (2015). https://doi.org/10.1016/j.compeleceng.2015.08.009
8. Ziółko, B., Ziółko, M.: Time durations of phonemes in Polish language for speech and speaker recognition. In: Human Language Technology, Challenges for Computer Science and Linguistics. Lecture Notes in Computer Science, vol. 6562, pp. 105–114. Springer (2011)
9. Czyżewski, A., Kostek, B., Bratoszewski, P., Kotus, J., Szykulski, M.: An audio-visual corpus for multimodal automatic speech recognition. J. Intell. Inf. Syst. **1**, 1–27 (2017). https://doi.org/10.1007/s10844-016-0438-z
10. Rosner, A., Kostek, B.: Automatic music genre classification based on musical instrument track separation. J. Intell. Inf. Syst. **5** (2017). https://doi.org/10.1007/s10844-017-0464-5
11. Korvel, G., Kostek, B.: Examining feature vector for phoneme recognition. In: Proceeding of IEEE International Symposium on Signal Processing and Information Technology, ISSPIT 2017. Bilbao, Spain (2017)
12. Korvel, G., Kostek, B.: Voiceless stop consonant modelling and synthesis framework based on MISO dynamic system. Arch. Acoust. **42**(3), 375–383 (2017). https://doi.org/10.1515/aoa-2017-0039
13. Plewa, M., Kostek, B.: Music mood visualization using self-organizing maps. Arch. Acoust. **40**(4), 513–525 (2015). https://doi.org/10.1515/aoa-2015-0051

14. Kostek, B., Kupryjanow, A., Zwan, P., Jiang, W., Raś, Z., Wojnarski, M., Swietlicka, J.: Report of the ISMIS 2011 contest: music information retrieval. Found. Intell. Syst. 715–724 (2011)
15. Gold, B., Morgan, N., Ellis, D.: Speech and Audio Signal Processing: Processing and Perception of Speech and Music, 2nd edn, 688 pp. Wiley, Inc., (2011)
16. Prabhu, K.M.M.: Window Functions and Their Applications in Signal Processing. CRC Press (2013)
17. Heinzel, G., Rudiger, A., Schilling, R.: Spectrum and spectral density estimation by the discrete Fourier transform (DFT), including a comprehensive list of window functions and some new flat-top windows. Internal Report, Max-Planck-Institut fur Gravitations physik, Hannover (2002)
18. Gillet, O., Richard, G.: Automatic transcription of drum loops. In: IEEE International Conference on Acoustics, Speech, and Signal Processing, (ICASSP '04) (2004)
19. Hyungsuk, K., Heo, S.W.: Time-domain calculation of spectral centroid from backscattered ultrasound signals. IEEE Trans. Ultrason. Ferroelectr. Freq. Control **59**(6) (2012)
20. Hyoung-Gook, K., Moreau, N., Sikora, T.: MPEG-7 Audio and Beyond: Audio Content Indexing and Retrieval. Wiley, Hoboken (2005)
21. Manjunath, B.S., Salembier, P., Sikora T.: Introduction to MPEG-7: Multimedia Content Description Interface. Wiley (2002)
22. Ma, Y., Nishihara, A.: Efficient voice activity detection algorithm using long-term spectral flatness measure. EURASIP J. Audio, Speech, Music Process 1–18 (2013)
23. Moattar, M.H., Homayounpour, M.M.: A simple but efficient real-time voice activity detection algorithm. In: 17th European Signal Processing Conference (EUSIPCO 2009). Glasgow, Scotland, Aug 24–28 (2009)
24. Mermelstein, P.: Distance measures for speech recognition, psychological and instrumental. Pattern Recognition and Artificial Intelligence, pp. 374–388. Academic, New York (1976)
25. Logan, B.: Mel frequency cepstral coefficients for music modeling. In: Proceedings of 1st International Symposium on Music Information Retrieval (ISMIR). Plymouth, Massachusetts, USA (2000)
26. Nijhawan, G., Soni, M.K.: Speaker recognition using MFCC and vector quantisation. J. Recent Trends Eng. Technol. **11**(1), 211–218 (2014)
27. Wang, Y., Lawlor, B.: Speaker recognition based on MFCC and BP neural networks. In: 28th Irish Signals and Systems Conference (2017)
28. Ahmad, K.S., Thosar, A.S., Nirmal, J.H., Pande, V.S.: A unique approach in text independent speaker recognition using MFCC feature sets and probabilistic neural network. In: 2015 Eighth International Conference on Advances in Pattern Recognition (ICAPR), 4–7 Jan 2015, pp. 1–6 (2015)
29. Dennis, J., Tran, H.D., Li, H.: Spectrogram image feature for sound event classification in mismatched conditions. Signal Process. Lett. IEEE **18**(2), 130–133 (2011)
30. Leonard, F.: Phase spectrogram and frequency spectrogram as new diagnostic tools. Mech. Syst. Signal Process. **21**(1), 125–137 (2007)
31. Lawrence, J.R., Borden, G.J., Harris K.S.: Speech Science Primer: Physiology, Acoustics, and Perception of Speech, 6th edn, 334 pp. Lippincott Williams & Wilkins (2011)
32. Steuer, R., Daub, C.O., Selbig, J., Kurths, J.: Measuring distances between variables by mutual information. In: Innovations in Classification, Data Science, and Information Systems, pp. 81–90 (2005)
33. Pohjalainen, J., Rasanen, O., Kadioglu, S.: Feature selection methods and their combinations in high-dimensional classification of speaker likability, intelligibility and personality traits. Comput. Speech Lang. **29**(1), 145–171 (2015)
34. Manocha, S., Girolami, M.A.: An empirical analysis of the probabilistic K-nearest neighbour classifier. Pattern Recogn. Lett. **28**, 1818–1824 (2007)
35. Palaniappan, R., Sundaraj, K., Sundaraj, S.: A comparative study of the SVM and k-nn machine learning algorithms for the diagnosis of respiratory pathologies using pulmonary acoustic signals. BMC Bioinf. **15**, 1–8 (2014)

36. Czyżewski, A., Piotrowska, M., Kostek, B.: Analysis of allophones based on audio signal recordings and parameterization. J. Acoust. Soc. Am. **141**(5), 3521 (2017). https://doi.org/10.1121/1.4987415
37. Kostek, B., Piotrowska, M., Czyżewski, A.: Comparative study of self-organizing maps versus subjective evaluation of quality of allophone pronunciation for nonnative english speakers. In: 143rd Audio Engineering Society Convention, Preprint 9847. New York (2017)
38. Han, J., Kamber, M., Pei, J.: Data mining: concepts and techniques. In: The Morgan Kaufmann Series in Data Management Systems, 2nd edn, 761 pp. Morgan Kaufmann (2006)
39. Kingma, P.D., Ba, J.L.: ADAM: a method for stochastic optimization. In: International Conference on Learning Representations, ICLR 2015 (2015). https://arxiv.org/pdf/1412.6980.pdf. Accessed Jan 2018
40. Keras library Keras Documentation Website. http://keras.io. Accessed Jan 2018
41. TensorFlow library. TensorFlow Documentation Website. https://www.tensorflow.org/. Accessed Jan 2018
42. TIMIT: Acoustic-Phonetic Continuous Speech Corpus. https://catalog.ldc.upenn.edu/ldc93s1. Accessed Jan 2018

Part II
Data Analytics in Social Studies and Social Interactions

Chapter 7
Trends on Sentiment Analysis over Social Networks: Pre-processing Ramifications, Stand-Alone Classifiers and Ensemble Averaging

Christos Troussas, Akrivi Krouska and Maria Virvou

Abstract Technology advancements gave birth to social networks during the last decade. Many people tend to increasingly use them in order to share their personal opinion on current topics of their everyday life as well as express their emotions about situations in which they are interested. Hence, the emotions that are expressed in social networks can be positive, negative or neutral. To this direction, the analysis of people's sentiments has drawn the attention of many scientists worldwide and offers a fertile ground for increasing research. Furthermore, a social network that is adopted by an ever growing percentage of people is Twitter. Twitter is an online news and social networking service where users post and interact with messages; such messages conceal people's feelings and sentiments. Therefore, Twitter can be seen as a source of information and holds a vast amount of data that can be exploited for sentiment analysis research. In view of above, the purpose of this paper is to provide a guideline for the decision of optimal pre-processing techniques and classifiers for sentiment analysis over Twitter. In this context, three well-known Twitter datasets (OMD, HCR and STS-Gold) were used and a set of experiments was conducted. In particular, firstly, an extended comparison of sentiment polarity classification methods for Twitter text and the role of text preprocessing in sentiment analysis are discussed in depth. Secondly, four well-known learning-based classifiers (Naive Bayes, Support Vector Machine, k-Nearest Neighbors and C4.5) have been evaluated based on confusion matrices. Thirdly, the most common ensemble methods (Bagging, Boosting, Stacking and Voting) are examined and compared to base classifiers' results. Finally, a case study concerning the application of Twitter sentiment analysis in an e-learning context is presented. The main result of the utilization of the Twitter-based learning application is that the exploitation of students' emotional

C. Troussas · A. Krouska (✉) · M. Virvou
Software Engineering Laboratory, Department of Informatics, University of Piraeus, Piraeus, Greece
e-mail: akrouska@unipi.gr

C. Troussas
e-mail: ctrouss@unipi.gr

M. Virvou
e-mail: mvirvou@unipi.gr

G. A. Tsihrintzis et al. (eds.), *Machine Learning Paradigms*, Intelligent Systems Reference Library 149, https://doi.org/10.1007/978-3-319-94030-4_7

states can be used to enhance adaptivity in the learning content as well as deliver recommendations about activities and provide personalized assistance. Concerning data pre-processing, the experimental results demonstrate that feature selection and representation can affect the classification performance positively. Regarding the selection of the proper classifier, the superiority of Naive Bayes and Support Vector Machine, regardless of datasets, is proved, while the use of ensembles of multiple base classifiers can improve the accuracy of Twitter sentiment analysis.

Keywords Sentiment analysis · Data preprocessing · Learning machines Ensembles · Polarity detection · Twitter

7.1 Introduction

Over the last few years, the wide spread use of world wide web in social contexts has changed the form of use of the internet by allowing users to interact and collaborate with each other in a social media dialogue, as creators of user-generated content in a virtual community. An inherent tool of social web is microblogging, which allows users to exchange small elements of content such as short sentences, individual images, or video links. Twitter is the most widespread[1] social networking microblogging service with more than 320 million active users and approximately 500 million tweets per day.[2] Tweets are short messages (no longer than 140 characters), that people share to show aspects of their everyday life, personal opinion about different topics, such economy, environment health care and so on. Twitter offers the possibility to users to post tweets in a convenient and rapid way. Hence, this vast amount of data renders Twitter a useful tool for sentiment analysis.

Sentiment analysis refers to the use of natural language processing, statistics and machine learning methods to identify public sentiment towards various issues in a text unit.[3] Several methods and approaches have been proposed to automatically detect sentiment content and have been categorized based on various attributes. Hence, there are three main classification levels in sentiment analysis: (a) document-level, considering the whole document a basic information unit, (b) sentence-level, identifying whether the sentence is subjective or objective and determining the sentiment expressed in subjective sentences, and (c) aspect-level, classifying the sentiment towards different aspects of text entities [1]. There is no fundamental difference between document and sentence level classifications, because sentences are just short documents, and in literature they are also referred as context-level, whereas aspect-level is mentioned as concept-level classification [2]. Moreover, the sentiment classification techniques can be divided into: (a) machine learning approach

[1] http://www.alexa.com/siteinfo/twitter.com—http://mywptips.com/top-microblogging-sites-list/.

[2] https://about.twitter.com/company—http://www.internetlivestats.com/.

[3] https://en.wikipedia.org/wiki/Sentiment_analysis.

(supervised), using a variety of features and labeled data for training sentiment classifiers, (b) lexicon based approach (unsupervised), relying on lists of words with predetermined emotional weight, and (c) hybrid approach, combining both techniques [3]. According to the opinion dimensions that the sentiment analysis methods attempt to measure, a taxonomy is: (a) polarity, categorized as positive, negative or neutral using lexical resources composed of positive and negative words, (b) strength, indicating the intensity of sentiment according to lists of opinion words with strength scores, and (c) emotion, extracting emotion or mood states based on a list of expressions [4]. This chapter focuses on context-level classification using machine learning approached to identify polarity.

Sentiment analysis over Twitter is the task of classifying tweets based on feeling that the user intended to transmit, utilizing the aforementioned approaches. Twitter sentiment analysis is a growing research area with significant applications [5]. For instance, the extracting information is invaluable to companies, organizations and governments alike, in order to evaluate human reaction on their services and products [6–8]. In the educational context, e-learning systems can incorporate student emotional state to the user model and provide adaptive content, recommendations about activities or personalized assistance. Such a case is CoMoLE, an adaptive e-learning system that uses students' sentiment and proposes to learner motivational activities (e.g. games, simulations, etc.) when detecting a negative emotion. Moreover, extracting learner sentiment towards an ongoing course can act as feedback for the teacher [9].

Sentiment analysis over Twitter is extremely valuable for both industry and academia and is becoming increasingly important for web mining. Consequently, different data pre-processing techniques have been emerged, a wide range of sentiment classification algorithms has been developed and novel features and hybrid methods have been researched for more efficient and accurate results [10].

Concerning data pre-processing techniques, tweets faces several new challenges due to the typical short length and irregular structure of such content. Hence, the data preprocessing is a crucial step in sentiment analysis, since selecting the appropriate preprocessing methods, the correctly classified instances can be increased [11]. This chapter tackles with the extended comparison of sentiment polarity classification methods for Twitter text and the role of text pre-processing in sentiment analysis [12]. The preprocessing methods evaluated by the current research are three different data representations: unigram, bigrams and 1-to-3 grams, and two feature extraction filters: one based on information gain and the other based on Random Forest algorithm. Four well-known machine learning algorithms were selected for tweets classification, namely Naïve Bayes (NB), Support Vector Machine (SVM), k-Nearest Neighbors (KNN) and Decision Tree (C4.5), using tenfold cross-validation method. The experiment results demonstrating that with the feature selection and representation can affect the classification performance positively.

Regarding the determination on the appropriate algorithms to apply and combine for better outcomes, the current chapter, firstly, focuses on the comparative analysis of four well-known classifiers, namely Naive Bayes (NB), Support Vector Machine (SVM), k-Nearest Neighbors (KNN) and Decision Tree (C4.5). These classifiers were chosen as the most representative of machine learning and tested using percentage split test model. Secondly, it is examined the combination of multiple classifiers in order to obtain better predictive performance. Hence, the performance of the aforementioned learning-based classifiers is compared with the most common types of ensembles—Bagging, Boosting, Stacking and Voting [13]. In these experiments, a data preprocessing phase was performed based on the optimal option emerged from preprocessing techniques evaluation and percentage split test model was used.

Finally, a case study has been presented showing a testbed for our research concerning the sentiment analysis in Twitter, when used for educational purposes. More specifically, undergraduate students of the Department of Informatics in the University of Piraeus were prompted to use Twitter in the contexts of object-oriented programming tutoring. They were supported in utilizing Twitter modules which were tailored to tutoring systems, such as content delivery, exercise answering, advice giving and so on. Many tweets were posted and the main result of these was that digital learning through social networks can be further enhanced by reaping the benefits of sentiment analysis.

7.2 Research Methodology

The goal of the current research is to address the following research issues:

1. Which is the role of data preprocessing techniques on classification problems?
2. Which stand-alone classifier results better accuracy?
3. Does the ensemble of algorithms outperform the stand-alone ones?
4. Does the use of multiple datasets on different domains demonstrate different performance indicators?

In order to answer these issues, a set of experiments was conducted using different data preprocessing options and a range of well-known classifiers. All experiments were performed in three Twitter datasets: a dataset on the Obama-McCain Debate (OMD), one on Health Care Reform (HCR) and one with no particular topic focus, the Stanford Twitter Sentiment Gold Standard (STS-Gold) dataset.

Firstly, we examine the performance of several well-known learning-based classification algorithms using various preprocessing options. A weighting scheme, a stemming library and a stop-words removal list were applied to tweets. Afterwards, three different tokenization settings were chosen to be compared: unigram, bigram and 1-to-3-gram. For each of these options, feature extraction methods were applied in order to estimate if the elimination of poorly characterizing attributes can be useful to get better classification accuracy. The methods used were: one based on information gain and the other based on Random Forest classifier. The results of these experiments were used in order to perform the proper preprocessing options for the comparative analysis of following classifiers.

In regard to second research issue, it was chosen four representative and state-of-the-art machine learning algorithms which cover different classification approaches.

The classifiers are the Naïve Bayes (NB), Support Vector Machine (SVM), k-Nearest Neighbors (KNN) and Decision tree (C4.5). For the validation phase, two commonly used methods were implemented: percentage split and k-fold cross validation.

Towards the combination of sentiment analysis algorithms, we run the ensemble algorithms, namely Bagging, AdaboostM1, Stacking and Vote, using the four base classifiers, mentioned above. The results of these ensembles were compared to those proceeded from the previous analysis of stand-alone classifiers.

The evaluation of models used in this research was the confusion matrix, one of the most popular tools.[4] Its focus is on the predictive capability of a model rather than how fast the model takes to perform the classification, scalability, etc. The confusion matrix is represented by a matrix which each row represents the instances in a predicted class, while each column represents in an actual class. Various measures, such as error-rate, accuracy, specificity, sensitivity, and precision, and several advanced measures, such as ROC and Precision-Recall, are derived from the confusion matrix. One of the advantages of using this performance evaluation tool is that it can be easily found if the model is confusing two classes (i.e. commonly mislabeling one as another). The matrix also shows the accuracy of the classifier as the percentage of correctly classified patterns in a given class divided by the total number of patterns in that class. The overall (average) accuracy of the classifier is also evaluated by using the confusion matrix.[5]

The preprocessing settings and the learning-based algorithms were executed using Weka data mining package.[6] The outcomes of the implementation have been tabulated. Afterwards, a descriptive analysis has been conducted to answer to research issues (Fig. 7.1).

7.3 Twitter Datasets

Three well-known Twitter datasets were used in current experiments; a dataset on the Obama-McCain Debate (OMD) [14], one on Health Care Reform (HCR) [15] and one with randomly selected tweets, the Stanford Twitter Sentiment Gold Standard (STS-Gold) dataset [16]. The reason they have been chosen is that they are available on the Web with no charge and have been created by reputable universities for academic scope with a significant number of tweets. Moreover, they have been used in various researches [17]. Thus, these datasets are considered as reliable for our experiments. The statistics of the datasets are shown in Table 7.1.

Obama-McCain Debate (OMD). The Obama-McCain Debate (OMD) dataset was constructed from 3238 tweets crawled during the first U.S. presidential TV debate in September 2008 [14]. Sentiment labels were acquired by using Amazon

[4] http://rali.iro.umontreal.ca/rali/sites/default/files/publis/SokolovaLapalme-JIPM09.pdf.

[5] http://aimotion.blogspot.gr/2010/08/tools-for-machine-learning-performance.html.

[6] http://www.cs.waikato.ac.nz/ml/weka/index.html.

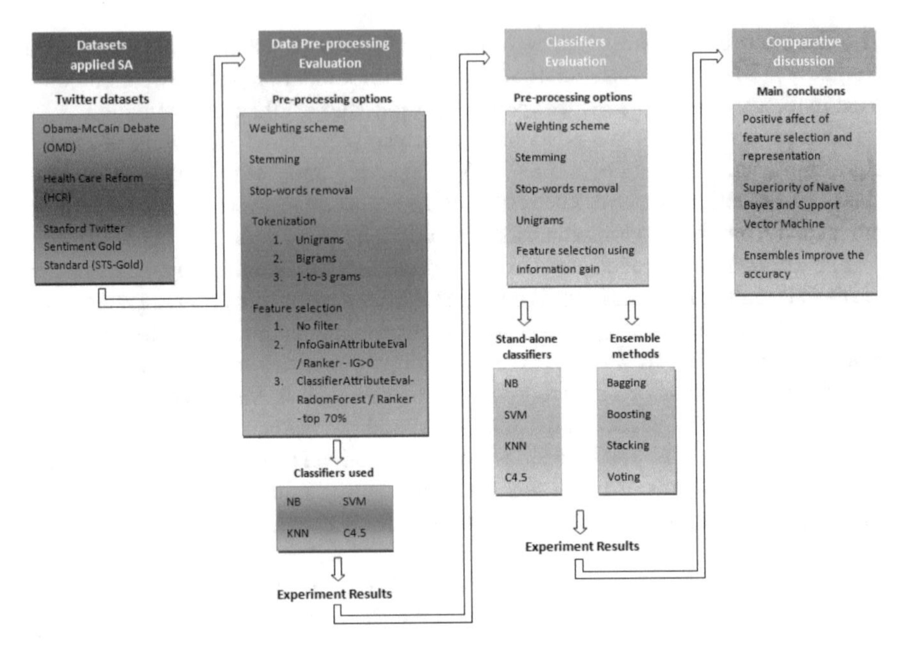

Fig. 7.1 Research methodology

Table 7.1 Statistics of the three Twitter datasets used

Dataset	Tweets	Positive	Negative
Obama-McCain Debate (OMD)	1904	709	1195
Health Care Reform (HCR)	1922	541	1381
Stanford Twitter Sentiment Gold Standard (STS-Gold)	2034	632	1402

Mechanical Turk. The set used in this paper consisted of 709 positive and 1195 negative, on which two-third of the voters had agreed.

Health Care Reform (HCR). The Health Care Reform (HCR) dataset was built by tweets with the hashtag #hcr (health care reform) in March 2010 [15]. A set of 2516 tweets was manually annotated by the authors with 5 labels: positive, negative, neutral, irrelevant, unsure. For this research, a subset of 1922 tweets was considered, excluding irrelevant, unsure and neutral labeled tweets. Hence, the final dataset included 541 positive and 1381 negative tweets.

Stanford Sentiment Gold Standard (STS-Gold). The STS-Gold dataset was created by selecting tweets from Stanford Twitter Sentiment Corpus 10 and contains independent sentiment labels for tweets and entities, supporting the evaluation of tweet-based as well as entity-based Twitter sentiment analysis models [16]. In current experiments, the set of 2034 tweets was used with 632 positive and 1402 negative ones.

7.4 Evaluation of Data Preprocessing Techniques

Preprocessing is a necessary data preparation step for sentiment classification. To perform the preprocessing in WEKA, we use the StringToWordVector filter. This filter allows the following configurations:

- **TF-IDF weighting scheme**: It is a standard approach to feature vector construction. TF-IDF stands for the "term frequency-inverse document frequency" and is a numerical statistic that reflects how important a word is to a document in a corpus.
- **Stemming**: Stemming algorithms work by removing the suffix of the word, according to some grammatical rules. In this study, we apply the Snowball stemmer library,[7] which is the most popular and standard approach.
- **Stop-words removal**: It is a technique that eliminates the frequent usage words which are meaningless and useless for the text classification. This reduces the corpus size without losing important information. The Rainbow list[8] is used for our experiments.
- **Tokenization**: This setting splits the documents into words/terms, constructing a word vector, known as bag-of-words. We propose NGramTokenizer to compare word unigram, bigram and 1-to-3-gram.

The above preprocessing generates a huge number of attributes, many of them being not relevant with classification. Hence, we apply the following operation:

- **Feature selection**: It is a process by which the number of attributes is decreased into a better subset which can bring highest accuracy. The benefits of performing this option on the data are the limitation of overfitting, the improvement of accuracy and the reduction in training time. Feature Selection methods can be classified as Filters and Wrappers. Filters are based on statistical tests, such as Infogain, Chisquare and CFS, while Wrappers use a learning algorithm to report the optimal subset of features. For this task, WEKA provides the AttributeSelection filter which allows to choose an attribute evaluation method and a search strategy. In this paper, we examine three options:

  a. No filter applied. We use all attributes created by StringToWordVector filter.
  b. InfoGainAttributeEval, which evaluates the worth of an attribute by measuring the information gain with respect to the class and we set the Ranker search method to select attributes with $IG > 0$, and
  c. ClassifierAttributeEval, which evaluates the worth of an attribute by using a user-specified classifier. We choose Random Forest as classifier and set the Ranker search method to select the top 70% attributes (Table 7.2).

In order to specify the optimal settings of the preprocessing techniques and the classifiers, we conducted a variety of experiments testing the options that would return more accurate results. The chosen preprocessing methods, described above,

[7] http://snowball.tartarus.org/.

[8] http://www.cs.cmu.edu/~mccallum/bow/rainbow/.

Table 7.2 Preprocessing techniques applied

Preprocessing technique	Applied option
Weighting scheme	TF-IDF
Stemming	Snowball stemmer
Stop-words removal	Rainbow list
Tokenization	1. Unigram 2. Bigram 3. 1-to-3-gram
Feature selection	1. All 2. InfoGainAttributeEval/Ranker—IG>0 3. ClassifierAttributeEval-RadomForest/Ranker—top 70%

Table 7.3 Number of attributes created by attribute selection option

Dataset	Attribute selection	N-gram		
		Unigram	Bigram	1-to-3-gram
OMD	All	2150	7400	2430
	IG>0	264	519	1074
	Top 70%	1500	5180	1680
HCR	All	3000	1835	2945
	IG>0	281	645	1280
	Top 70%	2100	1280	2060
STS-Gold	All	2990	8420	2115
	IG>0	252	354	720
	Top 70%	2090	5890	1480

was applied to the three Twitter datasets and for the classification, the Naïve Bayes Multinomial (NBM), nu-SVM type, KNN with k = 19 and default settings of C4.5 were chosen. For the validation phase, the tenfold cross validation method was used.

According to n-gram and attribute selection options, a different number of attributes was created based on which the classification was performed. Table 7.3 demonstrates that numbers. We observe that selecting the attributes with information gain upper than zero, the resultant number of them is decreased appreciably. For the second feature extraction, where the attributes are evaluated by Radom Forest algorithm, we choose approximately the 70% of attributes ranked as more worthy. An expected benefit of attribute selection is that algorithms train faster.

Table 7.4 demonstrates the performance of classifiers depending on the preprocessing methods applied. Regarding dataset representations, the behavior is not uniform. There is no representation that brings systematically better results in comparison with the others. In general, 1-to-3-grams perform better than the other representations, having a close competition with unigram.

Table 7.4 Classifiers' confusion matrices related with preprocessing techniques

N-gram	Attribute selection	Measures	Dataset											
			OMD				HCR				STS-Gold			
			Classifiers				Classifiers				Classifiers			
			NB	SVM	KNN	C4.5	NB	SVM	KNN	C4.5	NB	SVM	KNN	C4.5
Unigram	All	Acc. (%)	79.46	81.93	73.32	75.26	77.59	**83.15**	72.02	73.63	82.10	83.63	68.93	74.29
		Pr.	0.792	0.831	0.785	0.750	0.763	0.852	0.799	0.712	0.817	0.850	0.475	0.731
		R.	0.795	0.819	0.733	0.753	0.776	0.832	0.720	0.736	0.821	0.836	0.689	0.743
		FM.	0.793	0.809	0.689	0.742	0.763	0.808	0.605	0.714	0.814	0.821	0.562	0.711
	IG>0	Acc. (%)	86.61	**83.61**	**74.21**	**75.89**	85.34	79.04	**72.85**	**74.57**	88.35	**83.92**	**73.40**	**74.29**
		Pr.	0.868	0.860	0.735	0.762	0.853	0.839	0.815	0.731	0.882	0.852	0.776	0.734
		R.	0.866	0.836	0.691	0.759	0.851	0.799	0.771	0.751	0.883	0.839	0.734	0.743
		FM.	0.863	0.825	0.568	0.744	0.847	0.782	0.693	0.694	0.881	0.825	0.663	0.707
	Top 70%	Acc. (%)	**88.08**	83.25	73.58	74.79	**88.25**	82.53	72.02	73.27	**91.40**	83.68	68.93	74.04
		Pr.	0.880	0.852	0.748	0.748	0.884	0.853	0.799	0.707	0.914	0.854	0.475	0.732
		R.	0.881	0.832	0.736	0.748	0.882	0.825	0.720	0.733	0.914	0.837	0.689	0.740
		FM.	0.880	0.822	0.709	0.733	0.876	0.797	0.605	0.709	0.912	0.820	0.562	0.701
Bigrams	All	Acc. (%)	75.26	85.82	62.76	**69.22**	82.16	78.00	71.81	71.66	70.30	**85.05**	68.93	70.50
		Pr.	0.736	0.882	0.394	0.728	0.822	0.815	0.516	0.659	0.680	0.877	0.475	0.702
		R.	0.723	0.858	0.628	0.692	0.822	0.780	0.718	0.717	0.703	0.851	0.689	0.705
		FM.	0.692	0.849	0.484	0.632	0.822	0.728	0.600	0.635	0.624	0.834	0.562	0.616
	IG>0	Acc. (%)	**89.02**	82.88	**63.39**	68.28	**89.13**	**78.36**	71.81	71.92	**86.77**	80.63	**68.98**	70.21
		Pr.	0.908	0.848	0.415	0.731	0.895	0.834	0.516	0.672	0.874	0.836	0.683	0.697
		R.	0.898	0.829	0.632	0.691	0.891	0.784	0.718	0.719	0.868	0.806	0.690	0.702
		FM.	0.890	0.822	0.488	0.625	0.885	0.730	0.600	0.612	0.860	0.778	0.565	0.608

(continued)

Table 7.4 (continued)

N-gram	Attribute selection	Measures	Dataset											
			OMD				HCR				STS-Gold			
			Classifiers				Classifiers				Classifiers			
			NB	SVM	KNN	C4.5	NB	SVM	KNN	C4.5	NB	SVM	KNN	C4.5
	Top 70%	Acc. (%)	83.04	**87.08**	62.76	69.01	87.52	78.16	**71.82**	**72.07**	80.53	84.32	68.93	**70.55**
		Pr.	0.853	0.892	0.394	0.747	0.873	0.827	0.516	0.682	0.837	0.872	0.475	0.705
		R.	0.830	0.871	0.628	0.690	0.875	0.782	0.718	0.721	0.805	0.843	0.689	0.706
		FM.	0.818	0.863	0.484	0.621	0.871	0.728	0.600	0.619	0.776	0.825	0.562	0.616
1-to-3-grams	All	Acc. (%)	85.77	82.51	**68.86**	76.05	83.26	**77.54**	**72.13**	73.95	87.56	81.91	68.93	74.39
		Pr.	0.857	0.842	0.792	0.757	0.828	0.818	0.764	0.717	0.878	0.833	0.475	0.730
		R.	0.858	0.825	0.689	0.761	0.833	0.775	0.721	0.739	0.876	0.819	0.689	0.744
		FM.	0.857	0.814	0.608	0.754	0.830	0.719	0.609	0.720	0.877	0.800	0.562	0.718
	IG>0	Acc. (%)	**92.59**	**84.14**	66.02	**76.21**	**91.94**	76.76	71.97	**74.62**	**92.67**	**83.83**	69.12	74.73
		Pr.	0.928	0.872	0.780	0.767	0.919	0.824	0.798	0.729	0.926	0.867	0.735	0.742
		R.	0.926	0.841	0.660	0.762	0.919	0.768	0.720	0.746	0.927	0.838	0.691	0.747
		FM.	0.925	0.829	0.554	0.747	0.917	0.702	0.604	0.694	0.926	0.819	0.568	0.710
	Top 70%	Acc. (%)	88.34	83.88	66.44	75.37	87.94	77.17	71.92	73.90	90.81	82.74	**68.93**	**74.93**
		Pr.	0.883	0.860	0.771	0.754	0.879	0.826	0.797	0.717	0.907	0.849	0.475	0.741
		R.	0.883	0.839	0.664	0.754	0.875	0.782	0.719	0.738	0.908	0.827	0.689	0.749
		FM.	0.882	0.828	0.564	0.740	0.871	0.732	0.601	0.719	0.908	0.808	0.562	0.717

Table 7.5 Relative improvement in accuracy of classifiers depending on attribute selection options

Dataset		OMD		HCR		STS-Gold	
Attribute selection		IG>0	Top 70%	IG>0	Top 70%	IG>0	Top 70%
N-gram	Classifiers						
Unigram	NB	**+7.15**	**+8.62**	**+7.75**	**+10.66**	**+6.25**	**+9.30**
	SVM	+1.68	+1.32	−4.11	−0.62	+0.26	+0.05
	KNN	+0.89	+0.26	+0.83	0	+4.47	0
	C4.5	+0.63	−0.47	+0.94	−0.36	0	−0.25
Bigram	NB	**+13.76**	**+7.78**	**+6.97**	**+5.36**	**+16.47**	**+10.23**
	SVM	−2.94	+1.26	+0.36	+0.16	−4.42	−0.73
	KNN	+0.63	0	0	+0.01	+0.05	0
	C4.5	−0.94	−0.21	+0.26	+0.41	−0.29	+0.05
1-to-3-gram	NB	**+6.82**	+2.57	**+8.68**	+4.68	**+5.11**	+3.25
	SVM	+1.63	+1.37	−0.78	−0.37	+1.92	+0.83
	KNN	−2.84	−2.42	−0.16	−0.21	+0.19	0
	C4.5	+0.16	−0.68	+0.67	−0.05	+0.34	+0.54

Our evaluation results indicate that the attribute selection operation improves the performance of classification over selecting all attributes. This proceeds from the removal of redundant and irrelevant attributes from the datasets which can be misleading to modeling algorithms and result in overfitting. In Table 7.5, we observe that significant accuracy rates are obtained when applying the attribute selection based on information gain. Note particularly that in case of NB, the percentage of correctly classified instances is increased over 7 points. Moreover, the Random Forest algorithm as attribute selection classifier improves classification accuracy in comparison to using all attributes. Finally, in some experiment settings, there is no improvement in algorithms' performance by applying an attribute selection filter, but this is of insignificant value as the divergence is too low.

Figures 7.2, 7.3 and 7.4 illustrate the level of accuracy each classifier have achieved according to preprocessing technique applied to each dataset.

Evaluating the influence of dataset domain on preprocessing performance, we use three different datasets, one with no specific domain tweets and the others with a specific topic. The experiment results show that the effect of preprocessing techniques is the same regardless of the datasets.

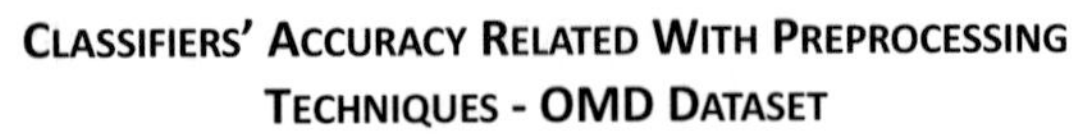

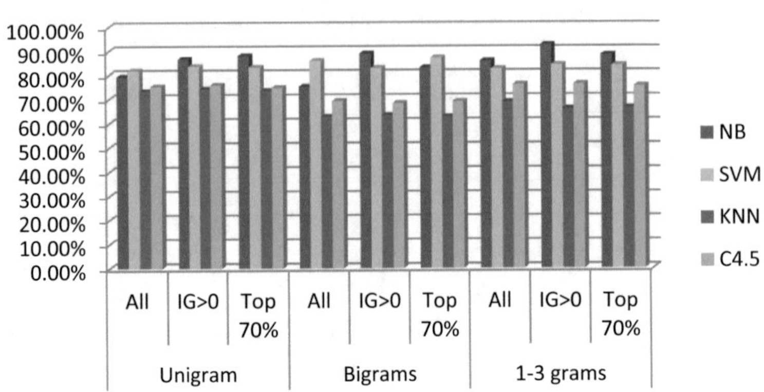

Fig. 7.2 Classifiers' accuracy related with preprocessing techniques on OMD dataset

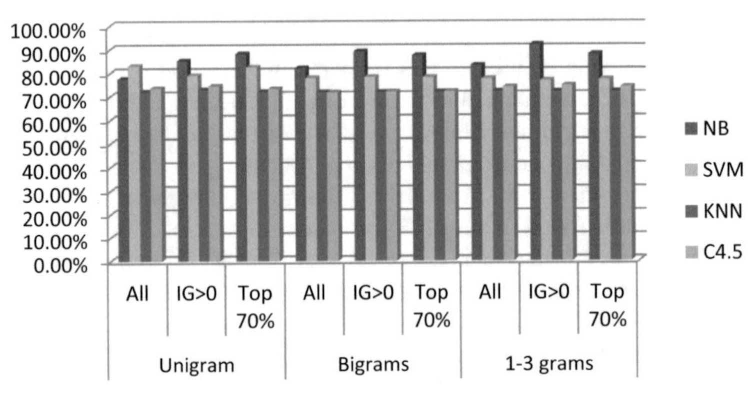

Fig. 7.3 Classifiers' accuracy related with preprocessing techniques on HCR dataset

7.5 Evaluation of Stand-Alone Classifiers

This section focuses on the comparative performance evaluation of different sentiment approaches. Therefore, it was chosen four representative and state-of-the-art machine learning algorithms, which are provided by Weka. Note particularly that the selected machine learning algorithms figured on the top 10 most influential data mining algorithms identified by the IEEE International Conference on Data Mining (ICDM) in December 2006, the 11 algorithms implemented by 11 Ants and the Oracle Data Mining (ODM) component.[9] Moreover, another parameter considered for

[9]https://www.quora.com/What-are-the-top-10-data-mining-or-machine-learning-algorithms.

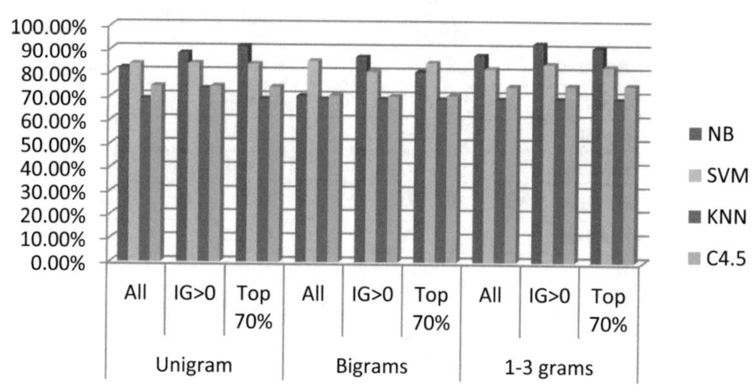

Fig. 7.4 Classifiers' accuracy related with preprocessing techniques on STS-Gold dataset

Table 7.6 Experiment classifiers

Classifier	Approach
Naïve Bayes (NB)	Probabilistic learning algorithm
Support Vector Machines (SVM)	Supervised learning model
k-Nearest Neighbor (KNN)	Instance-based learning algorithm
Logistic Regression (LR)	Regression model
C4.5	Decision tree

the algorithms' election was to cover different classification approaches. Table 7.6 shows the classifiers used.

Naïve Bayes (NB). Naive Bayes classifier is a probabilistic classifier based on applying Bayes' theorem with strong (naive) independence assumptions between the features.

Support Vector Machines (SVM). A Support Vector Machine (SVM) performs classification by finding the hyperplane that maximizes the margin between the two classes. The vectors (cases) that define the hyperplane are the support vectors.

k-Nearest Neighbors (KNN). The k-Nearest Neighbors algorithm is a instance-based learning, where a case is classified by a majority vote of its neighbors, with the case being assigned to the class most common amongst its k nearest neighbors measured by a distance function.

Decision tree (C4.5). C4.5 is an extension of earlier ID3 algorithm. C4.5 builds decision trees from a set of training data in the same way as ID3, using the concept of information entropy.

In order to perform the classification, a preliminary phase of text preprocessing and feature extraction is essential. Therefore, for each dataset, the tweets were transformed in a vector form by applying word tokenization, stemming and stop-words

Table 7.7 Classification results of machine learning algorithms

Classifiers	Measures	Datasets		
		OMD	HCR	STS-Gold
NB	Accuracy (%)	87.57	85.10	88.69
	Precision	0.876	0.848	0.886
	Recall	0.876	0.851	0.887
	F-Measure	0.873	0.841	0.884
SVM	Accuracy (%)	82.49	79.20	84.10
	Precision	0.836	0.820	0.849
	Recall	0.825	0.792	0.841
	F-Measure	0.814	0.743	0.826
KNN	Accuracy (%)	74.26	73.66	73.77
	Precision	0.773	0.740	0.766
	Recall	0.743	0.737	0.738
	F-Measure	0.704	0.631	0.660
C4.5	Accuracy (%)	74.96	74.35	73.44
	Precision	0.747	0.710	0.715
	Recall	0.750	0.744	0.734
	F-Measure	0.733	0.676	0.680

removal except emoticons. Moreover, the most worthy and relevant to the classification attributes were selected by measuring the information gain with respect to the class and rejecting them with information gain less than zero. With this option, emerged from the previous section of preprocessing evaluation, we can obtain better accuracy results and reduce training time. Afterwards, the algorithms were run.

Table 7.7 demonstrates the classification outcomes of the four machine learning algorithms used. The results show a close competition between NB and SVM, as they are more efficient than others, having precision rates over 0.8 approximately in all experiments with respective F-measure values, independently of dataset. This attests the fact that NB and SVM classifiers are widespread in sentiment analysis and the reason they are used in an abundance of such cases. On the other hand, the KNN and C4.5 return not so satisfied results, having their accuracy near 74%.

Regarding the recall, the proportion of positives that are correctly identified as such, in the majority of our experiments, the algorithms return values higher than 0.7 and near the precision rates. This results also satisfactory F-measure, which is the weighted harmonic mean of precision and recall. In particular, NB and SVM have around 0.8 recall as precision in all classifications of datasets. This means that the algorithms have high probability to avoid classifying false negatives.

In this study, each algorithm was applied in three different datasets. In STS-Gold, there is no specific domain, while the other datasets address specific topics. The results show that despite the fact that the algorithms' performance varies from one dataset to the other, the comparatively well-performed classifier is the same.

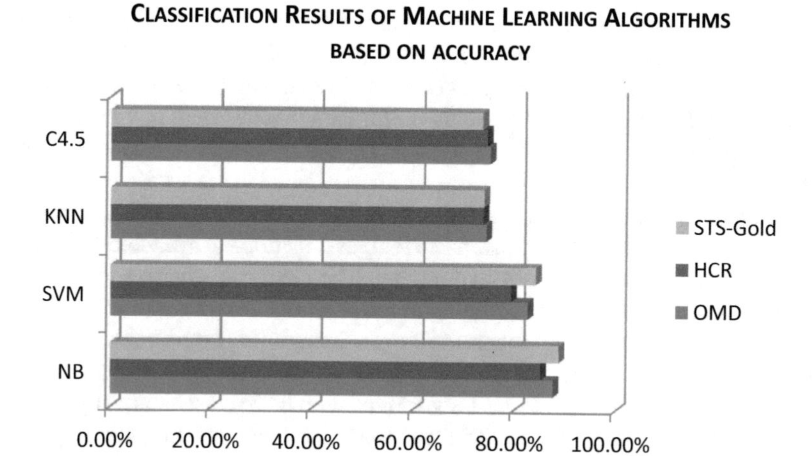

Fig. 7.5 Classification results of machine learning algorithms based on accuracy

Therefore, we conclude that NB and SVM algorithms are a reliable solution for sentiment analysis problems regardless of the dataset (Fig. 7.5).

The aforementioned results are confirmed also by the research in [18], where different preprocessing techniques were applied to the same datasets.

7.6 Evaluation of Ensemble Classifiers

Ensemble learning is the process by which multiple classifiers are combined in order to obtain higher accuracy than the individual classifiers. In this work, we examine four well-known ensemble learning techniques. A short description of each of them follows.

Bagging (**B**ootstrap **Agg**regation) generates random samples of the training dataset and applies the base learning algorithms on each sample. The results of these multiple classifiers are then combined using averaged or majority voting. This method reduces variance of unstable procedures (e.g. decision trees) and hence improves accuracy.

Boosting is an ensemble method that complementary models are generated by training each new model on previous models mis-classified. The procedure is repeated until a limit is reached in the number of models or accuracy. Although boosting outperforms bagging in some cases, it is more likely to over-fit the training data. AdaBoost is the most representative algorithm in this family. As weak classifiers, it is usually used rule based classifiers, one-two level trees, neural networks without hidden layers etc.

Stacking executes different learning algorithms on the training data and then uses a meta-classifier which takes the predictions of each classifier as additional inputs. This can lead to decrease in either bias or variance error depending on the combining learner used. Stacking typically yields performance better than any single one of the trained models. A single-layer logistic regression model is often used as the combiner.

Voting is a straight-forward adaptation of voting for distribution classifiers. Different combinations of probability estimates for classification are available, for instance average of probabilities, majority voting etc. The ensemble chooses the class that receives the largest total vote according to the combination rule.

In our experiments, we try to increase the efficiency of four well-known classifiers: Naïve Bayes, SVM, KNN and C4.5, using the above ensembles. In particular, we examine Bagging and AdaBoostM1 algorithms with these classifiers as base learner each time, Stacking algorithm with these four classifiers and Logistic Regression as the meta-classifier, and Vote algorithm again with these four classifiers using the average of probabilities and majority voting combination rule each time. In all implementations, the number of 10 interactions is performed for constructing the ensemble models, while the preprocessing options mentioned in the previous section were applied also. Table 7.8 shows the ensemble approaches used.

Table 7.9 demonstrates the overall picture of all ensemble approaches used. In regard to bagging and boosting methods, which use a base classifier trying to optimize its performance, the results show that they return more correctly classified instances than the stand-alone classifier and boosting outperforms bagging in most cases. Furthermore, these approaches yield better accuracy when using weak algorithms, such as KNN and C4.5, which is confirmed by literature.

Comparing stacking and voting, two methods that combine multiple base classifiers, to each individual base classifier, we observe that these ensembles outperform mainly the weak classifiers, KNN and C4.5. These ensembles may not necessarily improve the performance of the best classifier in the combination; however they certainly reduce the overall risk of making a mis-classification, as it is unlikely that all classifiers will make the same mistake. Table 7.10 represents the relative improvement in accuracy of ensembles in comparison with the stand-alone classifiers. It is worth noting that in some cases stacking and voting outperform KNN and C4.5 more than 10 points.

Evaluating the influence of dataset domain on ensemble learning, we use three different datasets, one with no specific domain tweets and the others with a specific topic. The experiment results show that the comparatively well-performed approach is the same regardless of the dataset.

Figure 7.6 illustrates the accuracy of tested ensembles on the selected datasets.

Table 7.8 Ensemble classifiers used

Ensemble type	Implementation
Bagging	Bagging with base classifier: 1. NB 2. SVM 3. KNN 4. C4.5
Boosting	AdaBoostM1 with base classifier: 1. NB 2. SVM 3. KNN 4. C4.5
Stacking	Stacking with 4 base classifiers: • NB • SVM • KNN • C4.5 and meta-classifier: 1. LR
Voting	Vote with 4 base classifiers: • NB • SVM • KNN • C4.5 and combination rule: 1. average of probabilities 2. majority voting

Table 7.9 Classification results of ensemble learning methods

Classifiers		Measures	Datasets		
			OMD	HCR	STS-Gold
NB	Simple	Accuracy (%)	87.57	85.10	88.69
		Precision	0.876	0.848	0.886
		Recall	0.876	0.851	0.887
		F-Measure	0.873	0.841	0.884
	Bagging	Accuracy (%)	**87.74**	**85.10**	88.69
		Precision	0.878	0.848	0.886
		Recall	0.877	0.851	0.887
		F-Measure	0.875	0.841	0.884
	AdaBoostM1	Accuracy (%)	86.34	83.54	**88.85**
		Precision	0.863	0.832	0.887
		Recall	0.863	0.835	0.889
		F-Measure	0.863	0.822	0.885

(continued)

Table 7.9 (continued)

Classifiers		Measures	Datasets		
			OMD	HCR	STS-Gold
SVM	Simple	Accuracy (%)	82.49	79.20	84.10
		Precision	0.836	0.820	0.849
		Recall	0.825	0.792	0.841
		F-Measure	0.814	0.743	0.826
	Bagging	Accuracy (%)	81.26	78.68	82.30
		Precision	0.836	0.835	0.842
		Recall	0.813	0.787	0.823
		F-Measure	0.796	0.729	0.799
	AdaBoostM1	Accuracy (%)	81.79	**82.67**	**87.70**
		Precision	0.816	0.820	0.877
		Recall	0.818	0.827	0.877
		F-Measure	0.816	0.814	0.872
KNN	Simple	Accuracy (%)	74.26	73.66	73.77
		Precision	0.773	0.740	0.766
		Recall	0.743	0.737	0.738
		F-Measure	0.704	0.631	0.660
	Bagging	Accuracy (%)	74.61	73.31	73.11
		Precision	0.791	0.671	0.754
		Recall	0.746	0.733	0.731
		F-Measure	0.704	0.623	0.647
	AdaBoostM1	Accuracy (%)	**75.13**	**75.04**	**80.00**
		Precision	0.745	0.722	0.800
		Recall	0.751	0.750	0.800
		F-Measure	0.745	0.695	0.778
C4.5	Simple	Accuracy (%)	74.96	74.35	73.44
		Precision	0.747	0.710	0.715
		Recall	0.750	0.744	0.734
		F-Measure	0.733	0.676	0.680
	Bagging	Accuracy (%)	**76.53**	75.22	73.11
		Precision	0.768	0.726	0.709
		Recall	0.765	0.752	0.731
		F-Measure	0.749	0.698	0.676
	AdaBoostM1	Accuracy (%)	74.96	**75.22**	**75.25**
		Precision	0.743	0.728	0.741
		Recall	0.750	0.752	0.752

(continued)

Table 7.9 (continued)

Classifiers		Measures	Datasets		
			OMD	HCR	STS-Gold
		F-Measure	0.741	0.693	0.712
Combined all	Stacking	Accuracy (%)	**87.57**	**85.10**	**89.02**
		Precision	0.875	0.848	0.890
		Recall	0.876	0.851	0.890
		F-Measure	0.874	0.841	0.886
	Vote-AP	Accuracy (%)	82.49	79.39	84.75
		Precision	0.837	0.803	0.855
		Recall	0.825	0.794	0.848
		F-Measure	0.813	0.753	0.834
	Vote-MV	Accuracy (%)	82.84	79.38	84.75
		Precision	0.840	0.803	0.855
		Recall	0.828	0.794	0.848
		F-Measure	0.818	0.753	0.834

Table 7.10 Relative improvement in accuracy of ensembles using stand-alone classifiers as baselines

Datasets	Ensembles	Base classifiers			
		NB	SVM	KNN	C4.5
OMD	Bagging	+0.17	−1.23	+0.35	+1.57
	AdaBoostM1	−1.23	−0.7	+0.87	0
	Stacking	0	**+5.08**	**+13.31**	**+12.61**
	Vote-AP	−5.08	0	**+8.23**	**+7.53**
	Vote-MV	−4.73	+0.35	**+8.58**	**+7.88**
HCR	Bagging	0	−0.52	−0.35	+0.87
	AdaBoostM1	−1.56	+3.47	+1.38	+0.87
	Stacking	0	**+5.90**	**+11.44**	**+10.75**
	Vote-AP	−5.71	+0.19	**+5.73**	**+5.04**
	Vote-MV	−5.72	+0.18	**+5.72**	**+5.03**
STS-Gold	Bagging	0	−1.80	−0.66	−0.33
	AdaBoostM1	+0.16	+3.60	**+6.23**	+1.81
	Stacking	+0.33	+4.92	**+15.25**	**+15.58**
	Vote-AP	−3.94	+0.65	**+10.98**	**+11.31**
	Vote-MV	−3.94	+0.65	**+10.98**	**+11.31**

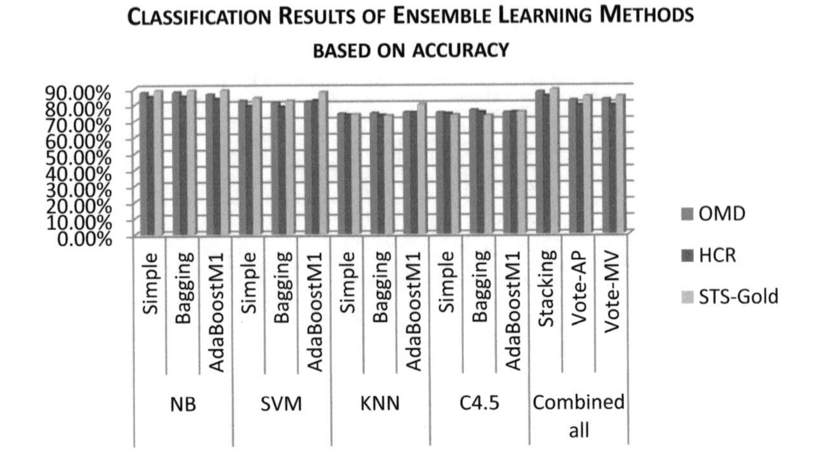

Fig. 7.6 Classification results of ensemble learning methods based on accuracy

7.7 The Case of Sentiment Analysis in Social e-Learning

As a testbed for our research, an e-learning module was created using Twitter. More specifically, Twitter served as the means to build a web-based course delivery platform and use it for supporting the teaching of the object-oriented language C# as part of undergraduate courses in the Department of Informatics of the University of Piraeus and assisting the undergraduate students of the aforementioned Department in the educational process.

Twitter was selected as an educational tool because of the fact that:

- Instructors can be connected to their students in a social context as well as in more a personal context.
- Interactions with the e-learning material and peers can take place outside the traditional class, since Twitter is ubiquitous and can be found in smart phones and laptops.
- Twitter allows a degree of adaptivity to the learners and as such tutoring can gain personalization.
- Twitter can be used to quickly connect to multimedia resources (e.g. YouTube) and offers effective ways of education.
- Twitter gives new opportunities to connect with other learning communities and new forms of educational material.
- The nature of Twitter itself is that it can support the rapid knowledge transmission.

The e-learning course delivery module uses the logical architecture of Twitter, as shown in Fig. 7.7.

The e-learning module was consisted of several important features of tutoring systems, such as the multimedia-based course delivery, exercise answering, personal

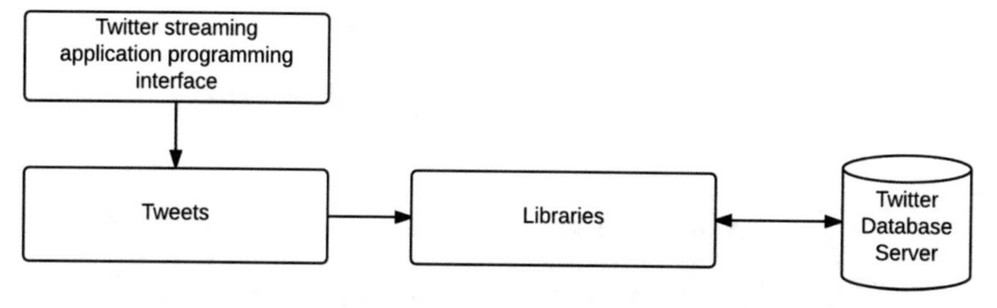

Fig. 7.7 Logical arcitecture of Thitter-based learning module

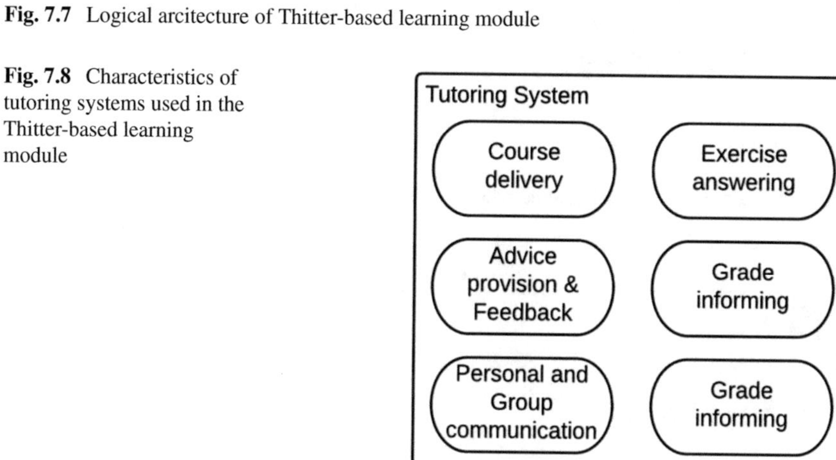

Fig. 7.8 Characteristics of tutoring systems used in the Thitter-based learning module

and group communication, advice provision and feedback, and grade informing, as shown in Fig. 7.8.

Following, several screenshots of the Twitter-based e-learning module are provided. Figure 7.9 illustrates the delivery of the learning material which can include multimedia. Figure 7.10 shows the communication between students and/or instructors with the use of text messaging. Figure 7.11 illustrates that the advice to students from instructors or peers or queries answering can take place with the use of hashtags, which can also serve as another means of communication. Figure 7.12 shows the exercise answering process using online voting. Furthermore, Fig. 7.13 shows the use of external links that can help users on finding additional learning material. Figure 7.14 illustrates how comments can serve as a way of assisting students by answering their queries or providing feedback to them. Figure 7.15 shows the uploading of exercises and Fig. 7.16 illustrates the delivery of grades to students.

The tweets that were exchanged by the 100 undergraduate students concern basically the teaching of the programming language C# and their number is 352. That means that the 100 undergraduate students exchanged 352 tweets related to the needs of the course.

With the use of two sentiment analysis approaches, i.e. the probabilistic learning algorithm Naive Bayes (NB) and the supervised learning model Support Vector

Fig. 7.9 Content delivery using multimedia

Fig. 7.10 Communication using text messaging between users

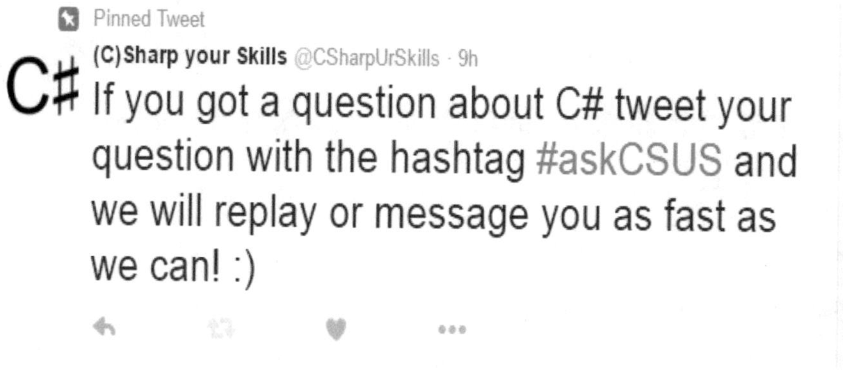
Pinned Tweet
(C)Sharp your Skills @CSharpUrSkills · 9h
C#
If you got a question about C# tweet your question with the hashtag #askCSUS and we will replay or message you as fast as we can! :)

Fig. 7.11 Instructor's advice and communication using hashtags

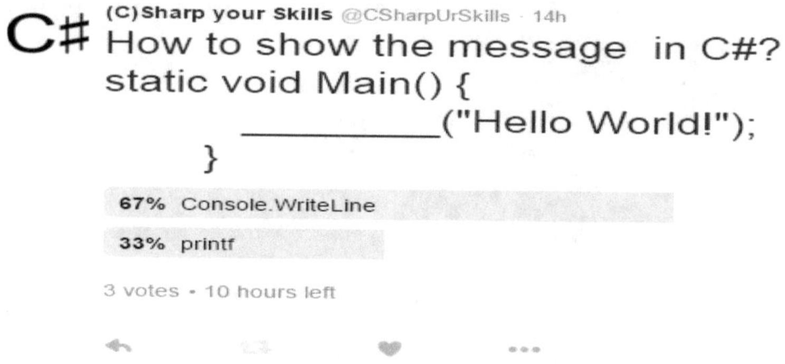
(C)Sharp your Skills @CSharpUrSkills · 14h
C#
How to show the message in C#?
static void Main() {
__________("Hello World!");
}
67% Console.WriteLine
33% printf
3 votes • 10 hours left

Fig. 7.12 Exercise answering using online voting

(C)Sharp your Skills @CSharpUrSkills · 14h
C#
Second Lesson!! The structure of a program written in C#!!
tutorialspoint.com/csharp/csharp_ …

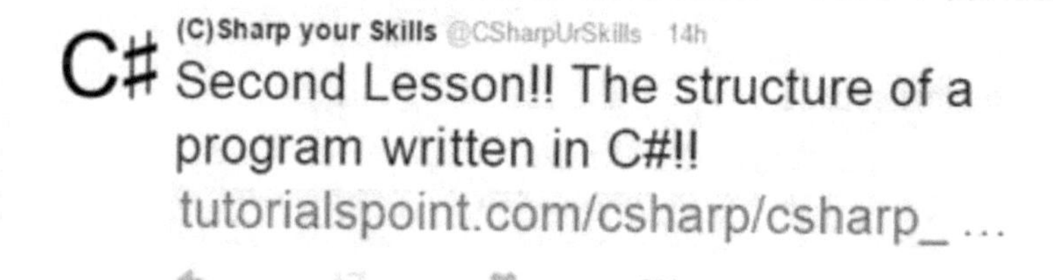

Fig. 7.13 Use of external links

Machines (SVM), students' tweets were classified in order to detect which of them are positive, negative or neutral. Both of these approaches showed remarkable results, as can be also proved by our experimental results, shown in previous sections.

The most important result is that the sentiment analysis in e-learning can serve as a powerful tool for instructors and can assist them on ameliorating the educational procedure. Specifically, instructors can acquire notion concerning the affective state

Fig. 7.14 Feedback to students using commenting

Fig. 7.15 Uploading of exercises

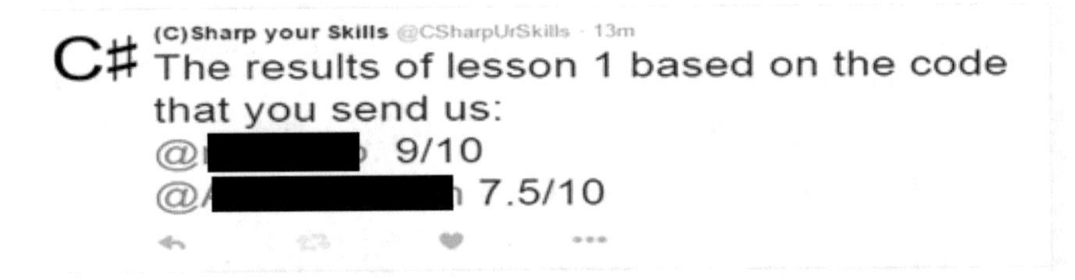

Fig. 7.16 Delivering students' grades

of students and can help them by delivering advice, changing the learning material and adapting it to students, motivating them and preventing them from quitting learning. As such, sentiment analysis can give input concerning the student's emotional states to student models so that the personalization in digital learning and the adaptivity to student's needs are further improved.

7.8 Conclusions and Future Work

Social networks have become a pool of data which can be appropriately used by sentiment analysis scientists. Therefore, they can elicit people's emotions and point of view for many different cases, such as commercial industry or e-learning. In this paper, a deep insight of sentiment analysis using different approaches is presented. More specifically, the application of pre-processing in analyzing people's emotions is described and its techniques are evaluated. Furthermore, an evaluation of standalone classifiers is shown. Also, the algorithms of ensemble averaging for sentiment analysis are examined; particularly, algorithmic techniques for the creation and combination of multiple models towards a desired output are presented and evaluated. Moreover, a case study for the sentiment analysis in a Twitter-based learning platform is given; the result showed that sentiment analysis plays a significant role in e-learning since instructors can acquire a clear representation of students' emotional states and as such they can adapt the learning content to students as well as individualized instruction can be promoted.

Future work includes the creation of a novel hybrid model using pre-processing techniques and ensemble classifiers for effective sentiment analysis. Also, the adjustment of this model in different cases, such as business marketing or public sector, is a future plan that may be proved beneficial for ameliorating the provision of services to individuals.

References

1. Yadav, S.K.: Sentiment analysis and classification: a survey. Int. J. Adv. Res. Comput. Sci. Manag. Stud. **3**(3), 113–121 (2015)
2. Saif, H., Fernandez, M., He, Y., Alani, H.: SentiCircles for contextual and conceptual semantic sentiment analysis of twitter. In: 11th International Conference on Semantic Web: Trends and Challenges (ESWC 2014), Crete, Greece (2014)
3. Medhat, W., Hassan, A., Korashy, H.: Sentiment analysis algorithms and applications: a survey. Ain Shams Eng. J. **5**(4), 1093–1113 (2014)
4. Bravo-Marquez, F., Mendoza, M., Poblete, B.: Meta-level sentiment models for big social data analysis. Knowl.-Based Syst. **69**(1), 86–99 (2014)
5. Martínez-Cámara, E., Martín-Valdivia, M.T., Ureña-López, L.A., Montejo-Ráez, A.R.: Sentiment analysis in twitter. Nat. Lang. Eng. **20**(1), 1–28 (2014)
6. Smailović, J., Grčar, M., Lavrač, N., Žnidaršič, M.: Stream-based active learning for sentiment analysis in the financial domain. Inf. Sci. **285**(1), 181–203 (2014)
7. Mostafa, M.M.: More than words: social networks' text mining for consumer brand sentiments. Expert Syst. Appl. **40**(10), 4241–4251 (2013)
8. Cheong, M., Lee, V.C.S.: A microblogging-based approach to terrorism informatics: exploration and chronicling civilian sentiment and response to terrorism events via twitter. Inf. Syst. Front. **13**(1), 45–59 (2011)
9. Ortigosa, A., Martín, J.M., Carro, R.M.: Sentiment analysis in facebook and its application to e-learning. Comput. Hum. Behav. **31**(1), 527–541 (2014)
10. Zhang, L., Ghosh, R., Dekhil, M., Hsu, M., Liu, B.: Combining lexicon-based and learning-based methods for twitter sentiment analysis. HP Laboratories Technical Report (89) (2011)

11. Haddi, E., Liu, X., Shi, Y.: The role of text pre-processing in sentiment analysis. Proc. Comput. Sci. **17**, 26–32 (2013)
12. Krouska, A., Troussas, C., Virvou, M.: The effect of preprocessing techniques on Twitter sentiment analysis. In: 2016 7th International Conference on Information, Intelligence, Systems and Applications (IISA), pp. 1–5. IEEE (2016)
13. Troussas, C., Krouska, A., Virvou, M.: Evaluation of ensemble-based sentiment classifiers for Twitter data. In: 2016 7th International Conference on Information, Intelligence, Systems and Applications (IISA), pp. 1–6. IEEE (2016)
14. Shamma, D., Kennedy, L., Churchill, E.: Tweet the Debates: Understanding Community Annotation of Uncollected Sources. ACM Multimedia, ACM (2009)
15. Speriosu, M., Sudan, N., Upadhyay, N., Baldridge, J.: Twitter polarity classification with label propagation over lexical links and the follower graph. In: Proceedings of the First Workshop on Unsupervised Methods in NLP, Edinburgh, Scotland (2011)
16. Saif, H., Fernez, M., He, Y., Alani, H.: Evaluation datasets for Twitter sentiment analysis: a survey and a new dataset, the STS-Gold. In: 1st International Workshop on Emotion and Sentiment in Social and Expressive Media: Approaches and Perspectives from AI (ESSEM 2013), Turin, Italy (2013)
17. Bravo-Marquez, F., Mendoza, M., Poblete, B.: Combining strengths, emotions and polarities for boosting twitter sentiment analysis. In: Proceedings of the 2nd International Workshop on Issues of Sentiment Discovery and Opinion Mining, WISDOM 2013 (2013). Da Silva, N.F., Hruschka, E.R., Hruschka, E.R., Jr.: Tweet sentiment analysis with classifier ensembles. Decis. Support Syst. **66**, 170–179 (2014)
18. Krouska, A., Troussas, C., Virvou, M.: Comparative evaluation of algorithms for sentiment analysis over social networking services. J. Univ. Comput. Sci. **23**(8), 755–768 (2017)

Chapter 8
Finding a Healthy Equilibrium of Geo-demographic Segments for a Telecom Business: Who Are Malicious Hot-Spotters?

J. Sidorova, O. Rosander, L. Skold, H. Grahn and L. Lundberg

Abstract In telecommunication business, a major investment goes into the infrastructure and its maintenance, while business revenues are proportional to how big, good, and well-balanced the customer base is. In our previous work we presented a data-driven analytic strategy based on combinatorial optimization and analysis of the historical mobility designed to quantify the desirability of different geo-demographic segments, and several segments were recommended for a partial reduction. Within a segment, clients are different. In order to enable intelligent reduction, we introduce the term infrastructure-stressing client and, using the proposed method, we reveal the list of the IDs of such clients. We also have developed a visualization tool to allow for manual checks: it shows how the client moved through a sequence of hot spots and was repeatedly served by critically loaded antennas. The code and the footprint matrix are available on the SourceForge.

Keywords Business intelligence · Combinatorial optimization · Fuzzy logic MOSAIC · Geo-demographic segments · Mobility data

8.1 Introduction

In the telecommunication industry, the lion's share of capital is spent on the infrastructure and its maintenance. The revenues are dependent on the size and the quality of the customer base: without a customer there is no business, yet satisfying everyone is simply not feasible, unless the right ones are chosen and others are let to go [1]

J. Sidorova (✉) · O. Rosander · H. Grahn · L. Lundberg
Department of Computer Science and Engineering, Blekinge Institute of Technology, Karlskrona, Sweden
e-mail: Julia.a.sidorova@gmail.com

L. Lundberg
e-mail: Lars.lundberg@bth.se

L. Skold
Telenor, Stockholm, Sweden

G. A. Tsihrintzis et al. (eds.), *Machine Learning Paradigms*, Intelligent Systems Reference Library 149, https://doi.org/10.1007/978-3-319-94030-4_8

or, in a less drastic manner, such clients can be put on more expensive personalized tariffs. The problem we concern ourselves with is how to find a balanced user portfolio in order to optimally exploit the infrastructure and get maximum benefit from the investments.

Methodology-wise, the literature in telecommunications research is abundant with optimization approaches formulated for the exploitation of telecommunication networks under the disguise of problems, which at a first glance one may consider unrelated to the topic, such as optimal location of cell towers, optimization of base stations deployment and so on, e.g. [2–5]. For example, the dual formulation of the optimal positioning of new cell towers turns out to be the problem of finding an optimal portfolio with user segments [6]. The theoretical formulation was extended with the use of historical data in [7], where a methodological framework was formulated as a classical resource allocation problem to calculate in a data-driven way the degree of desirability of a group of clients. The work addresses the problem of intelligent growth of the customer base, and yet some user groups were recommended for a partial reduction unless the infrastructure is upgraded, in order to keep the service reliable.

Another question is how to define and operationally single out different user groups. For example, such segments can be clustered from mobility data, e.g. [8]. Alternatively, the so called geo-demographic segmentations can be used, which both yield strong predictors of user behavior and are operational user descriptors. The client's home address is a surprisingly powerful predictor of client's behavior, because people of similar social status and lifestyle tend to live close together. Compared with conventional occupational measures of social class, postcode classifications typically achieve higher levels of discrimination, whether averaged across a random basket of behaviors recorded on the Target Group Index or surveys of citizen satisfaction with the provision of local authority services. One of the reasons that segmentation systems (MOSAIC, ACORN, and others) are so effective is that they are created by combining statistical averages for both census data and consumer spending data in pre-defined geographical units [9, 10]. For example, it was demonstrated that middle-class MOSAIC categories in the UK such as 'New Urban Colonists', 'Bungalow Retirement', 'Gentrified Villages' and 'Conservative Values', despite being very similar in terms of overall social status, nonetheless register widely different public attitudes and voting intentions; they support different kinds of charities and have preferences for different media as well as different forms of consumption. Geodemographic categories correlate with diabetes propensity [9], school students' performance [10], broadband access and availability [9], and so on. When acquiring new customers, industries rely increasingly on geodemographic segmentations to classify their markets. The localized versions of MOSAIC have been developed for a number of countries, including the USA and the EU. As a drawback it can be said that main geodemographic systems are in competition with each other and the exact details of the data and methods for generating lifestyles segments are never released [11] and, as a result, the specific variables or the derivations of these variables are unknown.

The contributions of this paper are as follows. As mentioned above, in our previous work [7] some segments were recommended for a partial reduction, but only the strategies for the segments to be boosted were discussed, leaving it undecided what to do with the other segments. To carry out the reduction in an intelligent way and generally to better understand the clients that can be described as hot spotters, we propose a notion of the infrastructure-stressing client and a method to reveal such clients based on their footprint on the system. We also have developed a visualization tool to allow for manual checks: it shows how the client moved through a sequence of hot spots and was repeatedly served by critically loaded antennas. (In our previous work [12], we also resorted to fuzzy logic and natural language processing, in order to better understand this stratum of the customer population.)

The rest of the paper is organized as follows. In Sect. 8.2, the data set is described. Section 8.3 explains the resource allocation formulation and the resulting linear programming system, which is at the heart of the proposed method. In Sect. 8.4, the notion of the Infrastructure-Stressing client is formulated and the proposed methodology is applied to find such users relying on the historical data. In Sect. 8.5, experimental results are reported. Finally, the conclusions are drawn in Sect. 8.6.

8.2 Geospatial and Geo-demographic Data

The study has been conducted on anonymized geospatial and geo-demographic data provided by a Scandinavian telecommunication operator. The data consist of CDRs (Call Detail Records) containing historical location data and calls made during one week in a midsized region in Sweden with more than one thousand radio cells. Several cells can be located on the same antenna. The cell density varies in different areas and is higher in city centers, compared to rural areas. The locations of 27,010 clients are registered together with which cells served them. The client's location is registered every five minutes. In the periods when the client does not generate any traffic, she does not make any impact on the infrastructure and such periods of inactivity are not relevant in the light of the resource allocation analysis and thus are not included in the present study. Every client in the database is labeled with her geo-demographic segment. The fields of the database used in this study are:

- the cell's ID with the information about the users, whom the cell served at different time points,
- the location coordinates of the cells,
- the time stamps of every event, which generated traffic, and the ID of the user who originated the event, and
- the MOSAIC geo-demographic segment for each client.

There are 14 MOSAIC segments present in the database; for their detailed description the reader is referred to [13].

8.3 The Combinatorial Optimization Module

The individual mobility patterns of different user segments sum up into the collective footprint, which the whole customer base produces on the infrastructure in a time-continuous manner. The desired property of such a collective footprint is that it does not exhibit skinny peaks and gaps in time. The closer to the optimal "even load" scenario, the better the infrastructure is exploited. The model's variables are the following.

Variables:

- clientSet: set with IDs of clients;
- *I:* the set with defined user segments {segment_1, …, segment_k};
- *D:* the mobility data for a region that for each user contain client's ID, client's geo-demographic segment, time stamps, when the client generated traffic, and which antenna served the client.
- S_i: the footprint by segment i, and the number of subscribers that belong to a geo-demographic segment *i* (will be discussed in the assumptions at the end of Section 3);
- $S_{i,t,j}$: the number of subscribers that belong to a geo-demographic segment *i*, at a time moment *t*, who are registered with a particular cell *j*;
- C_j: the capacity of cell *j* in terms of how many persons it can safely handle simultaneously;
- **x**: the vector with the scaling coefficients for the geo-demographic segments or other groups such as IS clients;
- x_{IS}: the coefficent for the IS segment from the vector **x**;
- $N_{t,j}$= number of users at cell j at time t.

The problem of finding an optimal combination of user segments, given that we want to maximize the overall number of users, who consume finite resources, belongs to a classical family of resource allocation problems. The formulation of our problem is as follows:

- *The vector* **x** *with the decision variables*

$$\boldsymbol{x} = \left\{ x_{\text{CC}},\ x_{\text{CA}},\ x_{\text{MJM}},\ x_{\text{QA}},\ x_{\text{T}},\ x_{\text{VA}} \right\}.$$

  The decision variables represent the scaling coefficients for each geo-demographic segment. In case of segmentation at Telenor (a Scandinavian operator for whom the study was carried out) they are: cost-aware (CA), modern John/Mary (MJM), quality aware (QA), traditional (T), value aware (VA), and corporate clients (CC). A scaling coefficient x_i is greater than 1, if the number of clients of a given geo-demographic segment is desired to be increased. For example, for the category in the customer base that is to be doubled $x_i = 2$. Similarly, if $x_i < 1$ for a geo-demographic segment, it means that the number of clients is to be reduced. The $x_i = 0$ indicates that the segment is absolutely unwanted in the clientele. By formulation **x** is non-negative.
- *The objective function* seeks to maximize the number of subscribers:

$$Maximize \quad \Sigma_{i \in \{\text{CC, CA, MJM, QA, T, VA}\}} S_i\, x_i \tag{8.1}$$

Table 8.1 The number of subscribers in each segment for all time slots and cells for the small example

	Cell 1		Cell 2	
Time slot	Segment 1	Segment 2	Segment 1	Segment 2
t_1	40	0	20	20
t_2	40	0	0	40
t_3	25	25	10	20

- *The restrictions*

$$for\ all\ j, t,\ \Sigma_{i \in \{\mathrm{CC, CA, MJM, QA, T, VA}\}} S_{i,t,j}\, x_i \le C_j \quad (8.2)$$

represent the observed number of persons in each user group at a particular time and served by a particular cell multiplied by the scaling coefficient. This value is required not to exceed the capacity of the cell C_j in terms of how many persons it can handle at a time. In other words the restriction says: if the historical number of users are scaled with a coefficient for their geo-demographic category, the cells should not be overloaded.

A consensus reached in the literature [14–16] is that the mobility pattern for the subscribers is predictable due to strong spatio-temporal regularity of human activities. Consequently, the increase in the number of subscribers in a given segment with a factor x will result in an increase of the load generated by the segment with a factor x for each time and cell.

The LP model is solved for the input data D and the set of segments I:

$$(\boldsymbol{x_I}, max\_obj_{I,D}) = combinatorial\_optimization\,(D, I). \quad (8.3)$$

The output is the vector with the optimal scaling coefficients $\boldsymbol{x_I}$ and the maximum value of the objective function.

Consider a small example with two cells, two subscriber segments and three time slots. The footprint values are shown in Table 8.1. The total number of subscribers in segment 1 is 60, and the total number of subscribers in segment 2 is 40 ($\mathbf{s} = (60, 40)^T$). The capacity of both radio cells is 200, i.e., $\mathbf{c} = (200, 200)^T$. The optimization problem becomes:

$$Maximize\ 60x_1 + 40x_2.$$

The LP problem has 6 restrictions:

$$\text{for } t_1,\ \text{cell } 1: 40x_1 \le 200,$$
$$\text{for } t_1,\ \text{cell } 2: 20x_1 + 20x_2 \le 200,$$
$$\text{for } t_2,\ \text{cell } 1: 40x_1 \le 200,$$

$$\text{for } t_2, \text{ cell } 2: 40x_2 \leq 200,$$
$$\text{for } t_3, \text{ cell } 1: 25x_1 + 25x_2 \leq 200,$$
$$\text{for } t_3, \text{ cell } 2: 10x_1 + 20x_2 \leq 200,$$
$$\boldsymbol{x} \geq \mathbf{0}.$$

Solving this LP problem yields the optimal $\mathbf{x} = (5, 3)^T$, corresponding to $\mathbf{s}^T\mathbf{x} = 420$.

Before we continue, let us discuss some implicitly made assumptions that may be not necessarily fair. Firstly, *all the clients generate the same revenue*. Concrete tariffs can be integrated in the form of the coefficients of the objective function. Let the tariff for the user category i be denoted with R_i. Then, the initial objective function from Eq. 8.1 is extended into

$$Maximize \quad \Sigma_{i \in \{CC, CA, MJM, QA, T, VA\}} R_i \; S_i \; x_i.$$

Secondly, *the impact on the network produced by different users is the same.* The calculation of the impact on the network can be refined taking into account the historical traffic. Let the traffic generated by the user group be T_i. The restrictions from Eq. 8.2 are modified into the footprint made on the network:

$$\Sigma_{i \in \{CC, CA, MJM, QA, T, VA\}} T_{i,t,j} \; x_i \leq C_j.$$

We can easily accommodate these clarifications in our system, but currently the relevant data is out of reach.

8.4 Infrastructure-Friendly and Stressing Clients

Despite geo-demographic segmentations are so powerful, there is an individual aspect to each client. In the light of the network capacity and the demand by the client population, there are infrastructure-stressing (IS) and friendly (IF) clients. Informally speaking, the former use the infrastructure in a taxing manner, such as always staying in the zones of high demand and the latter predominantly stay in the zones of low demand, where the cells tend to be idle. This individual grading of the clients is naturally combinable with the segment-based approach: first a decision is made which segment is to be reduced, and then it is revealed which clients in this particular segment are IS. The method to reveal the infrastructure-stressing clients from the database with historical mobility (formalized as Algorithm 1) uses the following heuristic.

Heuristic: From the list of the users ordered according to their relative mobility (how many other persons were served by the same antenna) at hot spots, the top 1% (most infrastructure-stressing) users are tentatively labelled being IS. (The 1% is an ad hoc value, where the intention is to cut thin homogeneous layers from the customer base.) To prove or disprove this characterisation, the scaling coefficient for the group is calculated via combinatorial optimization according to Eq. 8.3. As was discussed above, the scaling coefficient equal to 0 indicates no desirability to keep this group of users in the customer base, and in this case that group is confirmed to be infrastructure-stressing.

```
Input: data set D: <User_ID, time stamp t, cell j>.

[I. Characterize each user with respect to the relative
mobility.]
        for t: 1 to 2016 do {
            for j in cells{
               N_{t,j}= number of users at cell j at time t;
                          }
                            }
        for each user_ID {
          trajectory_ID = cell_t1, …, cell_t2016;
          relativeTrajectory_ID = N_{t1,j}, …, N_{t2016,j};
          sortedTrajectory_ID = sort_decreas_or.(relativeTrajectory_ID);
          topHotSpots_ID = Σ_{k=1..100} sortedTrajectory_ID[k];
          userTopHotSpots = <user_ID, topHotSpots_ID>
          }
        rankedUserList = sort_{decreasing_or_by 2nd field}(userTopHotSpots)

   [II. Initialization.]

        x_stressing = 0;
        setStressingUsers = ∅.

[III. Reveal the infrastructure-stressing clients.]

        While (x_stressing = 0) do {
         tentativeStressingUsers = head_1%(rankedUserList);
         setFriendlyUsers = bottom_1%(rankedUserList);
         otherUsers = rankedUserList −
         tentativeSetStressingUsers - setFriendlyUsers;

         [Confirm the tentative labeling via the tetris.]
         I = {stressing, medium, friendly};
         {x_stressing, x_medium, x_friendly} =
        combinatorial_optimization(I,D);

        IF (x_stressing = 0), THEN {
        tentativeSetStressingUsers = field userID from
        tentativeStressingUsers;
        setStressingUsers = setStressingUsers +
        tentativeSetStressingUsers

        D = D - D_stressing
        } end of While

  Output: setStressingUsers: <user_ID>.
```

Algorithm 1: Revealing the IDs of Infrastructure-Stressing clients based on their mobility relative to the location of other users.

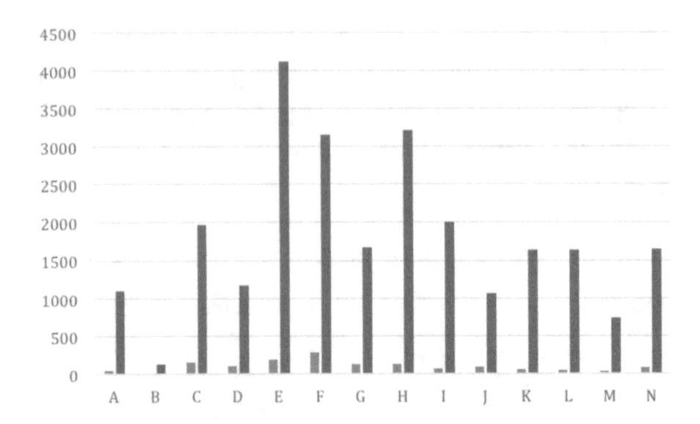

Fig. 8.1 The number of IS clients (blue) in different MOSAIC categories (red for the whole population)

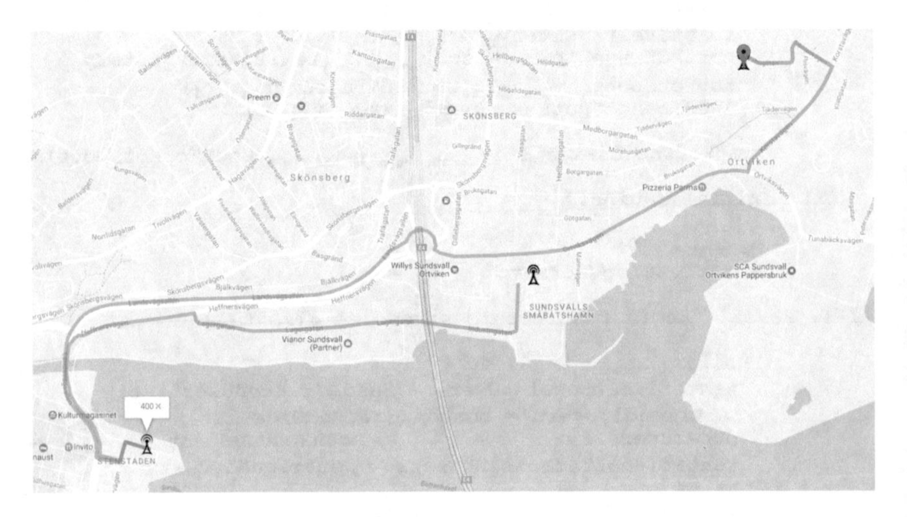

Fig. 8.2 Visualization for manual checks of the conclusions

8.5 Experiments

With the method formalized with Algorithm 1, the list with the UserIDs of IS clients has been obtained (those clients whose scaling coefficient $x = 0$). Seven percent of the customer base were revealed to be IS and they have turned out to be quite uniformly distributed across the geo-demographic segments, as depicted for the MOSAIC segments in Fig. 8.1 [12].

For further decision support, we have developed a visualization tool for manual checks of the conclusions: it shows how the client moved through a sequence of hot spots and was repeatedly served by critically loaded antennas, as it happened with the trajectory for the user in Fig. 8.2. Only the "critically loaded" antennas are depicted.

8.6 Conclusions

Our previous research suggested the necessity for the reduction in some geo-demographic segments (or expenses in a corresponding infrastructure upgrade to be able to provide everybody with telecommunication services in a reliable way). This paper suggests an intelligent reduction strategy. Based on a data-driven resource allocation formulation, we have proposed an algorithm to reveal the list of the so called infrastructure-stressing clients. Once the list is given, we have developed a visualization tool to allow for manual checks of the conclusions—per individual or in small packs.

The implications of the new EU legislation GDPR soon coming into force regarding projects like Tetris that involve the use of user's spatial data are addressed in [17].

Acknowledgements The experiments were run on the servers of the Future SOC Lab, Hasso Plattner Institute in Potsdam. This work is part of the research project "Scalable resource-efficient systems for big data analytics" funded by the Knowledge Foundation (grant: 20140032) in Sweden.

References

1. Haenlein, M., Kaplan, A.M.: Unprofitable customers and their management. Bus. Horiz. **52**(1), 89–97 (2009)
2. Tutschku, K.: Demand-based radio network planning of cellular mobile communication systems. In: Proceedings of the Seventeenth Annual Joint Conference of the IEEE Computer and Communications Societies, IEEE INFOCOM'98, vol. 3, pp. 1054–1061. IEEE (1998)
3. Tutschku, K., Tran-Gia, P.: Spatial traffic estimation and characterization for mobile communication network design. IEEE J. Sel. Areas Commun. **16**(5), 804–811 (1998)
4. Mathar, R., Niessen, T.: Optimum positioning of base stations for cellular radio networks. Wirel. Netw. **6**(6), 421–428 (2000)
5. González-Brevis, P., Gondzio, J., Fan, Y., Poor, H.V., Thompson, J., Krikidis, I., Chung, P.J.: Base station location optimization for minimal energy consumption in wireless networks. In: 2011 IEEE 73rd Vehicular Technology Conference (VTC Spring), pp. 1–5. IEEE (2011)
6. Sidorova, J., Rosander, O., Skold, L., Lundberg, L.: Data-driven solution to intelligent network updates for a telecom. Operator, Optim Eng, Accepted. https://rdcu.be/PkFM, Accessed on 28 May 2018
7. Sidorova, J., Skold, L., Rosander, O., Lundberg, L.: Recommendations for marketing campaigns in telecommunication business based on the footprint analysis. In: The 8th IEEE International Conference on Information, Intelligence, Systems and Applications, IISA, Cyprus, 27–31 Aug 2017
8. Sagar, S., Lundberg, L., Skold, L., Sidorova, J.: Trajectory segmentation for a recommendation module of a customer relationship management system. In: The 2017 International Symposium on Advances in Smart Big Data Processing (SBDP-2017)
9. Grubesic, T.H.: The geodemographic correlates of broadband access and availability in the United States. Telemat. Inform. **21**(4), 335–358 (2004)
10. Webber, R., Butler, T.: Classifying pupils by where they live: how well does this predict variations in their GCSE results? Urban Stud. **44**(7), 1229–1253 (2007)
11. Debenham, J., Clarke, G., Stillwell, J.: Extending geodemographic classification: a new regional prototype. Environ. Plan. A **35**(6), 1025–1050 (2003)

12. Podapati, S., Lundberg, L., Skold, L., Rosander, O., Sidorova, J.: Fuzzy recommendations in marketing campaigns. In: The 1st International Workshop on Data Science: Methodologies and Use-Cases (DaS 2017) at 21st European Conference on Advances in Databases and Information Systems (ADBIS 2017). Nicosia, Cyprus, 24 Sept 2017. LNCS, 28–30 August, Larnaca, Cyprus
13. InsightOne MOSAIC lifestyle classification for Sweden. http://insightone.se/en/mosaic-lifestyle/, Accessed on 15 Apr 2017
14. Song, C., Qu, Z., Blumm, N., Barabási, A.L.: Limits of predictability in human mobility. Science **327**(5968), 1018–1021 (2010)
15. Lu, X., Wetter, E., Bharti, N., Tatem, A.J., Bengtsson, L.: Approaching the limit of predictability in human mobility. Sci. Reports **3** (2013)
16. Naboulsi, D., Fiore, M., Ribot, S., Stanica, R.: Large-scale mobile traffic analysis: a survey. IEEE Commun. Surv. Tutor. **18**(1), 124–161 (2016)
17. Sidorova, J., Skold, L., Lundberg, L.: The concluding remarks about the tetris of big spatial data, report for the work carried on the HPI premises. HPI report. Spring (2018)

Part III
Data Analytics in Traffic, Computer and Power Networks

Chapter 9
Advanced Parametric Methods for Short-Term Traffic Forecasting in the Era of Big Data

George A. Gravvanis, Athanasios I. Salamanis and Christos K. Filelis-Papadopoulos

Abstract We live in the era of big data in all fields of activity and intensity. From econometrics and bioinformatics to robotics and aviation and from computational linguistics and social networks to traffic and transportation analytics, big data is the dominating factor of progress. Especially in the field of Intelligent Transportation Systems (ITS), the plethora of multisource traffic data has given a tremendous boost to the development of sophisticated systems for the confrontation of the several traffic related problems. One of the most challenging and at the same time crucial traffic related problems, which has significant impact in many ITS systems (e.g. Advanced Traveler Information Systems, multimodal routing systems, dynamic pricing systems, etc.), is the accurate and real-time traffic forecasting. The task of traffic forecasting, i.e. predicting the state of traffic in large scale urban and inter-urban networks within multiple intervals ahead in time, includes addressing several subproblems, like data acquisition from multiple sources (e.g. inductive loop detectors, moving vehicles, traffic cameras, etc.), preprocessing (outlier detection, missing data imputation, map-matching, etc.), integration and storage, design and development of complex algorithmic methods, overall network coverage of the forecasting results, performance issues, etc. In this chapter, the several state-of-the-art methods used in all aspects of the traffic forecasting problems are presented, with particular emphasis given on both the algorithmic and the efficiency aspects of the problem, in the light of the large amounts of available traffic data. In particular, the design of advanced traffic forecasting algorithms in large scale urban and inter-urban road networks are described along with their implementation and utilization on large amounts of real world traffic data.

G. A. Gravvanis (✉) · A. I. Salamanis · C. K. Filelis-Papadopoulos
Department of Electrical and Computer Engineering, School of Engineering, Democritus University of Thrace, University Campus Kimmeria, 67100 Xanthi, Greece
e-mail: ggravvan@ee.duth.gr

A. I. Salamanis
e-mail: asalaman@ee.duth.gr

C. K. Filelis-Papadopoulos
e-mail: cpapad@ee.duth.gr

G. A. Tsihrintzis et al. (eds.), *Machine Learning Paradigms*, Intelligent Systems Reference Library 149, https://doi.org/10.1007/978-3-319-94030-4_9

9.1 Introduction

The term Intelligent Transportation Systems (ITS) refers to the application of information and communication technologies in the field of transportation aiming to improve safety, efficiency and sustainability of the mobility of users and goods in the various transport networks. The ITS include several systems like Advanced Traveler Information Systems (ATIS), multimodal routing systems, fleet management systems, vehicle-to-vehicle communication systems and much more. In this rapidly growing field, many interesting research problems have arisen and one of the most interesting and important of them is the problem of traffic forecasting.

Traffic forecasting is the research problem of estimating the traffic state of a transport network element (e.g. a road segment) for one or more steps ahead in time. Traffic is quantified by an appropriate variable known as traffic descriptor. Such variables are the speed, flow, density, etc., that will be described in detail in the following sections. The selection of the appropriate traffic variable depends on the source of the traffic data. For instance, an inductive loop detector is a device that counts the number of vehicles passing from a certain point of the network and therefore it provides flow and density measurements. On the other hand, a GPS-enabled mobile phone carried by a driver provides measurements of the location and the instantaneous speed of the vehicle. When the data are collected and preprocessed (i.e. remove the outliers, match them to the network elements, etc.), an appropriate algorithm is used to forecast the traffic of the element of interest. Finally, when the forecasting horizon is up to one hour, then the problem is called short-term traffic forecasting and is one of the most well-studied cases.

Several algorithms for short-term traffic forecasting have been proposed in the relevant literature, with most of them being classified into one of the following categories:

- parametric algorithms
- non-parametric algorithms
- hybrid algorithms.

The parametric algorithms include a specific model whose structure is defined in advance and only the values of a set of parameters need to be estimated. Such algorithms are based on models like the Autoregressive Moving Average (ARMA) model, the Kalman filter, the Holt-Winters exponential smoothing model and the Generalized Autoregressive Conditional Heteroscedasticity (GARCH) model [5]. For instance, Yu et al. [31] proposed a double seasonal Holt-Winters traffic forecasting model to address the problem of the double seasonality in the traffic time series, namely the within-day and within-week seasonal cycle. Additionally, Guo et al. [8] introduced an adaptive Kalman filter that combined a Seasonal Autoregressive Integrated Moving Average (SARIMA) model with a GARCH model for short-term traffic forecasting. Zhang et al. [34] proposed a method in which the traffic time series were decomposed into three components and for each component traffic forecasting was performed using a different model, namely the spectral analysis method, the Autoregressive Integrated Moving Average (ARIMA) model and the Glosten-Jagannathan-Runkle GARCH (GJR-GARCH) model. Moreover, an autoregressive model that adapts itself to unpredictable events was used in [1] for predicting traffic for all the road segments of a network when traffic data for only few of them were known. Also, in [15] a SARIMA model for predicting traffic in the event of limited traffic data availability, was presented. Finally, Lv et al. [17] proposed a plane moving average model for traffic forecasting which could operate without the need for large traffic datasets.

On the other hand, the non-parametric algorithms do not presuppose a particular model structure and therefore both the exact model structure and its parameters need to be estimated using the available traffic data. The methods belonging to this category are based on models like the k-nearest neighbors (k-NN), the Artificial Neural Networks (ANN) and the Support Vector Regression (SVR). In this context, Habtemichael et al. [10] proposed an enhanced k-NN model for short-term traffic forecasting which utilized the weighted Euclidean distance as similarity measure, winsorization of neighbors to dampen the effects of dominant neighbors and rank exponent to aggregate the candidate values. Also, Zheng and Su [35] introduced a k-NN based approach for short-term traffic forecasting in which a linearly sewing principal component algorithm was used to control the undesirable impact of extreme values.

Moreover, several SVR-based methods have been suggested. Yao et al. [30] described a single-step SVR model for predicting traffic in the city of Foshan, China using GPS data from taxis, while Hu et al. [11] designed a method for short-term traffic forecasting which utilized SVR as prediction model and Particle Swarm Optimization (PSO) algorithm for searching for the optimal values of the SVR parameters. The authors paid special attention in the case of particles flying out of the search space by implementing three different strategies. Similarly, a SVR-based model with Gaussian loss function (G-SVR) that utilized a hybrid evolutionary algorithm for searching for the optimal values of the SVR parameters, has been proposed by Huang [12]. In particular, the evolutionary algorithm for searching optimal SVR parameters values (which is the main innovative feature of the work) consists of a chaotic map, a cloud model and a genetic algorithm and is abbreviated CCGA. In the same way, Li et al. [16] developed a G-SVR model, combined with a cat chaotic cloud PSO algorithm for SVR hyperparameters searching, for urban traffic flow forecasting.

Artificial neural networks have been extensively used for traffic forecasting [14, 36], due to their ability to model the non-linearities of the traffic dynamics. The state-of-the-art ANN-based methods for traffic forecasting have been influenced by the recent major breakthroughs in the data science and machine learning field and namely the deep learning approaches. In this context, one of the most studied and cited works is [18] by Lv et al., in which the authors described a deep architecture model with autoencoders as building blocks to represent traffic flow features for forecasting. Similarly, Huang et al. [13] presented a deep learning approach in which a deep belief network (DBN) was utilized for unsupervised learning of effective traffic features for forecasting. Also, Yang et al. [29] introduced a stacked autoencoder Levenberg-

Marquardt model aiming to improve traffic forecasting accuracy. Similar are the works presented in [19, 21, 22, 32, 33].

Finally, significant research effort has been dedicated on the development of hybrid traffic forecasting techniques which try to exploit the advantages of both the parametric and non-parametric approaches. For instance, Wang et al. [24] proposed a hybrid model for vehicle-type specific traffic forecasting in work zones. The model consisted of an empirical mode decomposition (EMD) component and an ARIMA component and was stated, by the authors, that it outperforms traditional forecasting models (e.g. ARIMA, Holt-Winters, ANN) in different scenarios. Additionally, a hybrid model that linearly combines elementary traffic forecasts from an SVR and a Box-Jenkins [5] model was presented in [2]. Another interesting approach was presented in [28]. In this approach, initially the traffic data sensors are grouped into clusters using the k-means clustering algorithm and then for each cluster an Autoregressive Moving Average with Exogenous inputs (ARMAX) model (with the clusters centroid as exogenous input) is fitted using the Recursive Least-Squares (RLS) algorithm and used for traffic forecasting. Additionally, Wang and Shi [25] proposed an SVR-based traffic forecasting model which utilizes the Chaos Wavelet Analysis to construct a new kernel function of the SVR that efficiently captures the non-stationary characteristics of the traffic data. Finally, Moretti et al. [20] presented an ensemble traffic forecasting model consisting of an ANN and a simple statistical approach, and stated that this model outperforms, in terms of forecasting accuracy, the methods it puts together.

9.2 Traffic Data

As in every machine learning application, the data plays a decisive role in traffic forecasting. Any algorithm, however well designed, cannot provide meaningful results if it is not combined with large amounts of appropriately preprocessed data. In this section, all the necessary information regarding traffic data is provided and described from the perspective of the traffic forecasting problem.

9.2.1 Traffic Network

Before start collecting and preprocessing traffic data and implementing traffic forecasting algorithms, one should first clearly define a structure for the examined traffic network. The various transportation networks (e.g. transit networks, railway networks, road networks, etc.) are represented by digital maps, designed either by governments and organizations or by individuals (e.g. OpenStreetMap). In the traffic forecasting framework, the network of interest is a road network. In this network, the elements are points representing road intersections, locations of traffic counters etc., and lines representing road segments. Each point is uniquely identified by a set of coordinates and each line by a pair of points. Such maps are usually encoded by Geographic Information Systems (GIS) file formats. There are two types of GIS file formats:

- *raster*, which is essentially a set of digital images,
- *vector*, in which the geographical features are represented as geometrical shapes.

In most cases of traffic forecasting applications, the vector format is used because of the ease of processing. The vector format contains three types of geometrical shapes: (a) point, (b) line or polyline and (c) polygon. A point expresses a single location in the network and can be uniquely identified by a set of coordinates. It can be used to represent the beginning or the end of a road segment, the location of a traffic counter, the location of a traffic light etc. Then, a polyline in a road network represents a road segment. Each line is described by a set of points indicating the beginning and the end of the line. Also, a polyline may contain additional features like the traffic direction, its length, the number of traffic lights installed on it, etc. The vector GIS files that contain polylines are the most important ones because, in most traffic forecasting applications, the elements for which traffic forecasting is required, is a road segment or a road. Finally, the polygons represent entire areas of the network. For instance, a traffic network may be segmented into zones in order to examine the different traffic patterns of each zone separately. In this case, each zone will be geographically represented by a separate polygon.

All the aforementioned geometrical shapes correspond to a row in a database which in turn corresponds to the particular vector file. For example, in a vector file containing all the roads of a specific traffic network there will be a database in which each row will correspond to a specific road of the network and each element of the row to a specific characteristic of this road. The are several different vector file formats designed and distributed by different vendors. Some of the most widely used vector GIS file formats representing road networks for traffic forecasting are the shapefile (.shp) developed by the GIS company Esri, the Keyhole Markup Language (.kml) primarily used by Google Earth and the OpenStreetMap (.osm) used by the homonymous crowdsourcing GIS project. The aforementioned file formats can be processed, independently of the traffic forecasting task, by specialized GIS software, which is either proprietary (e.g. ArcGIS) or open source (e.g. QGIS). Also, if the structure of the traffic network should be incorporated into the traffic forecasting algorithm (e.g. when using the spatiotemporal correlations between the networks elements) the researcher should define an appropriate representation of the network that would be efficiently stored into memory and processed by the traffic forecasting algorithm.

9.2.2 Traffic Descriptors

After the traffic network has been set up, the next step of traffic forecasting is to find a way to accurately describe and quantify traffic. A variable used for this purpose is called *traffic descriptor*. Each traffic descriptor describes traffic in a different way and it is associated with a distinct traffic meter. For example, a commonly used traffic descriptor is speed. From physics, speed is defined as the distance covered per time unit. In large traffic networks, it is difficult to track the speed of each individual vehicle and therefore metrics of average speed for specific elements of the network are defined. The two popular definitions of average speed in traffic networks are the time mean speed and the space mean speed.

The *time mean speed* is computed using the speed measurements of vehicles passing a specific reference point of the network, e.g. the location of an inductive loop detector. If n vehicles passed a reference point at a time period t, then the time mean speed v_t is given by the formula

$$v_t = \frac{1}{n} \sum_{i=1}^{n} v_i, \tag{9.1}$$

where v_i the instantaneous speed of vehicle i. The time mean speed is not a very accurate metric for describing the traffic of a network element, because the same vehicle can pass different points of the same network element at significantly different speeds.

On the other hand, the *space mean speed* is a more accurate average speed metric because it is computed using speed measurements taken from different points across the entire network element. These speed measurements come from different sources such as inductive loop detectors, traffic cameras, GPS probes, etc. The space mean speed is defined by the following formula

$$v_s = \left(\frac{1}{n} \sum_{i=1}^{n} \frac{1}{v_i} \right)^{-1}, \tag{9.2}$$

where n the number of instantaneous speed measurements taken across the entire network element and v_i a specific instantaneous speed measurement. The space mean speed is the harmonic mean of speed measurements. Except that it covers an entire network element (instead of just one reference point), the space mean speed is used more frequently than the time mean speed, because it is a better approximation to travel time [27], which is a variable used in routing applications. The travel time of an element of the network is the average time required for traversing the element. When the outcome of traffic forecasting is intended to be used by a rooting application, then it is more convenient to use the space mean speed as traffic descriptor, because it can be more easily interpreted as travel time.

Another widely used traffic descriptor is *density* (k), which is defined as the number of vehicles per unit length of an element of the network. For example, if at a certain time there are 250 vehicles at 10 kilometers (km) of a road, then the density of this road is 25 vehicles/km. There are two important values of density: (a) the critical density k_c and (b) the jam density k_j. The *critical density* is the maximum density under free flow conditions and the *jam density* the maximum density under congestion. The inverse of density is *spacing* (s) which is defined as the distance between the centers of two vehicles as shown in the following equation:

$$s = \frac{1}{k}. \tag{9.3}$$

For a network element of length L, if n vehicles are on the element at a certain time period T and $\bar{s}_t$ is the average spacing between the vehicles, then the following relationship holds:

$$k_{L,T} = \frac{n}{L} = \frac{1}{\bar{s}_t}. \tag{9.4}$$

Furthermore, *flow* (q) is another traffic descriptor which is defined as the number of vehicles traversing a reference point per unit of time. For instance, if 500 vehicles traversed a reference point of the network in 2 hours then the flow at this point is 250 vehicles/h. The flow is also defined as the product of speed with density as shown in the following equation:

$$q = v \cdot k. \tag{9.5}$$

The inverse of the flow is called *headway* (h) and it is defined as the time between the moments at which a vehicle i and a vehicle $i+1$, respectively, passed the reference point. If n vehicles traversed a reference point X at a time period T, then the following relationship applies:

$$q_{X,T} = \frac{n}{T} = \frac{1}{\bar{h}_x}, \tag{9.6}$$

where $\bar{h}_x$ is the average headway of the n vehicles.

The choice of the appropriate traffic descriptor depends on both the application that will use the results of the traffic forecasting algorithm (e.g. a routing application) and the available data sources of the examined traffic network. Some of the most frequently occurring traffic data sources are presented in the next section.

9.2.3 *Traffic Data Sources*

In order to quantify traffic, the values of the aforementioned traffic descriptors are measured using the appropriate equipment. There are conventional sensors which are part of the typical road infrastructure, such as on-road sensors and traffic cameras, and others that are installed on the vehicles and operate as moving sensors, such as the mobile phones of the drivers and the GPS receivers of taxis.

The most traditional way of measuring traffic is by installing electronic devices on the road in order to detect the presence of a vehicle. In this category belong devices such as the pneumatic tubes, the piezo-electric sensors and inductive loop detectors. These devices are usually embedded in the roadway and detect the presence of a vehicle that passes over them. For instance, when a vehicle passes over an inductive loop detector or stops within the loop, the inductance of the loop is decreased which activates the output of the electronic component of the detector which in turn sends a signal to the traffic signal controller indicating the presence of a vehicle. The inductive loop detectors and the piezo-electric sensors are used to permanently measure traffic in specific locations, while the pneumatic tubes are used just to sample traffic from specific locations of the network. Moreover, these devices detect vehicles whose masses are above a certain value. On one hand, this is a good characteristic because the devices do not produce many false positives, but on the other hand it means that small vehicles like bicycles, scooters and low-volume motorcycles will not be detected. The inductive loop detectors can be manually configured to detect small motorcycles. Finally, there are also off-road devices that emit energy such as radar waves or infrared beams to identify the presence of vehicles. These devices are mainly used in cases where only vehicle detection and not vehicle classification is required.

In addition to the sensors that identify the presence of a vehicle, there are also the *traffic cameras* for measuring traffic. A traffic camera is a conventional video camera which records vehicular traffic on a road of the network. They are typically installed on major roads like highways, and are connected to each other through optical fibers alongside or under the road. Their power supply comes mainly from the mains power in urban areas and, in some cases, they have a backup power supply system which operates using alternate forms of energy like solar energy. The traffic cameras produce a feed of low-resolution live video of the traffic conditions of a road which is transmitted to a traffic monitoring center. Specialized video processing software is used to extract traffic information from this live video, namely the number of vehicles that passed the location of the traffic camera at a certain time period (i.e. flow). It should be pointed out that the traffic cameras are distinct from the road safety cameras whose purpose is to enforce traffic regulations.

Alternatively, in recent years a new source of traffic data that does not require installation of hardware on the road network has been developed. This source consists of all mobile phones of the drivers. Nowadays, almost all drivers have mobile phones with them when driving and these phones can provide very useful data such as the location of the vehicle, its speed and the direction of travel at a specific time. This type of data is called *Floating Car Data* or *Floating Cellular Data (FCD)* and constitutes a valuable source of information for most ITS. Usually the FCD, contain tuples of the following form:

$$\{VehicleID, Timestamp, Location, Speed\}$$

where the $VehicleID$ is a unique identifier of the vehicle, the *Timestamp* is the time of the recording (based on a time reference point), the *Location* of vehicle is a set of coordinates (e.g. longitude, latitude, altitude) and the *Speed* is the instantaneous speed of the vehicle at the time of the recording. The FCD can be collected from the mobile network, where no special hardware is necessary, and also through the Global Positioning System (GPS) when the mobile phone is GPS-enabled. Essentially every switched-on mobile phone becomes an anonymous source of traffic information. The quality of the provided traffic information depends on the number of mobile phones. In large and congested urban traffic networks there are a lot of cars and therefore a lot of mobile phones and traffic probes resulting in more accurate traffic information. The advantages of these type of "sensors" over conventional traffic sensors are the following:

- No infrastructure/hardware is required on cars or alongside the roads.
- More coverage of the traffic network. Given the fact that the vehicles can move around the network, using FCD can result in a more complete picture of the traffic state of the network as opposed to the conventional traffic sensors that are installed at specific locations.
- Less expensive.
- Faster to set up.
- Less maintenance is required.

It should be pointed out that there is a major concern regarding the anonymity of the FCD because such data can easily be used for surveillance purposes. The companies that deploy and manage FCD services, provide assurances regarding the security of the data by stating that all data are anonymized for as long as they are kept in their servers.

9.3 Traffic Data Preprocessing

The available traffic data in their raw form are not suitable for processing by the traffic forecasting algorithms. Therefore, several steps of preprocessing, namely time series formulation, outlier detection, missing data imputation and map-matching, should be applied. In this section these preprocessing steps will be described.

9.3.1 Time Series Formulation

The traffic measurements coming from the various traffic data sources correspond to specific time instances or time periods. For example, a recording from three inductive loop detectors installed on a road segment with three lanes for a specific time period may have the form of the traffic count table shown in Table 9.1. Additionally, a set of traffic recordings from GPS-enabled mobile phones during a specific time period may have the form shown in Table 9.2.

As shown, each recording corresponds to a specific timestamp whose value is determined by a time reference point (e.g. Unix Epoch on January 1st, 1970 UTC). This data granularity may lead to difficulties in data management and processing. Therefore, the traffic data should be organized appropriately.

Table 9.1 Example of traffic count table from inductive loop detectors

Inductive loop detector id	Lane 1	Lane 2	Lane 3	Total
1	15	25	22	62
2	30	50	35	115
3	17	31	24	72

Table 9.2 Example of traffic recording from GPS-enabled mobile phones

Vehicle ID	Timestamp (ms)	Location (lat., lon.)	Speed (km/h)
35250210	1332115205000	(44.240315, −91.493619)	85
35250210	1332115207000	(44.240376, −91.493854)	72
45701997	1332115205000	(44.844847, −93.549069)	60
45868956	1332115207000	(44.920474, −93.447851)	65
85392424	1332115207000	(44.240304, −91.493768)	55

The most suitable way for organizing the traffic data for the task of traffic forecasting is the time series. Let r be a road segment whose traffic state in several steps ahead in time is required. Supposing that an inductive loop detector is installed on r and a set of traffic count values is available. These values cover a specific time period and are taken at a specific rate. For a total period of one day and a rate of one value per second, the set will contain 86,400 traffic values. A *traffic time series* that will correspond to r for this day can be constructed. The number of values in this time series depends on the desired granularity. For example, if it is decided that all the available traffic values will be represented, then the traffic time series will contain 86,400 values (highest possible granularity in this case). If it is decided that the time series will contain one value for each minute of the day, then it will contain 1,440 values, while if it is decided to contain one value for every five minutes of the day then it will contain 288 values. The selection of the appropriate granularity is the result of a compromise between the accuracy of representation of r's traffic state and the computational and memory requirements of the traffic forecasting algorithms.

Provided that the data granularity has been selected, the set of available traffic values has to be transformed into the values of the corresponding traffic time series. If t is the selected granularity, then the traffic values are mapped to t-minutes intervals of the day and one single value is computed that corresponds to this interval. Given n traffic values inside a t-minutes interval, the average of these values is computed as follows:

$$\bar{x}_t = \frac{1}{n}\sum_{i=1}^{n} x_i, \tag{9.7}$$

where x_i is the traffic count value for r at the time moment i inside t. If instead of traffic count values instantaneous speeds v_i are available, then the space mean speed of these values is computed.

The resulting value from Eq. (9.7) (or from the Eq. (9.1) in the case of speed values) corresponds to a specific interval of the traffic time series. If t is a multiple of one minute, the traffic time series will contain $1{,}440/t$ such values in total. In the context of traffic forecasting, it is customary to organize the traffic data into 1, 5 or 15 minutes intervals and consequently the corresponding traffic time series to have 1,440, 288 or 96 values respectively. If the total time period covered by the available traffic data exceeds one day (e.g. one month or one year), then either one large time series for the total period or a set of time series, where each one corresponds to a day of the total period, is formulated.

9.3.2 Outlier Detection

Outlier detection is the identification of observations in a dataset that do not conform to an expected pattern. The traffic time series are likely to contain outliers for reasons like measurement errors of the traffic counters or occurrence of abnormal events (car accidents, road works, special events like concerts and sport games, etc.). The identification of such anomalies is a very important process and may result in a significant increase in the accuracy of the traffic forecasting algorithms. It should be pointed out that in this section methods for detecting and not for handling outliers (e.g. replacing them with characteristic values, discarding them) are presented.

One of the simplest and most widely used methods for outlier detection is the *z-score* method. The z-score (or standard score) is a metric that indicates how many standard deviations a data point is away from the sample's mean. This metric can be used only assuming a Gaussian distribution of the data. If there are strong indications that the data are not normally distributed, then appropriate transformation should be applied to the data in order to use this metric. Supposing that the required assumptions are met, the z-score of a data point X_t is computed by the following formula:

$$z = \frac{X_t - \mu}{\sigma}, \tag{9.8}$$

where μ and σ are the mean and the standard deviation of the data respectively. If the z-score of the data point is greater than a threshold, then this point is considered outlier. The threshold is usually selected as two or three times the standard deviation of the data.

Another class of outlier detection methods are those that measure the deviation of a data point with respect to its neighbors. In this context, Breunig et al. [6] proposed an algorithm called *local outlier factor (LOF)* which is based on the concept of k-nearest neighbors of a data point X_t and the local density of this point. In this approach, a distance metric between two data points X_{t_1} and X_{t_2}, called *reachability distance (rd)*, is defined based on the following formula:

$$rd_k\left(X_{t_1}, X_{t_2}\right) = max\left\{d_k\left(X_{t_2}\right), d\left(X_{t_1}, X_{t_2}\right)\right\}, \tag{9.9}$$

where $d\left(X_{t_1}, X_{t_2}\right)$ is the distance between the data points X_{t_1} and X_{t_2}, and $d_k\left(X_{t_2}\right)$ the distance between X_{t_2} and its k-th nearest neighbor. This is the reachability distance of X_{t_1} from X_{t_2}. Then, the *local reachability density (lrd)* of X_{t_1} is defined as follows:

$$lrd\left(X_{t_1}\right) = \frac{1}{\left(\frac{\sum_{X_{t_2} \in N_k\left(X_{t_1}\right)} rd_k\left(X_{t_1}, X_{t_2}\right)}{\left|N_k\left(X_{t_1}\right)\right|}\right)}, \tag{9.10}$$

where $N_k\left(X_{t_1}\right)$ is the set of the k nearest neighbors of X_{t_1}. The local reachability density of X_{t_1} is the inverse of the average reachability distance of X_{t_1} from its k neighbors. Finally, the local outlier factor of X_{t_1} is defined as follows:

$$LOF_k\left(X_{t_1}\right) = \frac{\sum_{X_{t_2} \in N_k\left(X_{t_1}\right)} \frac{lrd\left(X_{t_2}\right)}{lrd\left(X_{t_1}\right)}}{\left|N_k\left(X_{t_1}\right)\right|}. \tag{9.11}$$

A value of LOF less than or equal to 1 indicates a dense region around X_{t_1} and thus it is not considered outlier. Conversely, a LOF value significantly greater than 1 indicates that the density of the data point X_{t_1} is substantially different from that of its k neighbors and therefore it is considered outlier. Other algorithms, operating similarly, are the clustering-based outlier detection methods such as the *Density-based spatial clustering of applications with noise (DBSCAN)* algorithm and its generalization the *Ordering points to identify the clustering structure (OPTICS)* algorithm.

There are also methods for outlier detection that require learning a specific model over the entire traffic time series. For instance, there are methods that require fitting an ARMA or an ARIMA model [5], on the traffic time series, and predicting each value based on its past values and some error terms. If the deviation between the real value of the traffic time series and the predicted value is greater than a threshold (or in other words it does not belong in a specific confidence interval) then the real value is considered outlier. In many cases this approach is combined with decomposing the time series into its components (seasonal, trend and residual) and training a different model for each component. Also, there are supervised machine learning methods for outlier detection (e.g. replicator neural networks, one-class support vector machines) which however require annotated outlier data.

The aforementioned methods either require the validity of many conditions, or they present extensive complexity. For these reasons, they are not the most effective ways for outlier detection, especially when this process must be performed automatically in real time as in the case of preprocessing traffic time series. Therefore, simpler but more efficient methods have been proposed. For instance, a simple method for checking if a data point X_t in a traffic time series is an outlier, is to compute its first difference $X_t' = X_t - X_{t-1}$ and check if it exceeds a threshold. If the time series has a seasonal component, then the same method can be used but with the seasonal difference $X_t' = X_t - X_{t-L}$ where L is the length of the season. Moreover, a very interesting approach for simple and automatic outlier detection is the one presented by Basu and Meckesheimer [3], called the *two-sided median* method. In this method, for a data point X_t a neighborhood with $2k$ points is constructed as follows:

$$\eta_t^{(k)} = \{X_{t-k}, \ldots, X_{t-1}, X_{t+1}, \ldots, X_{t+k}\}. \tag{9.12}$$

Then, the median $m_t^{(2k)}$ of this neighborhood is computed and compared with X_t. If the absolute value of the difference between $m_t^{(2k)}$ and X_t is greater than a threshold τ, then X_t is considered outlier. Additionally, the authors presented a slight variation of the two-sided median method called *one-sided median*. In this method, the neighborhood of the point X_t is defined only by its previous $2k$ points. Using the observed traffic time series and its first difference the following median values are computed:

$$m_t^X = median\,\{X_{t-2k}, \ldots, X_{t-1}\}, \tag{9.13}$$

$$m_t^{X'} = median\,\left\{X_{t-2k}', \ldots, X_{t-1}'\right\}. \tag{9.14}$$

Based on these medians, the following quantity is defined:

$$m_t^{(2k)} = m_t^X + k \cdot m_t^{X'}. \tag{9.15}$$

If the absolute value of the difference between the data point X_t and the quantity $m_t^{(2k)}$ is greater than a threshold τ, then X_t is considered outlier. In both methods, the threshold τ is a parameter selected by the user and controls the sensitivity of the methods against outliers.

9.3.3 Missing Data Imputation

The constructed traffic time series may have missing values. For instance, a traffic time series with 5 minutes granularity may have no values for the intervals 10:00–10:05 and 10:05–10:10 due to malfunction of the traffic counter (e.g. inductive loop detector). Missing data can cause problems in the process of traffic forecasting such as the introduction of substantial amount of bias at the forecasting results and difficulties in data management.

Several techniques for missing data imputation have been proposed, which have also been used in the context of traffic analytics. The most straightforward and easy to implement method is the *listwise deletion* (or *complete case*), in which when a traffic time series includes at least one missing value, it is deleted from the dataset. There is also a small variation of the method in which if the percentage of missing values (in relation to the total number of time series values) is greater than a threshold (e.g. 10%), then the time series is deleted from the dataset. If the values are not missing at random, meaning that there are only missing values at specific parts of the time series, these methods introduce bias. Although the not random missing values is the most common case in traffic analytics, the method of listwise deletion is widely used due to its simplicity and ease of development.

Another widely used data imputation technique is the *hot-deck imputation*. In this method, an existing value is selected randomly from the traffic time series and is used to fill the missing values. One type of hot-deck imputation is the *last observation carried forward (LOCF)* technique in which the existing value that is immediately prior to the missing value is used to impute it. Moreover, a characteristic value derived from the existing values can be used for imputation. For instance, the mean value of the traffic time series can be used to fill the missing ones. This technique does not change the mean of the time series which is an important property. Also, there is a variation of this technique in which the mean of the previous N existing values is used to fill a missing value. Similarly, if a set of traffic time series is available and there are missing values for a specific interval in some of them, then the mean of the existing values for this interval throughout all the time series can be used for imputation.

Finally, data imputation can be performed using *interpolation* methods. In mathematics, interpolation is defined as the process of estimating the values of a function in intermediate, unknown values of the independent variable given a finite set of the function's values in known values of the independent variable. The simplest interpolation method used for missing data imputation is the *piecewise constant interpolation*, which uses the nearest located existing value as imputation value. This technique is the same as the aforementioned LOCF technique and is a favorable choice due to its simplicity and reduced required computational work. Additionally, the *linear interpolation* is used for data imputation. In general, supposing two existing traffic data points $\left(t_1, X_{t_1}\right)$ and $\left(t_2, X_{t_2}\right)$ at intervals t_1 and t_2 respectively, the linear interpolant is defined by the following formula:

$$\frac{X_{t_m} - X_{t_1}}{t_m - t_1} = \frac{X_{t_2} - X_{t_1}}{t_2 - t_1} \Longleftrightarrow X_{t_m} = X_{t_1} + (t_m - t_1)\,\frac{X_{t_2} - X_{t_1}}{t_2 - t_1}, \tag{9.16}$$

where X_{t_m} is the missing value and t_m its interval in the traffic time series. The main advantage of the linear interpolation technique is that it is easy to interpret and implement, while its main disadvantage is the error, which is proportional to the square of the distance between the existing traffic data points. Finally, there are also other interpolation methods, like higher order *polynomial* and *spline* interpolation, in which the interpolants are polynomials with degree higher than one. These interpolation methods have several positive characteristics (e.g. smaller interpolation errors and thus smoother interpolants) but they are not used in practice as widely as the linear interpolation method mainly due to their computational complexity.

9.3.4 Map-Matching

As already mentioned, one typical source of traffic data is the inductive loop detectors. These sensors measure and record traffic values that correspond to specific points of the network. However, usually the element of interest in the traffic forecasting problem is a road segment (or a road). Therefore, a process for matching the measured traffic data to the road segments of the network is required. Several advanced map-matching algorithms have been proposed in the relevant literature like [4, 9, 26]. In this section, two basic techniques, namely a *brute-force* approach and a *Particle-In-Cell (PIC)* approach will be presented. In both cases, the coordinates of the inductive loop detectors and the endpoints of the road segments are considered known. The distance between an inductive loop detector and a road segment of the network is the distance between a point and a line segment in the N-dimensional space, where N is the number of coordinates of the points of the network (i.e. 2 or 3). This distance can be computed analytically using typical algorithms from the field of computational geometry.

The *brute-force* map-matching technique requires the computation of distances between all the inductive loop detectors and all the road segments of the network. In particular, in order to match a specific inductive loop detector to a road segment, its distances from all the road segments are computed and it is matched to the road segment with the minimum distance. This process is repeated until all the loop detectors are matched. This is far from optimal especially for large scale traffic networks. For instance, in a traffic network with 10^3 inductive loop detectors and 10^4 road segments, the distance between a point and a line-segment should be computed 10^7 times in order to match all the inductive loop detectors to the road segments of the network using the brute-force approach.

On the other hand, in the *PIC* approach initially a mesh is constructed that contains the whole network. This mesh is usually square (although there are cases that it may not be) and the dimension of the sides of its cells is selected in such a way that a specific criterion is met. Usually, this criterion has to do with the number of network elements that each cell will contain. For example, the dimension of the cells can be selected in such a way that the number of road segments in each cell to be, at most, 5.

Provided that the mesh has been constructed, the inductive loop detectors and the road segments should be matched to the mesh. The process of matching an inductive loop detector, i.e. a specific point of the network, in the mesh is straightforward. In particular, starting from the lower left cell, the mesh is searched (either by rows or by columns) until the cell containing the point is found. In the special case that the point lies on the common edge of two neighboring cells, then it is considered to belong to both cells.

On the other hand, the process of matching a road segment on the mesh is more complicated. Based on the selection of the cell dimension, the most likely event is the entire segment to belong to a cell. Nevertheless, there is also the case that one road segment traverses more than one cells either horizontally, vertically or diagonally. In this case, the cells traversed by the road segment are identified by computing the intersections between their edges and the road segment, and it is considered that this road segment belongs to all these cells.

After both the inductive loop detectors and the road segments have been matched to the mesh, the final step is to match the former to the latter. For this purpose, the distances between an inductive loop detector and the road segments *that belong to the same cell* are computed and the road segment with the minimum distance is selected. This process is repeated until all the inductive loop detectors are matched to the road segments of the network.

The PIC map-matching technique is far more efficient than the brute-force technique. In particular, if n is the number of inductive loop detectors, N the number of road segments and m the mean number of road segments in each cell of the mesh, then the time complexity of the brute-force approach is $O(nN)$ and the time complexity of the PIC technique is $O(nm)$. If the number of cells in the mesh is large, then the following relationship holds:

$$m \ll N, \tag{9.17}$$

which indicates that the PIC technique is far more efficient than the brute-force technique.

9.4 Parametric Short-Term Traffic Forecasting

The *parametric short-term traffic forecasting algorithms* is an important class of traffic forecasting algorithms which has been successfully used in many ITS applications. Most of them are based on the classic Box-Jenkins model [5] and their training process is usually performed by the least-squares method. In this section, the details of some advanced parametric short-term traffic forecasting algorithms, from a supervised learning perspective, are presented.

9.4.1 Autoregressive Moving Average (ARMA)

The *Autoregressive Integrated Moving Average (ARMA)* model describes a stationary stochastic process in terms of two polynomials, namely the autoregressive part and the moving average part. For a traffic time series X_t, the ARMA(p, q) model is defined as follows [5]:

$$X_t - \phi_1 X_{t-1} - \cdots - \phi_p X_{t-p} = \varepsilon_t + \theta_1 \varepsilon_{t-1} + \cdots + \theta_q \varepsilon_{t-q}, \tag{9.18}$$

where p is the order of the autoregressive part, q the order of the moving average part and ε_t the (random) error terms of X_t, which are considered identically distributed with normal distribution of zero mean and constant variance σ_ε^2. The above equation can be rewritten as [5]:

$$\left(1 - \sum_{i=1}^{p} \phi_i L^i\right) X_t = \left(1 + \sum_{i=1}^{q} \theta_i L^i\right) \varepsilon_t, \tag{9.19}$$

where L is the *lag operator* for which it holds:

$$L^i X_t = X_{t-i}. \tag{9.20}$$

A more concise representation of Eq. (9.19) is the following:

$$\phi\,(L)\,X_t = \theta\,(L)\,\varepsilon_t, \tag{9.21}$$

or

$$\frac{\phi\,(L)}{\theta\,(L)} X_t = \varepsilon_t, \tag{9.22}$$

where

$$\phi\,(L) = 1 - \sum_{i=1}^{p} \phi_i L^i, \tag{9.23}$$

$$\theta\,(L) = 1 + \sum_{i=1}^{q} \theta_i L^i. \tag{9.24}$$

The error terms ε_t are considered independent and identically distributed random variables sampled from a normal distribution with zero mean and constant variance σ_ε^2. It is very important that these conditions apply, since failure to do so will fundamentally change the model.

9.4.2 Autoregressive Integrated Moving Average (ARIMA)

The *Autoregressive Integrated Moving Average (ARIMA)* model is a generalization of the ARMA model and it is the model on which the more advanced parametric short-term traffic forecasting algorithms are based [5]. The ARIMA model can be used when the traffic time series in question is stationary. Therefore, if characteristics of non-stationarity are shown, a *differencing* step should be applied one or more times in order to eliminate them. The differencing step in traffic time series in usually implemented with *backward differences*. In particular, if X_t is the traffic time series, the first order backward difference is defined by the following equation:

$$\nabla X_t = X_t - X_{t-1}, \tag{9.25}$$

where ∇ the *backward difference operator*. For higher order differences, the backward difference operator is applied recursively. For instance, for order 2:

$$\nabla_t^2 X_t = X_t - 2X_{t-1} + X_{t-2}. \tag{9.26}$$

In general, for order d,

$$\nabla_t^d X_t = \sum_{k=0}^{d} (-1)^k \binom{d}{k} X_{t-k}. \tag{9.27}$$

In most cases, the non-stationary traffic time series can become stationary with just 1st order backward differences.

As already mentioned, the ARMA(p, q) model is defined by Eq. (9.19). Supposing that the polynomial $\phi(L)$ has a unit root of multiplicity d, it can be expressed as follows [5]:

$$\phi(L) = 1 - \sum_{i=1}^{p} \phi_i L^i = \left(1 - \sum_{i=1}^{p-d} \phi_i L^i\right)(1-L)^d. \tag{9.28}$$

Based on this polynomial factorization, an ARIMA(p, d, q) model, in which p has been replaced by $p - d$, is expressed by the following equation [5]:

$$\left(1 - \sum_{i=1}^{p} \phi_i L^i\right)(1-L)^d X_t = \left(1 + \sum_{i=1}^{q} \theta_i L^i\right)\varepsilon_t, \tag{9.29}$$

where p the order of the autoregressive part, d the degree of differencing (i.e. the order of backward differences required to make the traffic time series stationary) and q the order of the moving average part.

The appropriate model that should be used for a specific traffic time series (e.g. plain AR, MA or ARMA model) as well as the optimal values of orders p and q are usually identified by the visual inspection of the *Autocorrelation Function (ACF)* and the *Partial Autocorrelation Function (PACF)* plots. The ACF describes the similarity between the values of a time series as a function of the lag between them. The PACF is similar to the ACF but it does not take into consideration the linear dependence between the values in the intermediate lags. The ACF and PACF plot depict the form of these functions against the number of lags. In general, if ACF has an exponentially decreasing appearance and PACF is zero at a specific lag $p+1$, then a plain autoregressive model of order p (AR(p)) is suitable for representing the time series and forecasting its future values. On the other hand, if ACF becomes zero in lag $q+1$ and PACF decreases exponentially, then a plain moving average model of order q (MA(q)) should be used. If none of the previous cases apply, an ARMA(p, q) model is the most suitable choice. In this case, the optimal values of p and q are difficult to estimate through the visual inspection of the ACF and PACF plots and therefore advanced information criteria methods (e.g. Bayesian information criterion, Akaike information criterion) are utilized.

One widely used parametric model for short-term traffic forecasting is the ARIMA(p, 1, 0) model which is described by the following equation [5]:

$$X_t^{(1)} = \phi_1 X_{t-1}^{(1)} + \cdots + \phi_p X_{t-p}^{(1)}, \tag{9.30}$$

where

$$X_t^{(1)} = \nabla_t X_t = X_t - X_{t-1}. \tag{9.31}$$

For this model, the appropriate values of the ϕ parameters are estimated through a learning process which is usually the *least squares regression*.

9.4.2.1 Supervised Learning of the ARIMA Model—The Least Squares Problem

The process of estimating the ϕ and θ parameters is the learning process of the ARIMA model and it is performed in a supervised way [5]. One of the most widely used parametric models for short-term traffic forecasting is the ARIMA(p, 1, 0) model. The learning process of this model will be described here.

The ARIMA(p, 1, 0) model is described by Eq. (9.30). This is as linear equation with $X_t^{(1)}$ being the dependent variable and $X_{t-1}^{(1)}, \ldots, X_{t-p}^{(1)}$ the independent variables. Supposing that a traffic time series of size M is available, the following system of linear equations can be constructed:

$$\begin{aligned} X_{p+1}^{(1)} &= \phi_1 X_p^{(1)} + \cdots + \phi_p X_1^{(1)} \\ X_{p+2}^{(1)} &= \phi_1 X_{p+1}^{(1)} + \cdots + \phi_p X_2^{(1)} \\ &\vdots \\ &\vdots \\ &\vdots \\ X_M^{(1)} &= \phi_1 X_{M-1}^{(1)} + \cdots + \phi_p X_{M-p}^{(1)}, \end{aligned} \tag{9.32}$$

or in matrix form:

$$\begin{bmatrix} X_p^{(1)} & \dots & X_1^{(1)} \\ X_{p+1}^{(1)} & \dots & X_2^{(1)} \\ \vdots & \ddots & \vdots \\ \vdots & \ddots & \vdots \\ \vdots & \ddots & \vdots \\ X_{M-1}^{(1)} & \dots & X_{M-p}^{(1)} \end{bmatrix} \begin{bmatrix} \phi_1 \\ \vdots \\ \phi_p \end{bmatrix} = \begin{bmatrix} X_{p+1}^{(1)} \\ X_{p+2}^{(1)} \\ \vdots \\ \vdots \\ \vdots \\ X_M^{(1)} \end{bmatrix}. \tag{9.33}$$

This system can be rewritten in the form:

$$A\phi = b, \tag{9.34}$$

where

$$A = \begin{bmatrix} X_p^{(1)} & \dots & X_1^{(1)} \\ X_{p+1}^{(1)} & \dots & X_2^{(1)} \\ \vdots & \ddots & \vdots \\ \vdots & \ddots & \vdots \\ \vdots & \ddots & \vdots \\ X_{M-1}^{(1)} & \dots & X_{M-p}^{(1)} \end{bmatrix}, \phi = \begin{bmatrix} \phi_1 \\ \vdots \\ \phi_p \end{bmatrix}, b = \begin{bmatrix} X_{p+1}^{(1)} \\ X_{p+2}^{(1)} \\ \vdots \\ \vdots \\ \vdots \\ X_M^{(1)} \end{bmatrix}. \tag{9.35}$$

This system contains $(M - p - 1)$ equations and p unknowns. Supposing that N traffic time series of size M are available and a single ARIMA(p, 1, 0) model should be constructed, then the corresponding system will have $N \cdot (M - p - 1)$ equations and p unknowns. A single ARIMA(p, d, 0) model that will be constructed from the N traffic time series will have $N \cdot (M - p - d)$ equations and p unknowns [5].

In the context of short-term traffic forecasting problem, the system (9.34) is *overdetermined*, meaning that it has more equations than unknowns. Such systems are solved approximately, where the best approximation is the one that minimizes the square of the difference between the actual and the approximate solution. This quantity, which is called the *residual*, is described as follows:

$$r = b - A\phi. \tag{9.36}$$

Thus, the following optimization problem arises:

$$\hat{\phi} = \arg\min_{\phi} \|b - A\phi\|^2. \tag{9.37}$$

This problem is called the *least squares problem*. The function

$$S(\phi) = \|b - A\phi\|^2, \tag{9.38}$$

is the objective function of this optimization problem and can be rewritten as follows:

$$S(\phi) = \|b - A\phi\|^2$$
$$= b^T b - b^T A\phi - \phi^T A^T b + \phi^T A^T A\phi. \quad (9.39)$$

The quantity $\phi^T A^T b$ has dimension 1×1 so it is scalar and it is equal to its transpose for which is applies:

$$\left(\phi^T A^T b\right)^T = b^T \left(A^T\right)^T \left(\phi^T\right)^T = b^T A\phi. \quad (9.40)$$

Thus, the objective function S becomes:

$$S(\phi) = b^T b - \left(\phi^T A^T b\right)^T - \phi^T A^T b + \phi^T A^T A\phi$$
$$= b^T b - 2\phi^T A^T b + \phi^T A^T A\phi. \quad (9.41)$$

In order to minimize the objective function S, its first derivative with respect to ϕ should be set to zero. Hence:

$$\frac{dS}{d\phi} = -A^T b + A^T A\phi = 0. \quad (9.42)$$

Based on the Eq. (9.42), the *normal equations* of the least squares problem are derived:

$$A^T A\phi = A^T b. \quad (9.43)$$

The least squares problem has a unique solution, if the matrix A has *full column rank*, given by

$$\hat{\phi} = \left(A^T A\right)^{-1} A^T b. \quad (9.44)$$

where the matrix $A^+ = \left(A^T A\right)^{-1} A^T$ is the Moore-Penrose pseudoinverse of A.

The solution of the least squares problem can be computed by inverting the matrix $A^T A$, but this is not computationally efficient, especially for large matrices. A more efficient way for solving the least squares problem is by computing the *Cholesky decomposition* of the $A^T A$ matrix. If this matrix is well-conditioned and positive definite, then its Cholesky decomposition is given by the following equation:

$$A^T A = R^T R. \quad (9.45)$$

After the decomposition of the $A^T A$ matrix, the normal equations (9.43) will become:

$$R^T R\phi = A^T b. \quad (9.46)$$

The solution of the above system is obtained in two steps. A forward substitution step, solving for z:

$$R^T z = A^T b, \tag{9.47}$$

and a back-substitution step, solving for ϕ:

$$R\phi = z. \tag{9.48}$$

Another way for solving the least squares problem is by decomposing the matrix A. For instance, using the *Singular Value Decomposition (SVD)*, A is decomposed as follows:

$$A = U\Sigma V^T, \tag{9.49}$$

where U_{nxn} orthogonal matrix whose columns are the left singular vectors of A, V_{pxp} orthogonal matrix whose columns are the right singular vectors of A and Σ_{nxp} diagonal matrix whose elements in the main diagonal are the singular values of matrix A. Based on this decomposition, the solution of the least squares problem will be:

$$\hat{\phi} = V\Sigma^{+}U^T b, \tag{9.50}$$

where Σ^{+} is the Moore-Penrose pseudoinverse of Σ, which can be obtained easily by inverting the non-zero elements of Σ.

Additionally, instead of using the SVD of A, the *QR decomposition* of A can be used in the following form:

$$A = Q\begin{pmatrix} R \\ 0 \end{pmatrix}, \tag{9.51}$$

where Q_{nxn} orthogonal matrix and R_{pxp} upper triangular matrix with $r_{ii} > 0$. The Eq. (9.36) left multiplied by Q^T gives:

$$\begin{aligned} Q^T r &= Q^T(b - A\phi) \\ &= Q^T b - Q^T A\phi \\ &= Q^T b - \left(Q^T Q\right)\begin{pmatrix} R \\ 0 \end{pmatrix}\phi = Q^T b - I\begin{pmatrix} R \\ 0 \end{pmatrix}\phi = \begin{bmatrix} \left(Q^T b\right)_p - R\phi \\ \left(Q^T b\right)_{n-p} \end{bmatrix}. \end{aligned} \tag{9.52}$$

The quantity $\|r\|^2$ is minimized when:

$$\begin{aligned} \left(Q^T b\right)_p - R\phi = 0 \Leftrightarrow \\ R\phi = \left(Q^T b\right)_p. \end{aligned} \tag{9.53}$$

This equation can be easily solved for $\hat{\phi}$ because R is an upper triangular matrix.

9.4.3 Space-Time ARIMA (STARIMA)

An extension of the ARIMA model is the *Space-Time ARIMA (STARIMA)* model which takes into account the spatiotemporal correlations between the roads segments of the network. The general seasonal STARIMA model is described by the following equation:

$$\phi_{p,\lambda}(L)\,\Phi_{P,\Lambda}(L)\,(1-L)^d\left(1-L^S\right)^D X_t = \theta_{q,m}(L)\,\Theta_{Q,M}\left(L^S\right)\varepsilon_t, \tag{9.54}$$

where

$$\phi_{p,\lambda}(L) = I - \sum_{k=1}^{p}\sum_{l=0}^{\lambda_k}\phi_{kl}W_l L^k, \tag{9.55}$$

$$\Phi_{P,\Lambda}\left(L^S\right) = I - \sum_{k=1}^{P}\sum_{l=0}^{\Lambda_k}\Phi_{kl}W_l L^{kS}, \tag{9.56}$$

$$\theta_{q,m}(L) = I - \sum_{k=1}^{q}\sum_{l=0}^{m_k}\theta_{kl}W_l L^k, \tag{9.57}$$

$$\Theta_{Q,M}\left(L^S\right) = I - \sum_{k=1}^{Q}\sum_{l=0}^{M_k}\Theta_{kl}W_l L^{kS}. \tag{9.58}$$

The indexes k, l represent the temporal and the spatial lag, respectively. The $\Phi_{k,l}$ and $\phi_{k,l}$ are the seasonal and non-seasonal autoregressive parameters, respectively, and similarly $\Theta_{k,l}$ and $\theta_{k,l}$ are the seasonal and non-seasonal moving average parameters. The P, p are the autoregressive orders and Q, q the moving average orders of the seasonal and non-seasonal components of the model respectively. The Λ_k and λ_k are the seasonal and non-seasonal spatial orders of the kth autoregressive term, M_k and m_k the seasonal and non-seasonal spatial orders of the kth moving average term and D, d the number of seasonal and non-seasonal degrees of differencing. Finally, ε_t are the (random) error terms of X_t, identically distributed with normal distribution of zero mean and constant variance σ_ε^2.

Supposing a time series X_t that does not exhibit seasonal characteristics (e.g. a traffic time series that corresponds to one day), the STARIMA(p, d, q) model is described by the following simplified equation:

$$\phi_{p,\lambda}(L)\,(1-L)^d X_t = \theta_{q,m}(L)\,\varepsilon_t, \tag{9.59}$$

or equivalently for $d = 1$

$$\phi_{p,\lambda}(L)\,X_t^{(1)} = \theta_{q,m}(L)\,\varepsilon_t. \tag{9.60}$$

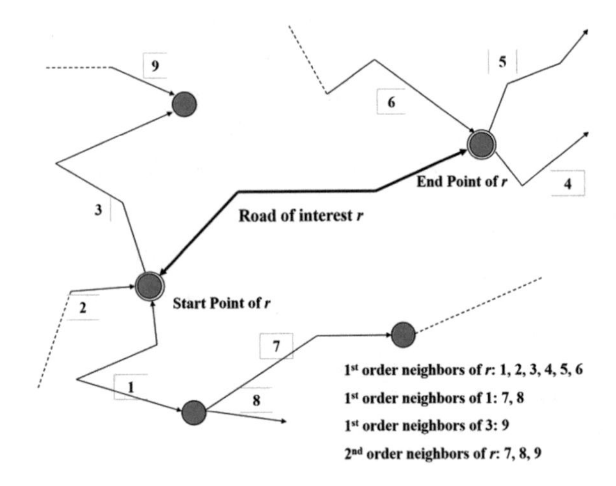

Fig. 9.1 Adjacency definition based on the concept of importing and exporting traffic from a road

The spatiotemporal correlations between the roads of the network are expressed through the W_l $N \times N$ matrix where N is the number of roads in the traffic network. This matrix is defined based on the selected definition of adjacency for the examined traffic network. For instance, the element w_{ij}, where i, j refer to two separate roads, may represent either the number of intermediate roads between i and j or the amount of available public transportation means connecting them or the number of toll stations between them, etc. [23]. The index l in W_l denotes the adjacency order.

An interesting definition of adjacency is based on the concept of importing or exporting traffic from a road. In particular, for a road i, a spatial 1st order neighbor is a road j that begins from or ends to the start or the end node of i and imports or exports traffic from i. Subsequently, the 2nd order neighbors of i are the 1st order neighbors of its 1st order neighbors, its 3rd order neighbors are the 1st order neighbors of its 2nd order neighbors and so on. Two roads i and j are kth order neighbors if and only if at least one of the $k-1$th order neighbors of i, is 1st order neighbor of j. A road i is 0th order neighbor with itself. This definition of adjacency is depicted in Fig. 9.1. According to the above description, the adjacency matrix W_l is defined as follows:

$$W_{l,ij} = \begin{cases} 1, & if\ i, j\ l\text{th}\ order\ neighbors \\ 0, & otherwise \end{cases}. \tag{9.61}$$

Each W_l can be constructed from W_{l-1} as follows:

$$W_{l,ij} = \begin{cases} 1, if\ \exists k : W_{l-1,ik} = W_{1,jk} = 1 \\ 0,\ otherwise \end{cases}. \tag{9.62}$$

The general STARIMA model is non-linear and a non-linear optimization technique (like the Levenberg-Marquardt algorithm) is required in order to estimate the ϕ and θ parameters. In the usual case that $q = 0$, $d = 1$ and $l = 1$ and assuming the 1st order neighbors of a road r are the same for all autoregressive terms, the corresponding simplified model is described by the following equation:

$$\begin{aligned} X_{r,t}^{(1)} &= \phi_1 X_{r,t-1}^{(1)} + \cdots + \phi_p X_{r,t-p}^{(1)} \\ &+\beta_{11} X_{neigh_1,t-1}^{(1)} + \cdots + \beta_{1k} X_{neigh_k,t-1}^{(1)} \\ &\vdots \\ &+\beta_{p1} X_{neigh_1,t-p}^{(1)} + \cdots + \beta_{pk} X_{neigh_k,t-p}^{(1)} \end{aligned} \quad (9.63)$$

The contribution of all 1st order neighbors for each autoregressive term m can be computed as follows:

$$X_{neigh_{av},t-m}^{(1)} = \frac{1}{k}\sum_{i=1}^{k} X_{neigh_i,t-m}^{(1)}. \quad (9.64)$$

Based on the Eq. (9.64), the Eq. (9.63) can take its final form:

$$X_{r,t}^{(1)} = \phi_1 X_{r,t-1}^{(1)} + \cdots + \phi_p X_{r,t-p}^{(1)} + \gamma_1 X_{neigh_{av},t-1}^{(1)} + \cdots + \gamma_p X_{neigh_{av},t-p}^{(1)}. \quad (9.65)$$

As shown, the above equation is linear. Therefore, the learning process of this STARIMA model, i.e. the estimation of the ϕ and β parameters, can be performed by solving the least squares problem as in the case of the ARIMA model.

9.4.4 *Lag-STARIMA*

Another parametric model that has been successfully used for short-term traffic forecasting is the *Lag-STARIMA* model [7]. The Lag-STARIMA model is a variant of the STARIMA model, which again takes into account the spatiotemporal correlations between the roads of the network, but it uses an adjacency definition that is not based on the topology of the network but solely on the traffic time series of the roads.

The model is based on the identification of the traffic time series whose values affect those of the examined traffic time series. For instance, the speed value for a road at the current time interval may be affected from the speed values of its spatial local neighbors at the current interval, but also from the speed values of other (probably distant) roads of the whole network at previous time intervals. This happens because the effects of traffic at a specific road require some time intervals to be propagated at the whole network. These relations between the traffic values of the roads of the whole network can be captured using correlation metrics.

One of the most widely used metrics for discovering correlations between time series is the *Pearson Correlation Coefficient (PCC)*. Given two time series X_t and Y_t, PCC is defined as follows:

$$PCC_{XY} = \frac{E\left[(X_t - \mu_X)(Y_t - \mu_Y)\right]}{\sigma_X \sigma_Y}, \quad (9.66)$$

where μ_X, μ_Y are the mean values and σ_X, σ_Y the standard deviations of the time series X_t and Y_t respectively. The PCC fails to capture the effect of lag at the correlation between the time series. For example, the speed value of a road r_1 at the time interval t may not be well correlated with the speed values of the roads r_2 and r_3 at the same interval. But supposing that the vehicles travel in the direction $r_3 \rightarrow r_2 \rightarrow r_1$ and that it takes one time interval for vehicles to travel from r_3 to r_2 and one time interval to travel from r_2 to r_1, the speed value of r_1 at interval t will be well correlated with the speed value r_2 at interval $t-1$ and with the speed value of r_3 at interval $t-2$. Such correlations can be captured using the *Cross Correlation Coefficient (CCC)* defined as follows:

$$CCC_{XY}(l) = \frac{E\left[(X_t - \mu_X)(Y_{t+l} - \mu_Y)\right]}{\sigma_X \sigma_Y}, \tag{9.67}$$

where $l < 0$ is the lag between the correlated time series. The metric used in the Lag-STARIMA model for estimating the correlations between the traffic time series of the roads of the network is derived by the CCC and described by the following equation:

$$CoD_{XY}(l) = 100 \cdot \left[\frac{E\left[(X_t - \mu_X)(Y_{t+l} - \mu_Y)\right]}{\sigma_X \sigma_Y}\right]^2. \tag{9.68}$$

This metric is called *Coefficient of Determination (CoD)* and it expresses the effect that one time series has to the other at a specific lag l. Defining as M_l, the set of m largest values of $CoD_{XY}(l)$, the weight matrix W_l is defined as follows:

$$W_{l,ij} = \begin{cases} 1, & \textit{if } CoD_{XY}(l) \in M_l \\ 0, & \textit{otherwise} \end{cases}, \tag{9.69}$$

where i, j two traffic time series corresponding to two roads of the traffic network. As it is understood, the lag l here has the same meaning with the order of neighbors described in the previous section.

Using the same assumptions as in the case of the STARIMA model (i.e. $q = 0$, $d = 1$, $l = 1$ and that for a road r the most well correlated roads for lag 1 are the same for all autoregressive terms) the Lag-STARIMA(p, 1, 0) with $l = 1$ model is defined by the following equation:

$$\begin{aligned} X_{r,t}^{(1)} &= \phi_1 X_{r,t-1}^{(1)} + \cdots + \phi_p X_{r,t-p}^{(1)} + \\ &+ \delta_1 X_{tr\_neigh_{av},t-1}^{(1)} + \cdots + \delta_p X_{tr\_neigh_{av},t-p}^{(1)} \end{aligned}, \tag{9.70}$$

where the index *tr_neigh* is used to distinguish between the neighbors estimated with spatial definition and the neighbors estimated using the CoD coefficient. Again, the equation describing the model is linear and therefore the learning process of this Lag-STARIMA model is done by solving the least squares problem.

9.4.5 Graph-Based Lag-STARIMA (GBLS)

The Lag-STARIMA model is based on the estimation of traffic correlations between the roads of the network as already mentioned. The straightforward approach (*brute-force*) for the identification of the most well correlated roads with a road of interest r, requires $N - 1$ CoD computations, where N is the total number of roads in the network. This process takes place N times for all the roads of the traffic network and therefore, the total number of CoD computations for all the roads of the network is $N \times (N - 1)$. The computational complexity of the process that implements these computations is $O(N^2)$. For large traffic networks, where N is in the order of hundreds or even thousands, the required time for these computations can be prohibitive, especially for applications that need (close to) real time short-term traffic forecasting.

In order to deal with this problem, Salamanis et al. [23] proposed a graph-based method for reducing the computational requirements of the CoD evaluations. In this approach, initially the traffic network was represented in the form of a directed graph based on the definition of the *in* and *out* roads. The in roads of a road of interest are those that end to its start or end node, whereas the out roads are those that start from its start or end node. For instance, in Fig. 9.1 the in roads or r are 1, 2 and 6 while the out roads are 3, 4 and 5. The in roads import traffic to the road of interest and the out roads export traffic. When a road has only one direction then it has only in or only out roads, whereas when it is bidirectional it has both.

Based on this definition, a finite directed graph is constructed in which each node represents a road of the network and each edge a direct in or out connection. More formally, two nodes n_1, n_2 in the graph are connected with an edge if and only if n_2 is an in or out road of n_1 or vice versa. The adjacency matrix of the graph is constructed based on the following formula:

$$A_{ij} = \begin{cases} 1, & \textit{if } j \textit{ out road of } i \\ 0, & \textit{otherwise} \end{cases} . \tag{9.71}$$

An example of an adjacency matrix of such graph is

$$A = \begin{bmatrix} 0\,1\,0\,1\,0\,0 \\ 1\,0\,0\,0\,0\,1 \\ 1\,1\,0\,0\,0\,1 \\ 0\,0\,0\,0\,1\,0 \\ 0\,0\,1\,0\,0\,1 \\ 0\,1\,1\,0\,0\,0 \end{bmatrix} . \tag{9.72}$$

The non-zero elements of the adjacency matrix in row i represent the out connections of i and those at column j the in connections of j. The matrix is not (in general) symmetric, meaning that if $A_{ij} = 1$, that is road j is an out road of road i, then this does not (necessarily) mean that $A_{ji} = 1$, that is road i is an out road of j. The matrix will be symmetric if and only if all the roads of the network are bidirectional. Also,

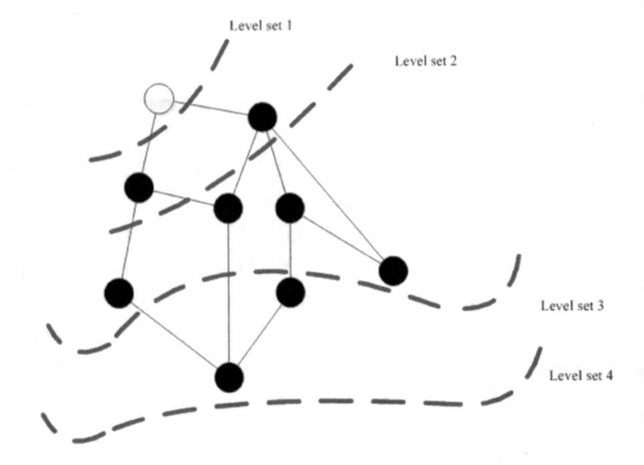

Fig. 9.2 Process of accumulating the vertices of an adjacency class

the matrix is sparse, which reduces the memory requirements for storing it. Finally, a road is not considered as in or out road of itself and therefore:

$$A_{ii} = 0, \forall i \in [1, N]. \tag{9.73}$$

Following the formation of the graph, the neighbors of a vertex at a given distance d can be found by accumulation of all the neighboring vertices up that distance. These neighbors define an *adjacency class* whose level is $d-1$ [23]. The adjacency class of level 0 (i.e. distance 1) of a specific vertex, includes only the vertex itself. The vertices of an adjacency class of a specific vertex are accumulated using a variant of the *Breadth First Search (BFS)* graph traversal algorithm. The process of accumulating the vertices is shown in Fig. 9.2.

The aforementioned variant of the BFS algorithm is *Modified Breadth First Search (MBFS)* is described in Algorithm 1. The algorithm operates on the rows of the adjacency matrix A, meaning that it accumulates vertices with respect to the out roads. The accumulation of vertices with respect to the in roads requires operation on the columns of the A matrix or alternatively the rows of its transpose A^T. Finally, if both the in and the out roads are required, then the $A + A^T$ matrix is utilized. In this case, the $A + A^T$ matrix is not explicitly stored but the rows of A and A^T are used to determine all the neighbors of the vertex. Thus, in this case, step 6 of the algorithm should be replaced by $B = B \cup \{A_u\} \cup \left\{A_u^T\right\}$ [23].

Algorithm 1 ModifiedBFS(u,A,l_s)

```
1: N = {u}
2: Mark u as visited
3: for i = 2 to l_s do
4:     B = {}
5:     for v ∈ N do
6:         B = B ∪ A_v
7:     N = N ∪ B
```

In order to avoid duplicate entries at step 6 of the MBFS algorithm, instead of accumulation of all neighbors and sorting, a structure containing a dense vector and a list is used. This dense vector is of size equal to the total number of vertices (roads) with all its entries initially set to zero. The list is used to retain the positions of the non-zero elements in the dense vector. A vertex is admitted the set B only if its respective position in the dense vector is zero and does not belong to the set N (step 7) during a previous iteration [23]. The implementation of the set N is similar to that of set B. When the algorithm finishes, the l_s level neighbors of the vertex u are copied to a vector with size equal to the number of neighbors. The computational complexity of the MBFS algorithm [23] is given by:

$$O\left(N \cdot \left(\frac{nnz\,(A)}{N}\right)^{p}\right), \tag{9.74}$$

where $nnz(A)$ is the number of non-zero elements of the adjacency matrix A and p the level of the adjacency class until which neighboring vertices are selected. For the adjacency matrix A, it holds:

$$nnz\,(A) \geq N. \tag{9.75}$$

This inequality holds because each road of the network should have at least 1 in or out road. If there is a road with 0 in or out roads, it means that this road is cut off from the rest of the network and the corresponding graph has an additional connected component. As a single coherent graph with just 1 connected component is considered, this cannot be the case. Hence, the computational complexity of MBFS is:

$$O\,(aN)\,,\ a \geq 1. \tag{9.76}$$

This is a significant improvement compared to the complexity of the naive approach which is $O(N^2)$. This improvement is depicted in Figs. 9.3, 9.4, 9.5 and 9.6. In particular, supposing four traffic networks with 100, 200, 300 and 400 roads respectively and one traffic time series of size 288 per road, the CoD computations are performed using the naive (brute-force) and the MBFS method (with 3 different levels). As shown, the performance of the MBFS method is, in all cases, substantially improved compared to the brute-force method. The data used for producing the results shown in the above figures correspond to the traffic network of the state of

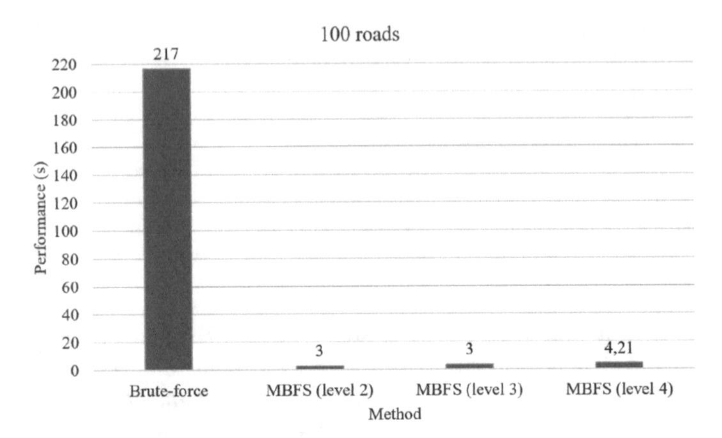

Fig. 9.3 Performance of the CoD computations using the brute-force and the MBFS method in a traffic network with 100 roads

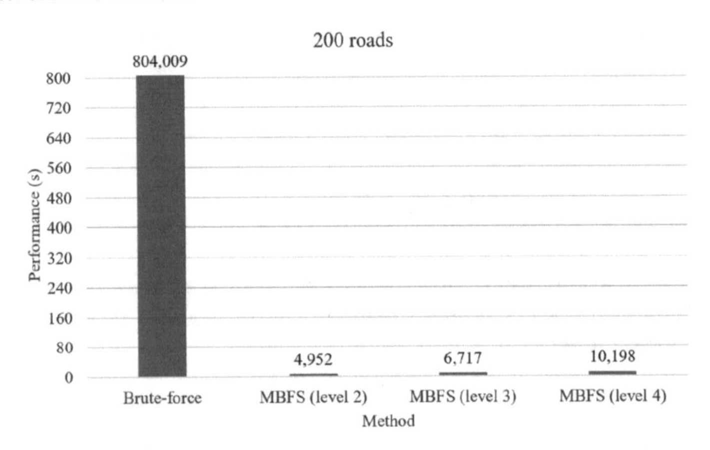

Fig. 9.4 Performance of the CoD computations using the brute-force and the MBFS method in a traffic network with 200 roads

California, USA and are available through the online repository: *Caltrans Performance Measurement System (PeMS)*.

The memory requirements of the MBFS algorithm, are limited to two dense vectors of size equal to the total number of vertices and two lists. Furthermore, as the adjacency matrices are sparse, they are stored using the Compressed Sparse Row storage format. As the correlations between the roads of the network are computed using this method, the Lag-STARIMA model is used for conducting the predictions. The described methodology for efficiently computing the correlations between roads of large traffic networks combined with the Lag-STARIMA model results in a new parametric algorithm for short-term traffic forecasting, namely the *Graph-Based Lag-STARIMA (GBLS)* whose learning process is, as in the case of the Lag-STARIMA model, the solution of the least squares problem.

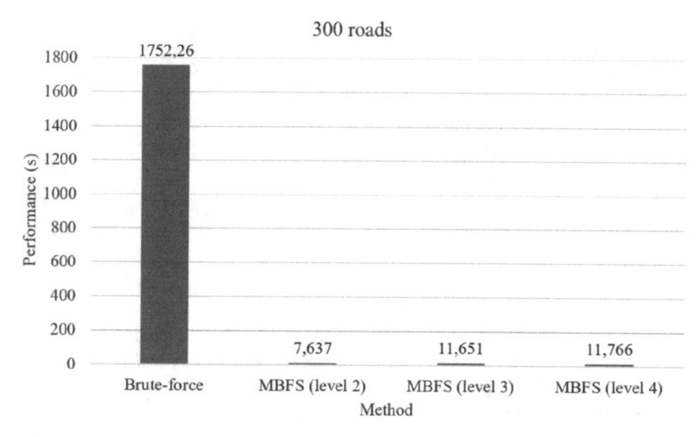

Fig. 9.5 Performance of the CoD computations using the brute-force and the MBFS method in a traffic network with 300 roads

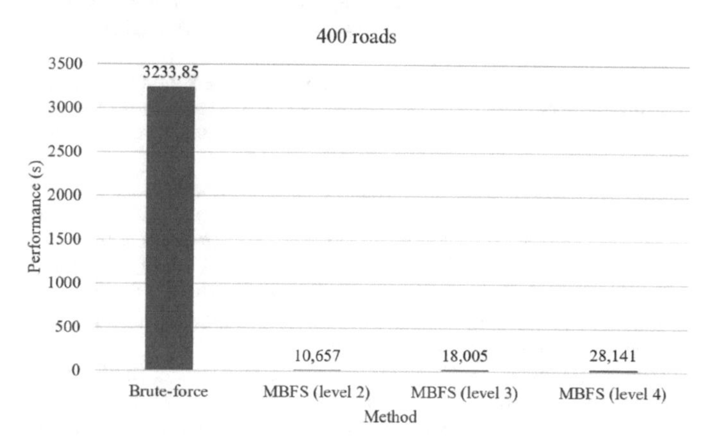

Fig. 9.6 Performance of the CoD computations using the brute-force and the MBFS method in a traffic network with 400 roads

References

1. Abadi, A., Rajabioun, T., Ioannou, P.A.: Traffic flow prediction for road transportation networks with limited traffic data. IEEE Trans. Intell. Transp. Syst. **16**(2), 653–662 (2015)
2. Agafonov, A., Myasnikov, V.: Traffic flow forecasting algorithm based on combination of adaptive elementary predictors. Commun. Comput. Inf. Sci. **542**, 163–174 (2015)
3. Basu, S., Meckesheimer, M.: Automatic outlier detection for time series: an application to sensor data. Knowl. Inf. Syst. **11**(2), 137–154 (2007)
4. Bekhor, S., Lotan, T., Gitelman, V., Morik, S.: Free-flow travel speed analysis and monitoring at the national level using global positioning system measurements. J. Transp. Eng. **139**(12), 1235–1243 (2013)
5. Box, G.E.P., Jenkins, G.M., Reinsel, G.C.: Time Series Analysis: Forecasting & Control, 4th edn. Wiley (2008)

6. Breunig, M.M., Kriegel, H.P., Ng, R.T., Sander, J.: LOF: identifying density-based local outliers. In: Proceedings 2000 ACM Sigmod International Conference Management Data, pp. 1–12 (2000)
7. Diamantopoulos, T., Kehagias, D., Konig, F. G., Tzovaras, D.: Investigating the effect of global metrics in travel time forecasting. In: IEEE Conference on Intelligent Transportation Systems, Proceedings, ITSC, pp. 412-417 (2013)
8. Guo, J., Huang, W., Williams, B.M.: Adaptive Kalman filter approach for stochastic short-term traffic flow rate prediction and uncertainty quantification. Transp. Res. Part C Emerg. Technol. **43**, 50–64 (2014)
9. Gustafsson, F., et al.: Particle filters for positioning, navigation, and tracking. IEEE Trans. Signal Process. **50**(2), 425–437 (2002)
10. Habtemichael, F.G., Cetin, M.: Short-term traffic flow rate forecasting based on identifying similar traffic patterns. Transp. Res. Part C Emerg. Technol. **66**, 61–78 (2016)
11. Hu, W., Yan, L., Liu, K., Wang, H.: A Short-term traffic flow forecasting method based on the hybrid PSO-SVR. Neural Process. Lett. **43**(1), 155–172 (2016)
12. Huang, M.L.: Intersection traffic flow forecasting based on GSVR with a new hybrid evolutionary algorithm. Neurocomputing **147**, 343–349 (2015)
13. Huang, W., Song, G., Hong, H., Xie, K.: Deep architecture for traffic flow prediction: deep belief networks with multitask learning. IEEE Trans. Intell. Transp. Syst. **15**(5), 2191–2201 (2014)
14. Kumar, K., Parida, M., Katiyar, V.K.: Short term traffic flow prediction for a non urban highway using artificial neural network. Proc. Soc. Behav. Sci. **104**, 755–764 (2013)
15. Kumar, S.V., Vanajakshi, L.: Short-term traffic flow prediction using seasonal ARIMA model with limited input data. Eur. Transp. Res. Rev. **7**, 3 (2015)
16. Li, M.W., Hong, W.C., Kang, H.G.: Urban traffic flow forecasting using Gauss-SVR with cat mapping, cloud model and PSO hybrid algorithm. Neurocomputing **99**, 230–240 (2013)
17. Lv, L., Chen, M., Liu, Y., Yu, X.: A plane moving average algorithm for short-term traffic flow prediction. In: Advances in Knowledge Discovery and Data Mining: 19th Pacific-Asia Conference, PAKDD 2015. Ho Chi Minh City, Vietnam, 19–22 May 2015, pp. 357–369, Proceedings, Part II; Cao, T., Lim, E.P., Zhou, Z.H., Ho, T.B., Cheung, D., Motoda, H. (eds.) Cham, Springer International Publishing (2015)
18. Lv, Y., Duan, Y., Kang, W., Li, Z., Wang, Y.: Traffic flow prediction with big data: a deep learning approach. IEEE Trans. Intell. Transp. Syst. **16**(2), 865–873 (2015)
19. Ma, X., Tao, Z., Wang, Y., Yu, H., Wang, Y.: Long short-term memory neural network for traffic speed prediction using remote microwave sensor data. Transp. Res. Part C Emerg. Technol. **54**, 187–197 (2015)
20. Moretti, F., Pizzuti, S., Panzieri, S., Annunziato, M.: Urban traffic flow forecasting through statistical and neural network bagging ensemble hybrid modeling. Neurocomputing **167**, 3–7 (2015)
21. Niu, X., Zhu, Y., Zhang, X.: DeepSense: a novel learning mechanism for traffic prediction with taxi GPS traces. In: 2014 IEEE Global Communications Conference, GLOBECOM 2014, pp. 2745–2750 (2014)
22. Polson, N., Sokolov, V.: Deep learning for short-term traffic flow prediction. Transp. Res. Part C Emerg. Technol. **79**, 1–17 (2017)
23. Salamanis, A., Kehagias, D.D., Filelis-Papadopoulos, C.K., Tzovaras, D., Gravvanis, G.A.: Managing Spatial graph dependencies in large volumes of traffic data for travel-time prediction. IEEE Trans. Intell. Transp. Syst. **17**(6), 1678–1687 (2016)
24. Wang, H., Liu, L., Dong, S., Qian, Z., Wei, H.: A novel work zone short-term vehicle-type specific traffic speed prediction model through the hybrid EMDARIMA framework. Transp. B **4**(3), 159–186 (2016)
25. Wang, J., Shi, Q.: Short-term traffic speed forecasting hybrid model based on chaos-wavelet analysis-support vector machine theory. Transp. Res. Part C Emerg. Technol. **27**, 219–232 (2013)

26. Wang, Z., Lu, M., Yuan, X., Zhang, J., Van De Wetering, H.: Visual traffic jam analysis based on trajectory data. IEEE Trans. Vis. Comput. Graph. **19**(12), 2159–2168 (2013)
27. Wojnarski, M., Gora, P., Szczuka, M., Nguyen, H.S., Swietlicka, J., Zeinalipour, D.: IEEE ICDM 2010 contest: TomTom traffic prediction for intelligent GPS navigation. In: Proceedings—IEEE International Conference on Data Mining, ICDM, pp. 1372–1376 (2010)
28. Wu, C.J., Schreiter, T., Horowitz, R.: Multiple-clustering ARMAX-based predictor and its application to freeway traffic flow prediction. In: Proceedings of the American Control Conference, pp. 4397–4403 (2014)
29. Yang, H.F., Dillon, T. S., Chen, Y.P.P.: Optimized structure of the traffic flow forecasting model with a deep learning approach. IEEE Trans. Neural Netw. Learn. Syst. 1–11 (2016)
30. Yao, B., et al.: Short-term traffic speed prediction for an urban corridor. Comput. Civ. Infrastruct. Eng. **32**(2), 154–169 (2017)
31. Yu, Y.: Short-term traffic forecasting using the double seasonal holt-winters method. In: CICTP 2016, pp. 397–407 (2016)
32. Yu, H., Wu, Z., Wang, S., Wang, Y., Ma, X.: Spatiotemporal recurrent convolutional networks for traffic prediction in transportation networks. Sensors (Switzerland) **17**(7) (2017)
33. Yu, R., Li, Y., Shahabi, C., Demiryurek, U., Liu, Y.: Deep learning: a generic approach for extreme condition traffic forecasting. In: Proceedings 2017 SIAM International Conference Data Mining, pp. 777–785 (2017)
34. Zhang, Y., Zhang, Y., Haghani, A.: A hybrid short-term traffic flow forecasting method based on spectral analysis and statistical volatility model. Transp. Res. Part C Emerg. Technol. **43**, 65–78 (2014)
35. Zheng, Z., Su, D.: Short-term traffic volume forecasting: a k-nearest neighbor approach enhanced by constrained linearly sewing principle component algorithm. Transp. Res. Part C Emerg. Technol. **43**, 143–157 (2014)
36. Zhu, J.Z., Cao, J.X., Zhu, Y.: Traffic volume forecasting based on radial basis function neural network with the consideration of traffic flows at the adjacent intersections. Transp. Res. Part C Emerg. Technol. **47**(2), 139–154 (2014)

Chapter 10
Network Traffic Analytics for Internet Service Providers—Application in Early Prediction of DDoS Attacks

Apostolos P. Leros and Antonios S. Andreatos

Abstract In this chapter an approach for modelling intra-values forecasts of a time-series Network Traffic using a mean reverting stochastic process (MRSP) is presented. An autoregressive model of order n, AR(n), formalized in state space, with its unobservable coefficients estimated by a Kalman filter using n past time series observations produces [AR(n)-KF] estimates, which constitute the mean reverting part of the process. A Brownian motion multiplied by a diffusion (or volatility) term constitutes the stochastic part of the process. The determinant and trace of the Kalman filter error covariance matrix multiplied by the process itself is used to capture the diffusion dynamics in the intra-values time-series. The proposed algorithm is designed especially for network traffic and it does not assume stationary data. The method was tested using real traffic data from GRnet concerning our institutional network. Experimental as well as simulation results based on real daily data from the GRnet IP traffic demonstrate the applicability of the model. The proposed MRSP algorithm was able to identify successfully unusual activities contained in the test datasets and produce proper warnings. Applications on real-time D/DoS bandwidth-flooding attack detection, are also presented.

Keywords Network traffic · Mean reverting stochastic process · Autoregressive model · State space · Kalman filter · Time-series prediction · Router bandwidth demand prediction · DDoS bandwidth-flooding attack detection

---

A. P. Leros
Department of Automation, School of Technological Applications, Technological Educational Institute of Sterea Ellada, 34400 Psachna, Evia, Greece
e-mail: lerosapostolos@gmail.com

A. S. Andreatos (✉)
Division of Computer Engineering and Information Science, Hellenic Air Force Academy, Dekeleia Air Force Base, 1010 Dekeleia, Attica, Greece
e-mail: antonios.andreatos@hafa.haf.gr; aandreatos@gmail.com

G. A. Tsihrintzis et al. (eds.), *Machine Learning Paradigms*, Intelligent Systems Reference Library 149, https://doi.org/10.1007/978-3-319-94030-4_10

10.1 Introduction

Today almost all homes, organizations and companies in the world are connected to the Internet, using millions of devices of all kinds. Bandwidth demand is exponentially growing in order to cover emerging needs of multimedia traffic, especially video; the arrival of the Internet of Things (IoT) will further complicate the situation. This trend pushes for the upgrade of infrastructure (e.g., Fiber-to-the-Home—FTTH) but this won't happen instantaneously. Therefore, Internet Service Providers (ISPs) with thousands of customers, usually of different priority, are in need of applying smart prediction algorithms in order to be able to serve their customer needs during the peak hours.

A main issue that Internet providers face is the availability of services. Distributed denial of service (DDoS) attacks against targeted servers may happen anytime, putting critical infrastructures out of order [8, 17]. In DDoS attacks, many agents cooperate to cause excessive load to a victim host, service or network. Over the years, DDoS attacks have increased in importance, number and strength, becoming a major problem. In a recent survey of network operators [3], DDoS was the most commonly identified 'significant threat' (76% of respondents). Furthermore, significant growth in size of attacks and in their sophistication is reported [3, 8, 26].

In this work real network traffic of an institutional LAN in "reduced resolution dataset" (RRD) format is processed. Bandwidth-flooding DoS/DDoS attacks producing excessive traffic aiming at exceeding the server's capability to process the requests will also be discussed. The traffic values are provided in 5 min intervals. A brief description of how the Fiber Ethernet Bandwidth is managed by providers via the institutional router is given in [29]. Hence, traffic from all sources is aggregated and the final volume is provided at the end of the interval (for each input port). Therefore, the router cannot discriminate whether excessive load within a single interval is caused by one or by multiple distributed agents. Under this perspective, bandwidth-flooding DoS and DDoS attacks are treated equally in our approach.

The proposed approach is a Mean Reverting Stochastic Processes (MRSP) consisting of a deterministic part and a stochastic part. The deterministic part is an AR(n) model with coefficients estimated with a Kalman filter, [AR(n)-KF]. The stochastic part is a Brownian motion multiplied by a diffusion (or volatility) term. The determinant and the trace of the Kalman filter error covariance matrix multiplied by the process itself is used to capture the diffusion dynamics in the intra-values time-series. The MRSP algorithm produces intra-values within the 5 min intervals and identifies successfully normal, as well as abnormal activities contained in the test datasets. Abnormal activities include sudden peaks in traffic (due to an attack or misuse), zero traffic rate (due to failures), and a nearly constant high peak (due to congestions or DDoS attacks) [22].

This chapter is organized as follows: Section 10.2 describes the procedure followed and reviews related work. The proposed algorithm is described and mathematically formulated in Sect. 10.3. Section 10.4 presents the structure of the MRSP and the parameter selection for simulating intra-values and DoS attacks using 5 min data

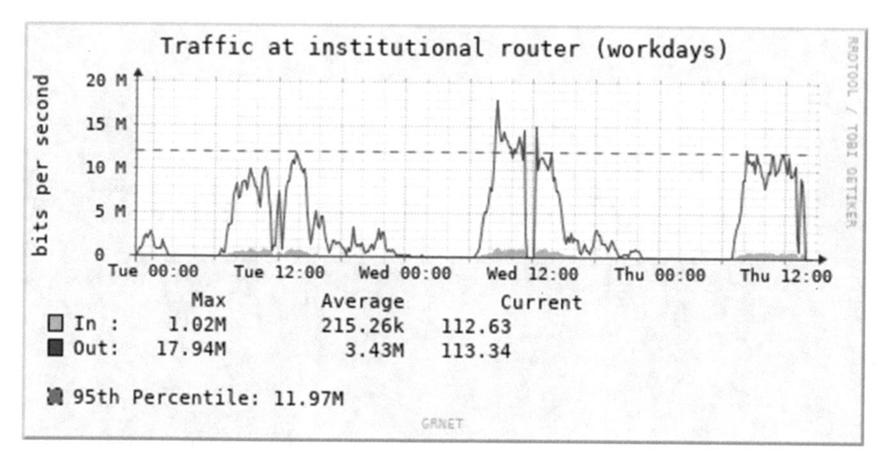

Fig. 10.1 A typical graph by the RRDTOOL (workdays)

intervals and AR(N)-Kalman filter predictions. Matlab simulations based on real network IP traffic obtained from GRnet are presented for the validation of the proposed modelling approach. Section 10.5 presents the performance of the MRSP algorithm on real data in RRD format. Next, some datasets were enhanced with artificial data realistically representing DDoS attacks and anomalies for testing purposes. Simulation results from these datasets demonstrate that the proposed algorithm detects any abnormal behaviour. Applications on Router Bandwidth Demand estimation [2, 15], as well as network traffic anomaly detection [20, 28] will be presented in Sect. 10.6. Section 10.7 concludes this chapter.

10.2 The Procedure Adopted

In this work we take the network traffic of an institutional LAN in rrd (reduced resolution dataset) format provided by default in 5 min intervals. Although this interval can be changed, this is the standard adjustment for Multi Router Traffic Grapher (MRTG) [23, 24] and it is used in most implementations. Therefore, the default sampling rate (5-min) was adopted in this work.

For visualizing long periods, MRTG does some averaging over a longer step. Weekly traffic is expressed in 30-min steps, monthly traffic in 2-h steps, and yearly traffic in 1-day steps. In order to preserve some information lost by averaging, the tool logs together the average and the maximum value for each step. They can both be used to produce forecasts, regarding the average or the maximum behaviour respectively [21].

A typical Network traffic produced by the popular and widely used RRDTOOL [25] is shown in Fig. 10.1.

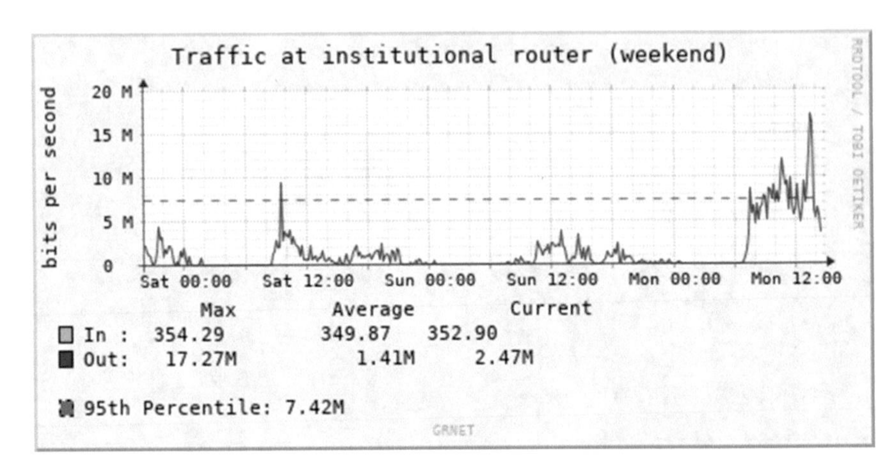

Fig. 10.2 Network traffic graphs for a weekend (left) and half workday (right)

In the graph of Fig. 10.1, the blue line represents the "outbound" traffic from the ISP to the institutional LAN. In other words, the blue curve represents the download traffic from the Internet towards the institution. On the other hand, the green shaded area, which is significantly smaller, represents the "inbound", traffic from the institutional LAN to the ISP (i.e., the upload traffic). The 24 hour time-line at the bottom is full of 5 min durations [29].

As we can observe in Fig. 10.2, network traffic varies significantly during workdays and weekends for the same institutional network.

The MRSP algorithm produces an artificial time-series of intra-values traffic (which is not provided by the router) within the 5 min intervals which may be used to:

1. Predict the network traffic for future intervals;
2. Detect a DDoS attack.

For the second item, simulated traffic data was generated, taking into account the characteristics of DDoS traffic from the bibliography, as well as our own experiments.

10.2.1 Related Work

Geva et al. [8] survey Bandwidth Distributed (BW) DDoS attacks and defenses. They claim that, while current BW-DDoS attacks employed relatively crude, inefficient, 'brute force' mechanisms, future attacks may be significantly more effective, hence, much more harmful.

Moussas et al. model network traffic using autoregressive moving-average classical models (ARMA, ARIMA, SARIMA) [19, 22]. Shu et al. [27], model Wireless

Traffic using SARIMA Models. Kuan Hoong et al. [12] use ARIMA to model bit torrent network traffic. However, these classical models require stationary time series.

Giannopoulos et al. [9] propose a Stochastic Model with an Adaptive Proportional Controller for the Evolution of User-Router Bandwidth Demand for Quality of Service. The stochastic behaviour of their modified MRSP process is implemented by a Bernoulli and a uniform distribution together (instead of the standard Brownian motion).

Giannopoulos et al. [10] propose a Model for the Evolution of Router Bandwidth Demand Using an Adaptive PID Controller for the production of intra-values within the 5-min intervals. They also use Bernoulli and a uniform distribution together for their modified MRSP process.

The current work uses an innovative MRSP model which is significantly different compared to Giannopoulos et al. [9, 10]. The proposed algorithm not only produces intra-values within the 5 min intervals but also, it estimates predictions for the next intervals, as presented below.

10.3 The Proposed Approach

In our approach the ARMA and ARIMA models were avoided because they assume stationary data (characterized by constant average and variance), which is not the case for network traffic. Another reason calling for this choice is that the aforementioned classical models need a lot of past (historical) data (>50) in order to produce predictions. A third reason for not selecting the ARMA and ARIMA models is that users' behaviour varies with time as some users are leaving while new users join, thus, long historical data do not matter so much.

The proposed approach uses an AR(n) model with coefficients estimated by a Kalman filter based on n past 5 min interval traffic values [5, 10]. The Kalman filter does not need a stationary time-series in order to make predictions—in contrast to AR, ARMA, ARIMA and SARIMA models. "It is clear also from the definition that neither $\{X_t\}$ nor $\{Y_t\}$ is necessarily stationary" [5, Chap. 8, p. 276].

The proposed algorithm uses an AR(n) model for predictions with n being 3–5 historical data values—a reasonable number for network traffic. Old data such as 50 historical values do not offer reliable information for the near future (e.g., the next 5–10 min), because users' behavior varies largely with time. The coefficients of the AR(n) model are estimated by a Kalman filter which uses an adaptive system noise covariance matrix based on the variance of the real measurements. These [AR(n)-KF] predictions then are used as the mean values the mean reverting stochastic process (MRSP) will revert to. The MRSP uses an adaptive strength coefficient for its deterministic part. The stochastic part consists of **three components**: an adaptive diffusion (or volatility) coefficient, based on the square root of the determinant and the trace of the Kalman filter error covariance (which, to the best of our knowledge, has not been used before); the adaptive diffusion is multiplied by the square root of the stochastic process itself according to the Cox-Ingersoll-Ross (CIR) model [7] (which is the second component), and a Brownian motion to capture the dynamics of the evolution from a starting point to reach the AR(n) predictions (which is the third component). This way the MRSP model produces a number of intra-values (e.g. 300) for every 5 min sample data value. Execution of the proposed algorithm on real traffic data proves its correctness and efficiency.

10.3.1 Mathematical Formulation

Description of the Mean Reverting Stochastic Process

A stochastic process S_t with general mean reverting behavior is formally described by the stochastic differential equation [4, 14]:

$$dS_t = A(t)[\mu(t) - S_t]dt + G(t, S_t)dB_t; \quad S_0 = \text{initial condition} \tag{10.3.1}$$

The deterministic part of (10.3.1) or so-called drift term is $A(t)[\mu(t) - S_t]$, with $A(t) > 0$ being the rate of reversion and $\mu(t)$ being the mean value (also called long term mean) around which the process tends to oscillate, and it may be a constant, a deterministic function of time, or even a stochastic process. The stochastic part is $G(t, S_t)dB_t$, with $G(t, S_t) > 0$ being the diffusion-coefficient and B_t being the Brownian motion (or Wiener or Wiener-Levy).

When $S_t > \mu(t)$, the drift term is negative and it results in a pull of the process S_t back down toward the equilibrium level. Conversely, when $S_t < \mu(t)$, the drift term is positive and it results in a pull of the process S_t back up to a higher equilibrium value. For the stochastic part, since B_t is a random variable composed of a sum of independent Gaussian increments, it is also Gaussian.

The solution of (10.3.1) is unique and takes the form [14]:

$$S_t = e^{-A(t)[t-t_0]}S_0 + \int_{t_0}^{t} e^{-A(s)[t-s]}[\mu(t)ds + G(s, S_t)dB_s]; \quad t_0 \le t \le T \tag{10.3.2}$$

which has the characteristic that fluctuates randomly, but tends to revert to some fundamental level $\mu(t)$ with some reversion behavior, which depends upon the choices of the speed of the reversion parameter $A(t) > 0$, and the nonrandom or random but continuous function $G(t, S_t) > 0$.

For simulation purposes, using the method of Euler–Maruyama [11], the discretization of (10.3.1) in an interval $t \in [t_0, T]$ with increments $i = 1, 2, \ldots, N$, for some integer N, is:

$$S_{i+1} = A(t_i)[\mu(T) - S_i]dt + G(t_i, S_i)\left[\sqrt{dt}N(0, 1)\right]; \quad S_{i0} = \text{initial condition} \tag{10.3.3}$$

where $N(0, 1)$ denotes the normal (Gaussian) process with zero mean and unit variance and the required step interval $dt = T/N$. The behavior of the discretized mean reverting stochastic process S_i remains the same as that for the continuous case S_t.

10.3.2 State Space Model—Autoregressive Model—Discrete-Time Kalman Filter

State space model: A linear, discrete-time, finite-dimensional system with noisy input and noisy output is described with the following state-space equations [1, 16]:

$$x_{k+1} = F_k x_k + w_k \tag{10.3.4}$$

$$z_k = y_k + v_k = H_k^T x_k + v_k \tag{10.3.5}$$

The unobserved variables of interest is the system state x_k at discrete-time $k \geq 0$. The system output is y_k, which usually is noisy. As such a noise process $\{v_k\}$ is added to it resulting in the observed measurement process $\{z_k\}$. The matrices F_k and H_k are of proper dimensions and known. The system input noise process $\{w_k\}$ and the measurement output noise process $\{v_k\}$, are independent and individually Gaussian white (uncorrelated from instant to instant and stationary) noise with zero mean and known covariance, i.e., $\{w_k\} \sim WN(0, Q_k\delta_{ks})$ and $\{v_k\} \sim WN(0, R_k\delta_{ks})$, where δ_{ks} denotes the Kronecker delta which is 1 for $k = s$ and zero otherwise. We assume that the initial state x_0 is a Gaussian random variable with known mean $E[x_0] = \bar{x}_0$ and known covariance $E\{[x_0 - \bar{x}_0][x_0 - \bar{x}_0]^T\} = \overline{P}_0$, and also that it is independent of w_k and v_k, for any k.

Autoregressive model: Now a model that expresses a univariate time-series system output y_k as a linear combination of past observations y_{k-n} (which are the observed measurements z_{k-n}) and a white noise v_k is referred to as an autoregressive of order n [or AR(n)] model and is given by the equation:

$$y_k = -a_k^{(1)} y_{k-1} - a_k^{(2)} y_{k-2} - \cdots - a_k^{(n)} y_{k-n} + v_k \tag{10.3.6}$$

The unknown set of parameters $\left\{a_k^{(1)}, \ldots, a_k^{(n)}\right\}$ are referred to as the AR(n) coefficients, which in case they are constant, with sufficient y_{k-n} observed measurements, can be found by solving a set of linear equations. Due to random errors in y_k though, it is more realistic to consider the coefficients $a_k^{(i)}$, $i = 1, 2, \ldots, n$, as being noisy. Thus, it is assumed that they are of the form $a_{k+1}^{(i)} = a_k^{(i)} + w_k^{(i)}$, where each $w_k^{(i)}$ is a zero mean, white, Gaussian random process, independent of $w_k^{(j)}$ for $i \neq j$, and also independent of each v_k.

Defining now all these unknown noisy AR(n) coefficients as an n-dimensional state vector $x_k = \left[x_k^{(1)}\ x_k^{(2)} \ldots x_k^{(n)}\right]^T \triangleq \left[a_k^{(1)}\ a_k^{(2)} \ldots a_k^{(n)}\right]^T$, and also defining

Table 10.1 Discrete-time Kalman filter equations

Time propagation (or prediction equations):	Measurement update (or correction) equations:
$\hat{x}_{k+1/k} = \hat{x}_{k/k}$ (10.3.10)	$K_{k+1} = P_{k+1/k} H_{k+1} \left[H_{k+1}^T P_{k+1/k} H_{k+1} + R_{k+1} \right]^{-1}$ (10.3.12)
$P_{k+1/k} = P_{k/k} + Q_k$ (10.3.11)	$\hat{x}_{k+1/k+1} = \hat{x}_{k+1/k} + K_{k+1} \left[z_{k+1} - H_{k+1}^T \hat{x}_{k+1/k} \right]$ (10.3.13)
	$P_{k+1/k+1} = \left(I - K_{k+1} H_{k+1}^T \right) P_{k+1/k} \left(I - K_{k+1} H_{k+1}^T \right)^T + K_{k+1} R_{k+1} K_{k+1}^T$ (10.3.14)

an n-dimensional, white, zero mean, Gaussian process $\{w_k\}$ as the vector process formed from all the $w_k^{(i)}$, we get the following system state equation:

$$x_{k+1} = x_k + w_k \tag{10.3.7}$$

Also, if we define the row vector of past observations:

$$H_k^T = \left[-y_{k-1} \ -y_{k-2} \ \cdots \ -y_{k-n} \right] = \left[-z_{k-1} \ -z_{k-2} \ \cdots \ -z_{k-n} \right] \tag{10.3.8}$$

then Eq. (10.3.6) with (10.3.8) becomes the observed state measurements equation:

$$y_k = H_k^T x_k + v_k \tag{10.3.9}$$

Thus, with the above definitions, the autoregressive of order n [or AR(n)] model has been transformed into a linear, discrete-time, finite-dimensional noisy input state space Eq. (10.3.7) with Eq. (10.3.9) being the noisy output.

Remark 1 In this state space formulation of the AR(n) model, it is not required for the time series data to be stationary, in contrast with the classical formulation case, where it is necessary for the time series data to be transformed by differencing or by removing trend and seasonal components before processing [5, 6].

Discrete-time Kalman filter: The one-step prediction problem now is to produce an estimate at time $k + 1$ of the system states $\hat{x}_{k+1/k+1}$ (which are the AR(n) coefficients) using n noisy measured time-series data $\{z_{k-1}, z_{k-2}, \ldots, z_{k-n}\}$, and from (10.3.9) the predicted y_{k+1} can be calculated. In this case, with the above definitions, the solution to the problem is given by the discrete-time Kalman filter recursive equations [1, 16] as listed in Table 10.1.
Where the matrices $R_k = E\left[v_k v_k^T\right]$, $Q_k = E\left[w_k w_k^T\right]$, and $I^{n \times n}$ is the $n \times n$ identity matrix (all 1's in the main diagonal and zeros elsewhere).

Equation (10.3.13) is initialized with $\hat{x}_{1/0}$ set equal to the vector of a priori estimates of the AR(n) coefficients. The one step predicted output is the term $y_{k+1} = H_{k+1}^T \hat{x}_{k+1/k}$ and the residual or so called innovation sequence is $r_{k+1} = \left[z_{k+1} - H_{k+1}^T \hat{x}_{k+1/k} \right]$.

Equation (10.3.14) is initialized with $P_{1/0}$ set equal to the a priori covariance matrix of the error in these coefficients. The matrix K_{k+1} is called the Kalman filter gain. Notice that the gain matrix K_{k+1} depends inversely on R_{k+1}—the larger the variance of the measurement error, the lower the weight is given to the measurement in making the forecast for the next period, given the current information set. The matrix Q_k describes the confidence in the system state Eq. (10.3.4). It can be estimated using the Maximum Likelihood Estimation method [1, 16], but usually is picked by simulations to be $Q_k = \gamma I^{n\times n}$, with γ being a positive scalar. The error covariance matrix $P_{k+1/k+1}$ depends on the measurements via K_{k+1}. As the k measurements are processed, it is desirable for the covariance to be $P_{k+1/k+1} \leq \rho_k I^{n\times n}$, where the positive scalar ρ_k approaches zero or a small number ρ as $k \to \infty$. Then, for almost all the measurements, the mean square parameter estimation error will approach zero, or some small quantity [1].

10.4 Structure and Parameters of the MRSP Algorithm

The structure of the previously described mean reverting stochastic process model and the [AR(n)-KF] predictions for simulating intra-values and denial of service (DoS) attacks within a 5 min time interval $t_i \in [0, \mathrm{T}]$ are graphically depicted in the Fig. 10.3. In this figure, the bottom part represents the measured time series data values $\{z_k, k = 1, 2, \ldots\}$. The part above it represents the [AR(n)-KF] predicted values y_k, each based on a set of n previous time series data values $\{z_1, z_2, \ldots, z_n\}$. The top part represents the mean reverting stochastic process (MRSP) generated intra-values S_i while the process evolves to reach the predicted values y_k. The squares at the top part at the end of each S_i process indicate the error between the MRSP predicted S_{kN} values at the end and the corresponding predicted values y_k.

The values of the parameters of the mean reverting stochastic model (10.3.3) chosen for simulation are as follows:

- For the time interval $t_i \in [0, \mathrm{T}] = 5\,\mathrm{min}$, we choose to divide it into N= 300 subintervals. This way we would have a total of $i = 1, 2, \ldots, N = 300$ increments within the 5 min interval and each increment would be of duration $dt = T/N = 1\,\mathrm{s}$.
- The strength of the mean reversion coefficient is calculated adaptively as $A(i) = adaptA\_scale \cdot [i \cdot dt \cdot |\mu(T) - S_i|] > 0$ for $i = 1, 2, \ldots, N = 300$ with $A(1)$ being some small positive number, e.g., $A(1) = 0.001$. The reason for picking this adaptive expression stems from the fact that at the beginning in the 5 min interval, since the process S_i starts from small (or almost zero) values we want large values for $A(i)$ to boost and directionally evolve the process towards the mean value $\mu(T)$. Close to the end of the interval though, as the process approaches the desired value $\mu(T)$, we want relatively small $A(i)$ values so that as not to overshoot the desired mean value $\mu(T)$. This behavior is captured by the expression $[i \cdot dt \cdot |\mu(T) - S_i|]$. That is, the difference $|\mu(T) - S_i|$ divided by dt into 300 increments (one second each) at the beginning is very large and as $i = 1, 2, \ldots$. N = 300 progresses along with the evolving S_i values towards the $\mu(T)$ value, the $A(i)$ values become smaller

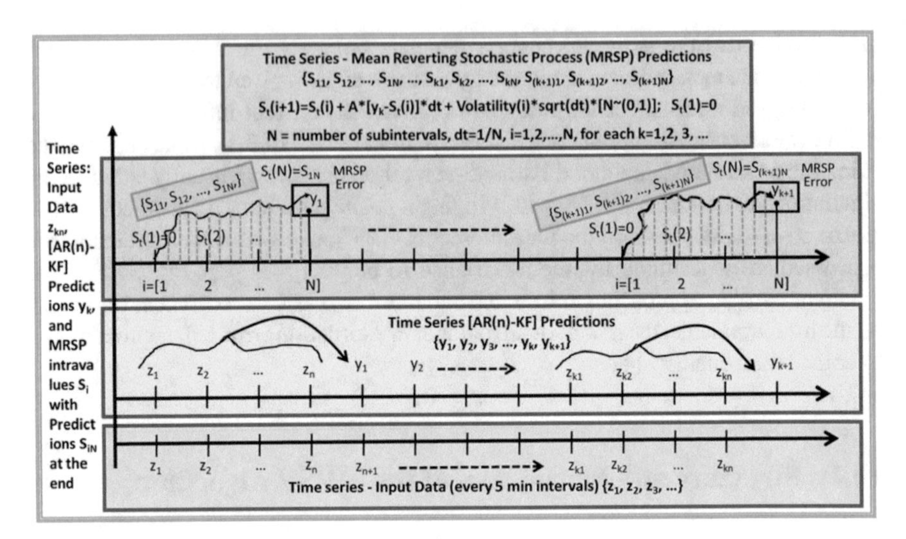

Fig. 10.3 Time series—input data z_k, [AR(n)-KF] predictions y_k, and mean reverting stochastic process (MRSP) intra-values $\mathrm{S_i}$ with predictions S_{iN} at the end

due to smaller deviations of S_i from $\mu(T)$. The additional factor $adaptA\_scale$ in the $A(i)$ expression is used to capture the dynamics produced by the order (or window-size) n of the autoregressive model, AR(n), as well as the number of historical measurements (or data points) needed for the Kalman filter covariance matrix P to reach a steady-state condition. That is, for the [AR(n)-KF] model this factor can take the form (a) $adaptA\_scale = 1.0/(n \cdot \text{dataPoints})$, or equivalently (b) $adaptA\_scale = [\text{Trace}P(k) + \text{Determinant}P(k)]/(n \cdot \text{dataPoints})$; $n + 1 \leq k \leq \text{dataPoints}$.

- The order of the autoregressive model was chosen to be AR(3), indicating regression over the past three (n = 3) 5 min values $\{z_{k-3}, z_{k-2}, z_{k-1}\}$. The choice n = 3 was determined by calculating the autocorrelation function with lags 20–50 points for a data set of 770 data points which gave a plot of a fast decaying exponential suggesting that for this data set an autoregressive model is appropriate, and also by calculating the partial autocorrelation function of the same data set which indicated 3–5 significant points [5].
- For the calculation of the long term mean $\mu(T)$ we use $\mu(T) = y_k = H_k^T \hat{x}_{k/k}$ with $H_k^T = \begin{bmatrix} -z_{k-1} & -z_{k-2} & \cdots & -z_{k-n} \end{bmatrix}$ being known from the measurements and $\hat{x}_{k/k}$ as being estimated by the discrete Kalman filter (Eqs. 10.3.10–10.3.14).
- The initial conditions S_1 of the MRSP model is chosen to be 1 bps which is the handshake (or "Keepalive") rate the router resets to after it provides the aggregate 5 min traffic value.
- In the MRSP algorithm, taking into consideration the physical aspects of the router, we set the predictions to the handshake value in case these are less than zero, i.e., if $y_k < 0$ then $y_k = handshake$. The same holds for the variable S_{i+1}, that is,

if $S_{i+1} < 0$ then $S_{i+1} = handshake$. In addition, if $S_{i+1} > C$ then $S_{i+1} = C$, where C is the capacity rate of the router, which in our case has the value C = 1 Gbps.

- The initial covariance matrix of the Kalman filter was chosen as $P_{1/0} = n \cdot dataPoints \cdot n^{(-n/4)} \cdot I^{n \times n}$, which by simulations reached a small steady-state value (good performance) with n values of 3–5 and a number of 20–40 data points.
- The covariance matrix Q_k describes the confidence in the system state Eq. (10.3.4); an increase in this matrix means that we trust less the process model and more the measurements. The traditional adaptation method proposed in Mohamed et al. [18] was used to adaptively calculate Q_k by using the residuals $r_{k+1} = \left[z_{k+1} - H_{k+1}^T \hat{x}_{k+1/k}\right]$ and the Kalman filter gain as $Q_k = K_{k+1}\left(\frac{1}{n} r_{k+1} r_{k+1}^T\right) K_{k+1}^T$.
- The measurement covariance R_k is set equal to the variance of the n observations vector $H_k^T = \left[-z_{k-1} \; -z_{k-2} \; \cdots \; -z_{k-n}\right]$, i.e., $R_k = \frac{1}{n-1} \sum_{i=1}^{n} (z_{k-i} - \mu_k)^2$, $\mu_k = \frac{1}{n} \sum_{i=1}^{n} z_{k-i}$.
- The initial state estimate was chosen as $\hat{x}_{1/0} = -0.35 * \left(\left[1 \; 1 \; \cdots \; 1\right]^{n \times 1}\right)^T$. This choice gave good plots for the initial errors at the beginning of the diagrams.
- The diffusion-coefficient (or diffusion function) as previously stated can be both deterministic and stochastic as well as a function of the process itself. Here it was chosen to be $G\left(k_{i+1/k}, S_i\right) = \sqrt{\left(tr\left[P_{k+1/k}\right] + \det\left[P_{k+1/k}\right]\right) S_i}$ at each instant $i = 1, 2, \ldots$. N $= 300$. This choice stems from the CIR model [7] which captures the mean reverting phenomenon and avoids the possibility of negative values for all values of $A(i) > 0$ and $\mu(T) > 0$ once the condition $2A(i)\mu(\mathrm{T}) > \left(\mathrm{Trace}\, P_{k+1/k} + \mathrm{Determinant}\, P_{k+1/k}\right)$ is satisfied. The stochastic part of this MRSP has the standard deviation $\sqrt{\left(\mathrm{Trace}\, P_{k+1/k} + \mathrm{Determinant}\, P_{k+1/k}\right)}$ and is proportional to $\sqrt{S_i}$. According to [7] this is significant because it states that as the short-rate increases, the standard deviation will decrease.

Remark 2 The choice in this work to use the Kalman filter error covariance matrix in the diffusion-coefficient provides information for the spread of error in the estimation of the AR(n) coefficients for the one 5 min ahead predicted value. That is, for the Kalman filter estimate $\hat{x}_{k+1/k}$ of the unobservable system state $x_{k+1/k}$, the covariance matrix of the error is $P_{k+1/k} = E\left\{\left(x_{k+1/k} - \hat{x}_{k+1/k}\right)\left(x_{k+1/k} - \hat{x}_{k+1/k}\right)^T\right\}$. Therefore, since the rows of this error covariance matrix span the error space of the AR(n) coefficients, different measures of this matrix can be used. Such measures giving constant diffusion for the whole 5 min interval include $\sqrt{\mathrm{Determinant}\left[P_{k+1/k}\right]} > 0$, $\sqrt{\mathrm{Trace}\left[P_{k+1/k}\right]} > 0$, $\sqrt{\mathrm{Eigenvalues}\left[P_{k+1/k}\right]} > 0$, $\sqrt{\mathrm{Norm}\left[P_{k+1/k}\right]} > 0$, or combinations of these. As the order though of the autoregressive model increases, so are these measures of the Kalman filter covariance matrix as well. Supportive to the thought of using the determinant of the Kalman filter covariance matrix $\sqrt{\mathrm{Determinant}\left[P_{k+1/k}\right]} > 0$ as an estimate of diffusion coefficient spanning the error space, is the known fact from linear algebra [13] that for a set of linearly independent vectors $u_1, u_2, \ldots, u_n$ in R^n, the absolute value of the determinant of the matrix M with rows $u_1, u_2, \ldots, u_n$ indicates the volume $V(\Pi) = |det(M)|$ of the solid parallelepiped $\Pi = \{a_1u_1 + a_2u_2 + \cdots + a_nu_n : 0 \leq a_i \leq 1 \ for \ i = 1, 2, \ldots, n\}$ formed by these vectors. When $n = 2$, Π is a parallelogram and $V(\Pi)$ denotes the area of Π. In general, $V(\Pi) = 0$ if and only if the vectors $u_1, u_2, \ldots, u_n$ are linearly dependent (i.e., if and only if the vectors do not form a coordinate system in R^n). Also the function Trace$\left[P_{k+1/k}\right] > 0$, gives another estimate of diffusion coefficient, since the trace is the sum of the diagonal elements of a matrix, and for the error covariance the trace is the sum of the mean square errors, which is a performance index for the Kalman filter. Similar diffusion coefficient measures provide the eigenvalues and the norm of the Kalman filter covariance matrix since the eigenvalues and the norm are interrelated for a positive definite matrix.

10.5 Results and Performance of the MRSP Algorithm

In the next series of figures the results obtained by Matlab simulation using real data are presented; the results demonstrate in details the operation of the MRSP algorithm and also prove that it works quite well.

In Fig. 10.4 the first two graphs show the stem plot and the continuous plot, respectively, of the total inbound time-series real data set consisting of 800 aggregate within a 5 min interval traffic values. For the largest part this inbound time-series real data set represents normal traffic which is less than 50 Mbps. The peak appearing from x = 161 to x = 169 reaching the value of 651 Mbps at x = 165 is inserted manually and simulates a traffic anomaly [22] which will be addressed in the next section. The peak appearing at x = 650 till x = 659 is inserted manually and simulates a DDoS attack which will be addressed in the next section as well.

The third and fourth graphs in the same figure show the stem plot and the continuous plot, respectively, of a sample of the total inbound time-series real data set consisting of 21 aggregate within a 5 min traffic values taken arbitrarily at times from 460 to 480 for testing the performance of the MRSP algorithm with normal traffic.

In Fig. 10.5 the top graph shows the stem plot of the 21 sample inbound traffic values. The middle part shows the corresponding [AR(3)-KF] prediction values. The bottom part shows the stem plot of the corresponding MRSP end-values. A direct comparison of the first two plots indicate that the [AR(3)-KF] prediction values are close to the real inbound data. The same holds for the MRSP end-values compared to the [AR(3)-KF] prediction values and the 21 sample inbound 5 min traffic values, indicating that the MRSP algorithm provides quite good results.

In Fig. 10.6 the top graph shows the stem plots for both the 21 sample inbound 5 min traffic values in blue squares and the corresponding [AR(3)-KF] prediction values in red circles. The middle part shows the stem plots for both the [AR(3)-KF] prediction values in red circles and for the corresponding MRSP end-values in blue squares. The bottom part shows the stem plots for both the MRSP end-values in blue squares and for the corresponding 21 sample inbound 5 min traffic values in

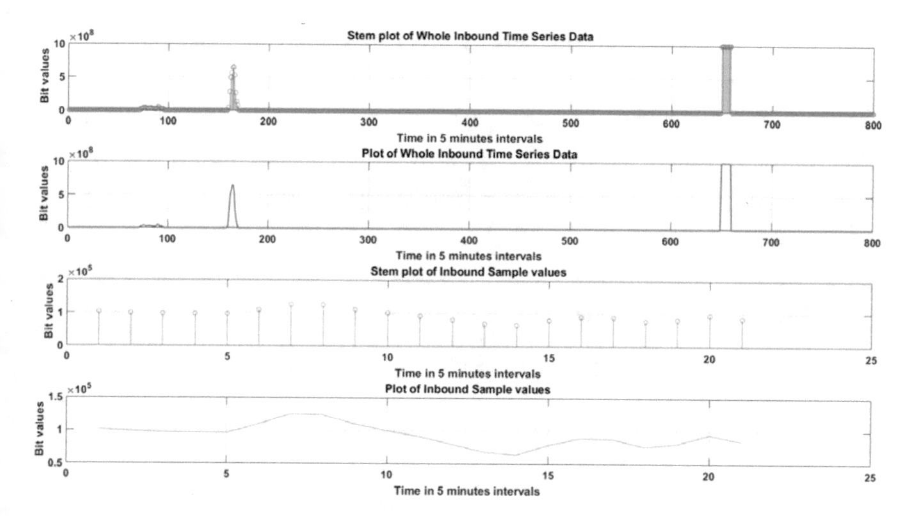

Fig. 10.4 The top first and second graphs show the stem plot and the continuous plot, respectively, of the total inbound time-series real data set consisting of 800 aggregate within a 5 min interval normal traffic values along with DDoS attack values from 650 to 659 and peak attack values from 160 to 169 intervals. The third and fourth graphs show the stem plot and the continuous plot, respectively, of a sample from the total inbound time-series real data set consisting of 21 aggregate within a 5 min normal traffic values at times from 460 to 480

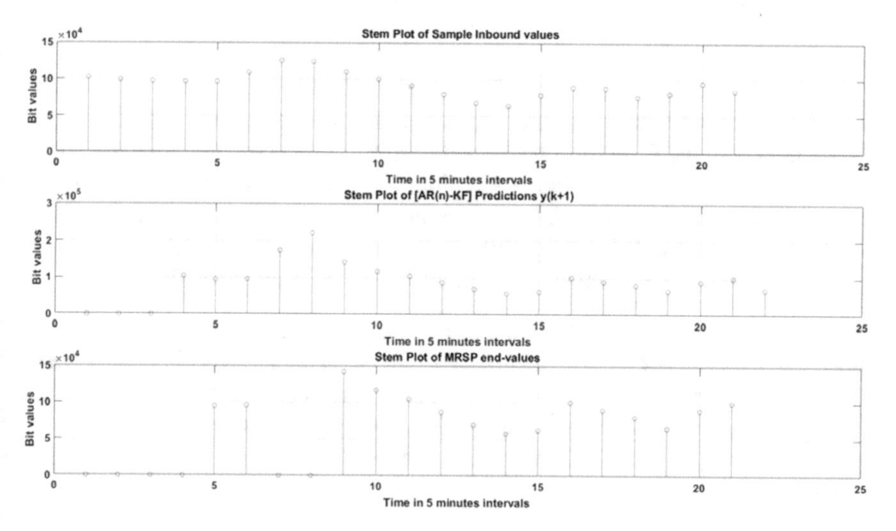

Fig. 10.5 The top part shows the stem plot of the 21 sample inbound 5 min traffic values. The middle part shows the stem plot of the corresponding [AR(3)-KF] prediction values. The bottom part shows the stem plot of the corresponding MRSP end-values

red circles. Again from these comparisons it is deduced that the MRSP algorithm provides quite good results, as it reaches steady-state after a few samples of 5 min traffic values at the beginning.

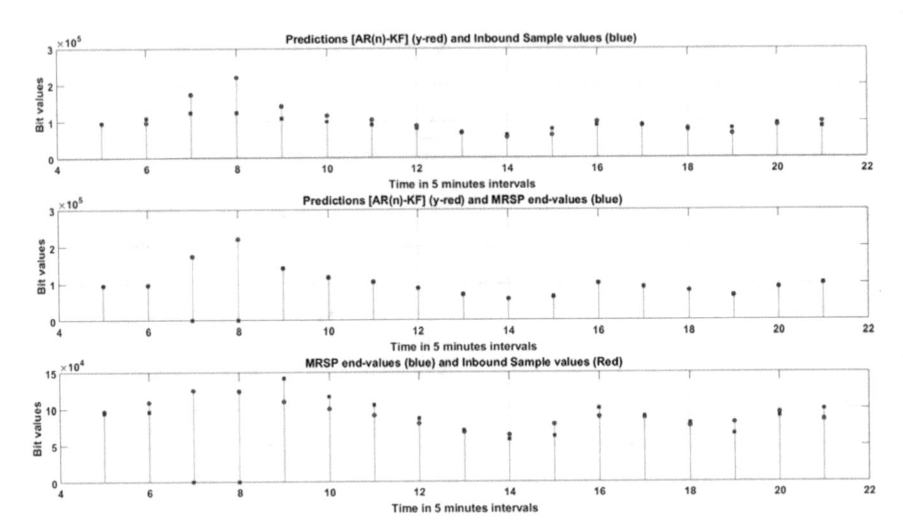

Fig. 10.6 The top part shows the stem plots for both the 21 sample inbound 5 min traffic values in blue squares and the corresponding [AR(3)-KF] prediction values in red circles. The middle part shows the stem plots for both the [AR(3)-KF] prediction values in red circles and for the corresponding MRSP end-values in blue squares. The bottom part shows the stem plots for both the MRSP end-values in blue squares and for the corresponding 21 sample inbound 5 min traffic values in red circles

In Fig. 10.7 the top graph shows the plot of the RMS error between the [AR(3)-KF] predictions and the 21 sample inbound 5 min traffic values. The middle part shows the plot of the RMS error between the [AR(3)-KF] predictions and the MRSP end-values. The bottom part shows the plot of the RMS error between the MRSP end-values and the 21 sample inbound 5 min traffic values. Again from this figure it is deduced that the MRSP algorithm provides quite good results, as it reaches steady-state after a few samples of 5 min traffic values at the beginning.

Remark 3 The RMS (Root Mean Square) error measures absolutely based on scale dependent measures of model performance, that is, RMS can only provide a relative comparison between different models. These deviance (or goodness-of-fit statistic for a statistical model) measures are zero if and only if the values are identical. For two data series $\{m_1(t),\ t = 1, 2, \ldots, N\}$ and $\{m_2(t),\ t = 1, 2, \ldots, N\}$, the RMS error measure is calculated as:

$$RMS = \frac{1}{N}\sqrt{\sum_{t=1}^{N}(m_{1t} - m_{2t})^2}.$$

The top part of Fig. 10.8 shows the plot of the MAPE error between the [AR(3)-KF] predictions and the 21 sample inbound 5 min traffic values. The middle part shows the plot of the MAPE error between the predictions of the [AR(n)-KF] model and the MRSP end-values. The bottom part shows the plot of the MAPE error between the MRSP end-values and the 21 sample inbound 5 min traffic values. Again from this figure it is deduced that the MRSP algorithm provides quite good results, as it reaches steady-state after a few samples of 5 min traffic values at the beginning.

Remark 4 The MAPE (Mean Absolute Percentage Error) measure between two data series is the average value of the absolute values of differences expressed in percentage terms. The data is considered to be in a relative scale if they are strictly positive and the importance of the difference is given by the ratio and not by the arithmetic. For two data series $\{m_1(t),\ t = 1, 2, \ldots, N\}$ and $\{m_2(t),\ t = 1, 2, \ldots, N\}$, the MAPE measure is calculated as: $MAPE = \frac{100}{N}\sum_{t=1}^{N}\left(\frac{|m_{1t}-m_{2t}|}{|m_{1t}|}\right)$. The MAPE measure cannot be determined if measured values are equal to zero and it tends to infinity if measurements are small or near zero. This is a typical behavior, when relative differences are considered.

In Fig. 10.9 the top part shows the stem plot of the directional evolution of the total deterministic MRSP values (consisting of 21 samples times 300 increments each). The middle part shows the stem plot of the volatility times the Brownian motion (or random up-down values) of the total stochastic values of the MRSP. The bottom part shows the stem plot of the sum all deterministic and all stochastic values of the MRSP. In this figure the end result is that the no matter what the deterministic or the stochastic values are the MRSP values always are bounded by the router specifications from above by the capacity value of 1 Gbps and from below by the handshake (Keepalive) value of 1 bps.

In Fig. 10.10 the top part shows the plot of the variance of all 21 sample inbound 5 min traffic values taken as a group of 3 at a time. The middle part shows the plot of the volatility [square root of (trace(P last) + det(P last) times the MRSP. The bottom part shows the plot of all the Brownian motion values for the MRSP. In this figure the large variance values in the top plot is noticed which in effect will adjust the Kalman filter gain to be small, thus giving much weight in the measurements rather in the state space model as desired.

In Fig. 10.11 the top part shows the plot of the determinants of all Kalman filter P(n $\times$ n) covariance matrices with n = 3. The second part shows the plot of the traces of all Kalman filter P(n $\times$ n) covariance matrices. The third part shows the plot of the sum of both the determinants and traces of all Kalman filter P(n $\times$ n) covariance matrices. The last bottom part shows the plot of the norms of all Kalman filter system noise Q(n $\times$ n) covariance matrices. From this figure, we can deduce that the Kalman filter performance is quite good (small P(n $\times$ n) and Q(n $\times$ n) covariances), as desired in steady-state after a few samples of 5 min traffic values at the beginning.

Figure 10.12 shows the evolution of the intra-values produced by the MRSP algorithm within a 5 min interval from Keepalive value to the [AR(3)-KF] prediction last-5 end-value indicated with the green square and the corresponding sample inbound last-5 value (or 16th 5 min traffic value out of 21) indicated with the magenta colour square. From this figure, we can deduce that the MRSP algorithm provides quite good results, since the end values at the increment 300 are relatively not that apart from each other.

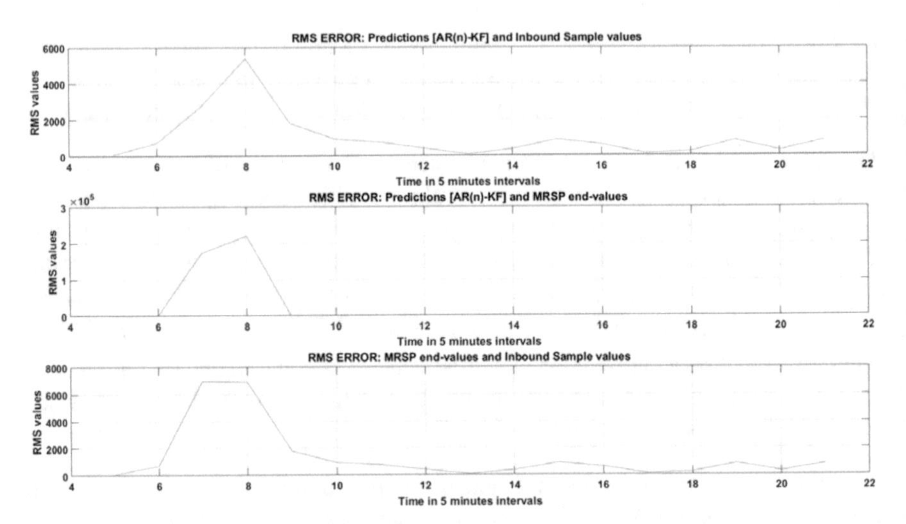

Fig. 10.7 The top part shows the plot of the RMS error between the [AR(3)-KF] predictions and the 21 sample inbound 5 min traffic values. The middle part shows the plot of the RMS error between the [AR(3)-KF] predictions and the MRSP end-values. The bottom part shows the plot of the RMS error between the MRSP end-values and the 21 sample inbound 5 min traffic values

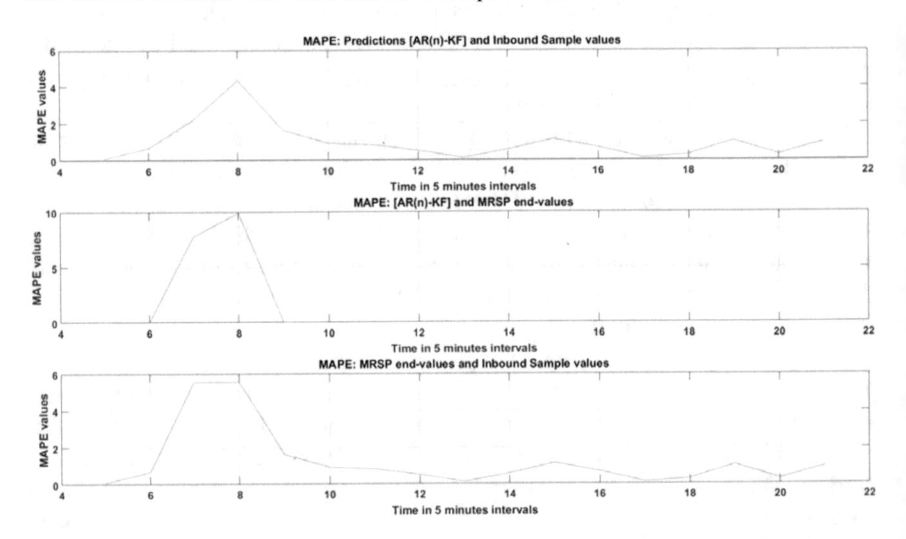

Fig. 10.8 The top part shows the plot of the MAPE error between the [AR(3)-KF] predictions and the 21 sample inbound 5 min traffic values. The middle part shows the plot of the MAPE error between the predictions of the [AR(n)-KF] model and the MRSP end-values. The bottom part shows the plot of the MAPE error between the MRSP end-values and the 21 sample inbound 5 min traffic values

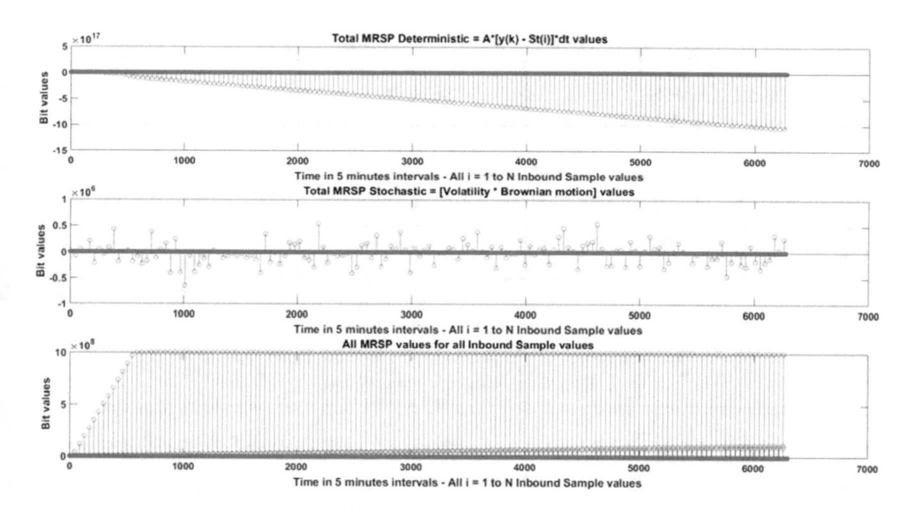

Fig. 10.9 The top part shows the stem plot of the directional evolution of the total deterministic MRSP values (consisting of 21 samples × 300 increments each). The middle part shows the stem plot of the volatility times the Brownian motion (or random up-down) of the total stochastic values of the MRSP. The bottom part shows the stem plot of all the sum deterministic and stochastic values of the MRSP

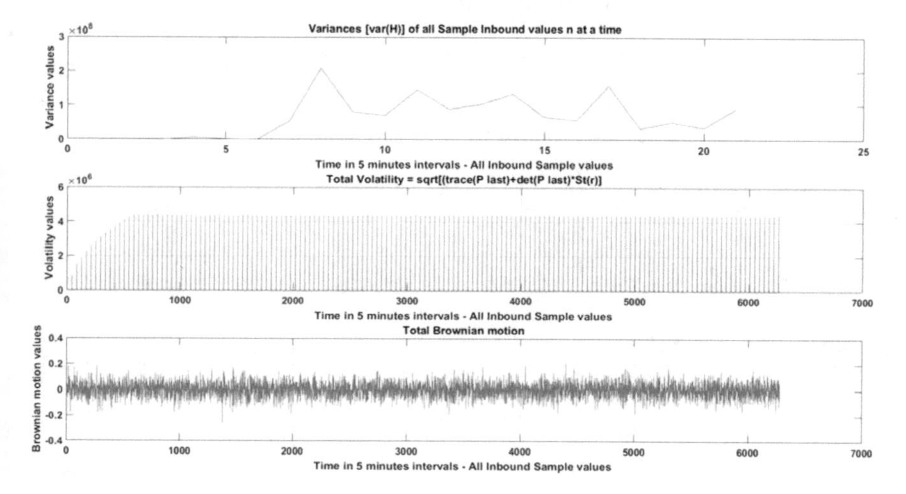

Fig. 10.10 The top part shows the plot of the variance of all sample inbound 5 min traffic values taken as a group of 3 at a time. The middle part shows the plot of the volatility [square root of (trace(P last) + det(P last) times the MRSP. The bottom part shows the plot of all the Brownian motion values for the MRSP

Note here that the plotted intra-values produced by the MRSP in Fig. 10.12 and subsequent corresponding figures is varying due to the stochastic part, although it looks like a smooth curve due to the scale of the y-axis.

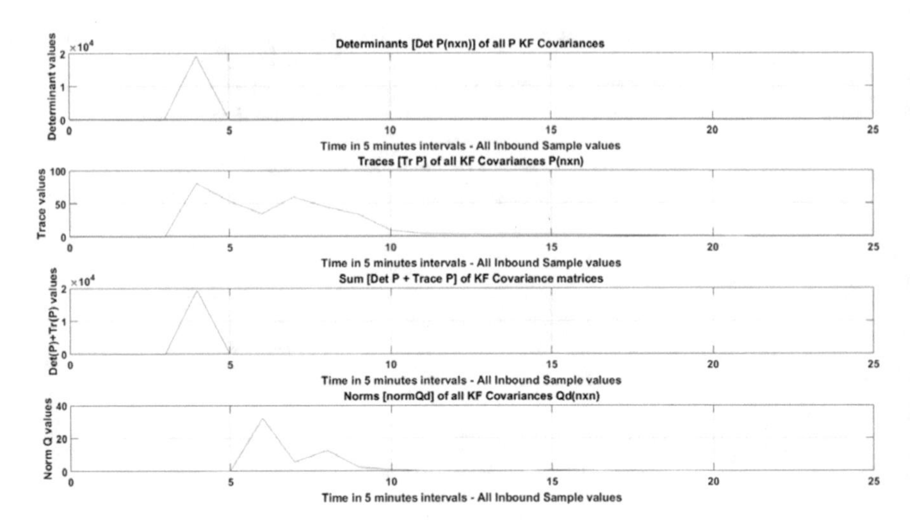

Fig. 10.11 The top part shows the plot of the determinants of all Kalman filter P(n × n) covariance matrices. The second part shows the plot of the traces of all Kalman filter P(n × n) covariance matrices. The third part shows the plot of the sum of both the determinants and traces of all Kalman filter P(n × n) covariance matrices. The last bottom part shows the plot of the norms of all Kalman filter system noise Q(n × n) covariance matrices

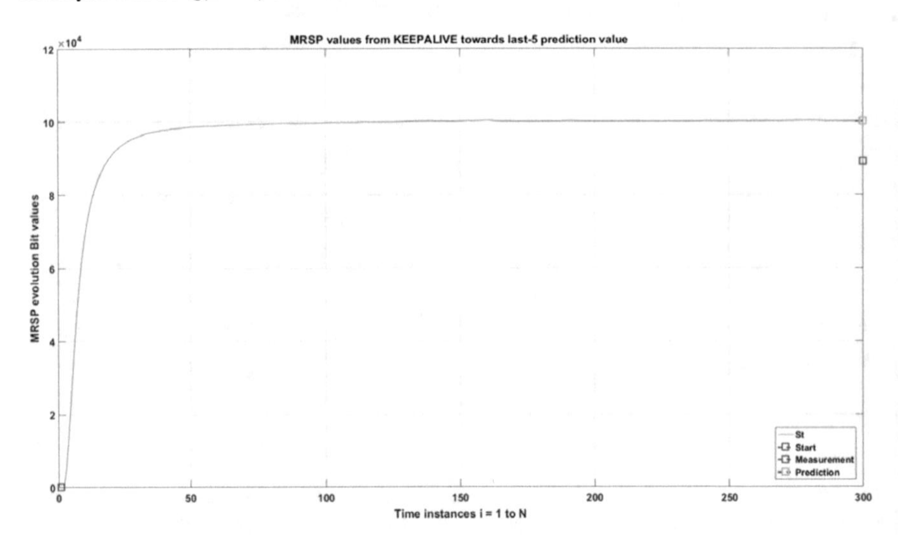

Fig. 10.12 Intra-values produced by the MRSP algorithm within a 5 min interval from Keepalive value to the MRSP prediction last-5 end-value indicated with the green square and the corresponding sample inbound last-5 value indicated with the magenta colour square

Figure 10.13 shows the evolution of the intra-values produced by the MRSP algorithm within a 5 min interval from Keepalive value to the [AR(3)-KF] prediction last-4 end-value indicated with the green square and the corresponding sample

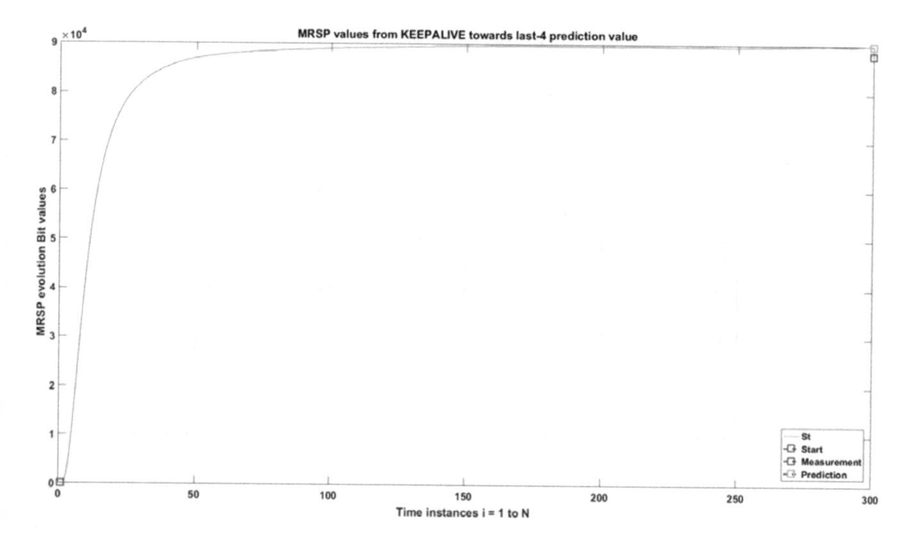

Fig. 10.13 Intra-values produced by the MRSP algorithm within a 5 min interval from Keepalive value to the MRSP prediction last-4 end-value indicated with the green square and the corresponding sample inbound last-4 value indicated with the magenta colour square

inbound last-5 value (or 17th 5 min traffic value out of 21) indicated with the magenta colour square. From this figure, we can deduce that the MRSP algorithm provides quite good results, since the end values at the increment 300 are relatively not that apart from each other.

Figure 10.14 shows the evolution of the intra-values produced by the MRSP algorithm within a 5 min interval from Keepalive value to the [AR(3)-KF] prediction last-3 end-value indicated with the green square and the corresponding sample inbound last-5 value (or 18th 5 min traffic value out of 21) indicated with the magenta colour square. From this figure, we can deduce that the MRSP algorithm provides quite good results, since the end values at the increment 300 are relatively not that apart from each other.

Figure 10.15 shows the evolution of the intra-values produced by the MRSP algorithm within a 5 min interval from Keepalive value to the [AR(3)-KF] prediction last-2 end-value indicated with the green square and the corresponding sample inbound last-5 value (or 19th 5 min traffic value out of 21) indicated with the magenta colour square. From this figure, we can deduce that the MRSP algorithm provides quite good results, since the end values at the increment 300 are relatively not that apart from each other.

Figure 10.16 shows the evolution of the intra-values produced by the MRSP algorithm within a 5 min interval from Keepalive value to the [AR(3)-KF] prediction last-1 end-value indicated with the green square and the corresponding sample inbound last-5 value (or 20th 5 min traffic value out of 21) indicated with the magenta colour square. From this figure, we can deduce that the MRSP algorithm provides

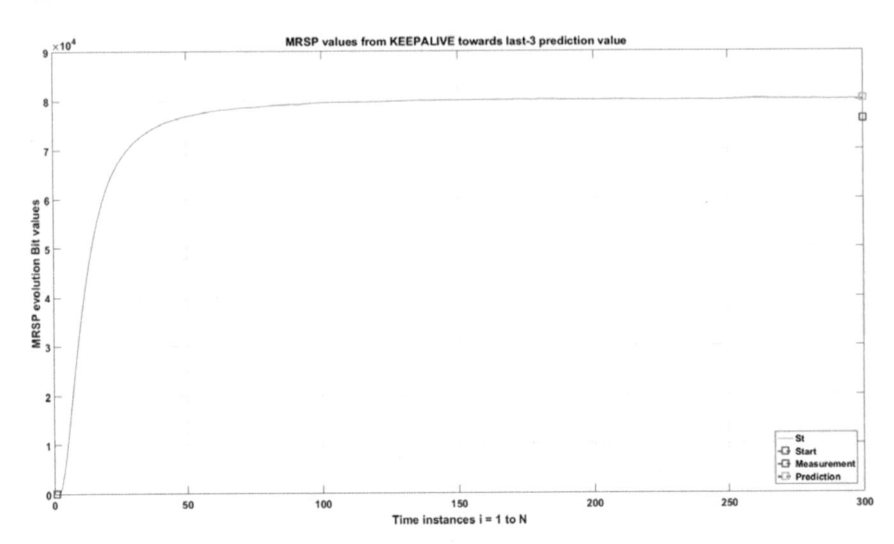

Fig. 10.14 Intra-values produced by the MRSP algorithm within a 5 min interval from Keepalive value to the MRSP prediction last-3 end-value indicated with the green square and the corresponding sample inbound last-3 value indicated with the magenta colour square

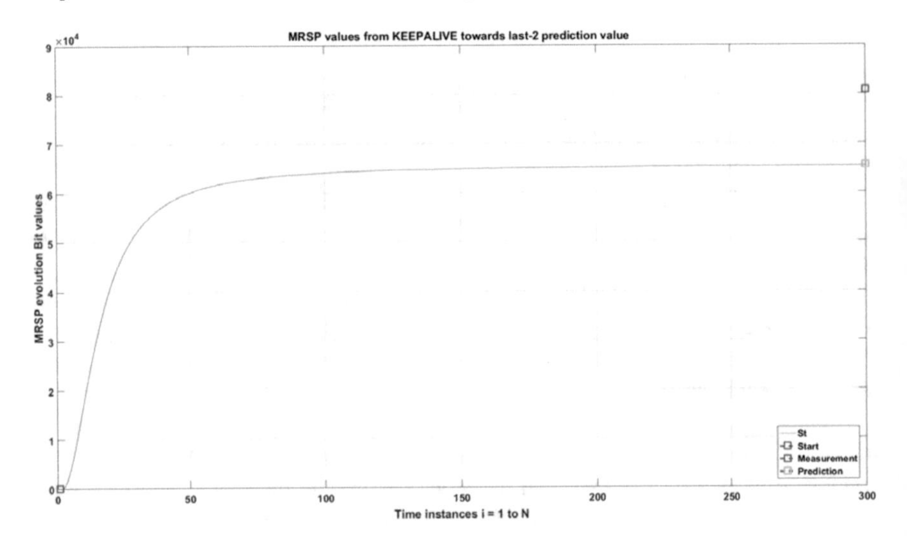

Fig. 10.15 Intra-values produced by the MRSP algorithm within a 5 min interval from Keepalive value to the MRSP prediction last-2 end-value indicated with the green square and the corresponding sample inbound last-2 value indicated with the magenta colour square

quite good results, since the end values at the increment 300 are relatively not that apart from each other.

Figure 10.17 shows the evolution of the intra-values produced by the MRSP algorithm within a 5 min interval from Keepalive value to the [AR(3)-KF] prediction

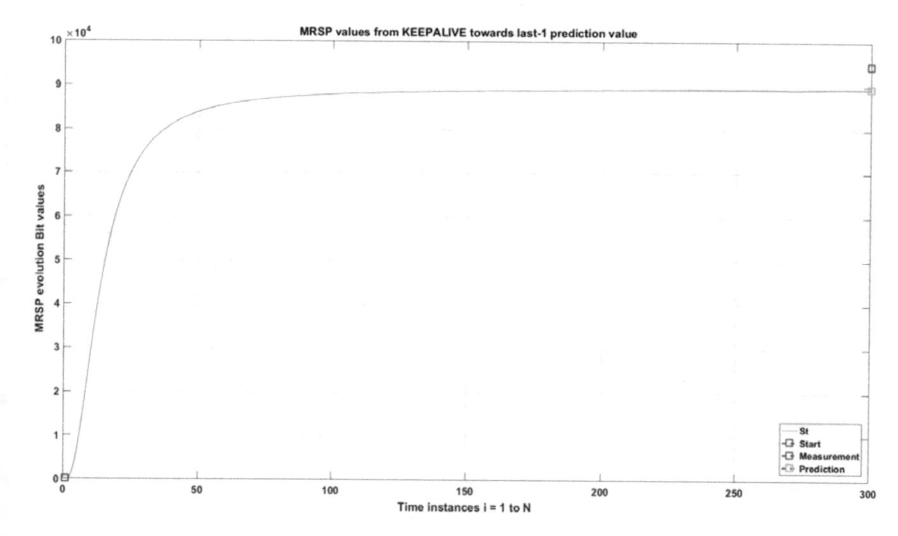

Fig. 10.16 Intra-values produced by the MRSP algorithm within a 5 min interval from Keepalive value to the MRSP prediction last-1 end-value indicated with the green square and the corresponding sample inbound last-1 value indicated with the magenta colour square

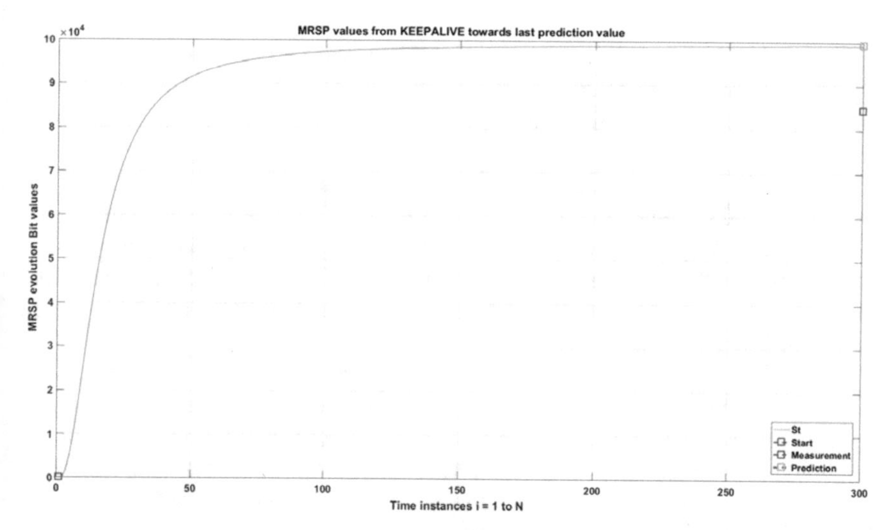

Fig. 10.17 Intra-values produced by the MRSP algorithm within a 5 min interval from Keepalive value to the MRSP prediction last end-value indicated with the green square and the corresponding sample inbound last value indicated with the magenta colour square

last end-value indicated with the green square and the corresponding sample inbound last value (or 21st 5 min traffic value out of 21) indicated with the magenta square. From this figure, we can deduce that the MRSP algorithm provides quite good results, since the end values at the increment 300 are relatively not that apart from each other.

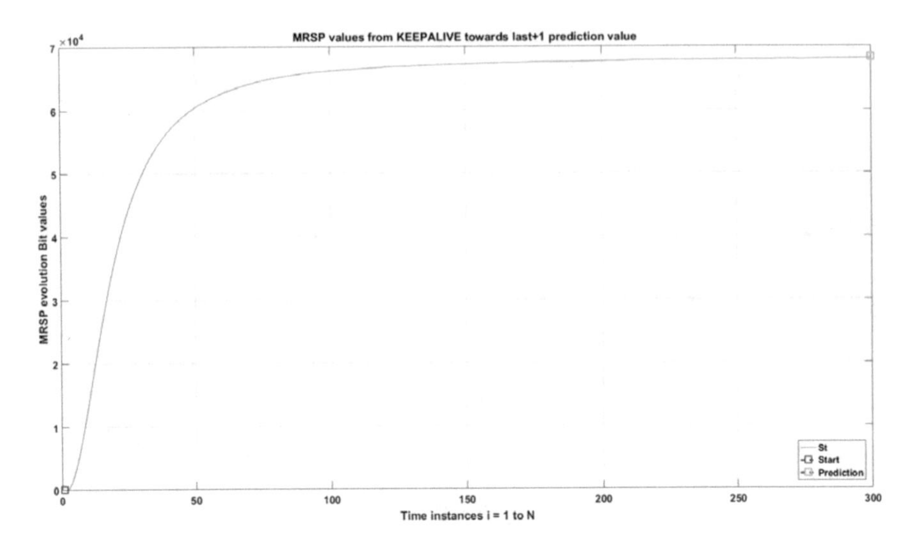

Fig. 10.18 Intra-values produced by the MRSP algorithm within a 5 min interval from Keepalive value to the MRSP prediction last + 1 end-value indicated with the green square

Figure 10.18 shows the evolution of the intra-values produced by the MRSP algorithm within a 5 min interval from Keepalive value to the [AR(3)-KF] prediction last + 1 (or 22nd 5 min traffic value out of 21) end-value indicated with the green square.

Figure 10.19 shows the adaptive MRSP coefficient values (or strength) A(i) for i = 1, 2, …, 300 increments within the last, last-1, last-2, and last-3 5 min interval out of 21. From this figure, we can deduce the adaptive behaviour of the values of A(i) as at the beginning of the interval starting from a small value rapidly increases to a high value and then slowly decays reaching a constant value at the end, as designed for the MRSP algorithm.

Figure 10.20 shows the adaptive MRSP coefficient values (or strength) A(i) for i = 1, 2, …, 300 increments within the last + 1 (or 22nd 5 min interval). From this figure, again we can deduce the adaptive behaviour of the values of A(i) as at the beginning of the interval starting from a small value rapidly increases to a high value and then slowly decays reaching a constant value at the end, as designed for the MRSP algorithm for normal 5 min traffic.

Figure 10.21 shows the adaptive MRSP volatility values for i = 1, 2, …, 300 increments within the last, last-1, last-2, and last-3 5 min interval out of 21. From this figure, we can deduce the adaptive behaviour of the volatility values as at the beginning of the interval starting from a small value increase to a high level reaching a constant value at the end, as designed for the MRSP algorithm for normal 5 min traffic.

Figure 10.22 shows the adaptive MRSP volatility values for i = 1, 2, …, 300 increments within the last + 1 (or 22nd) 5 min interval out of 21. From this figure, again we can deduce the adaptive behaviour of the volatility values as at the beginning

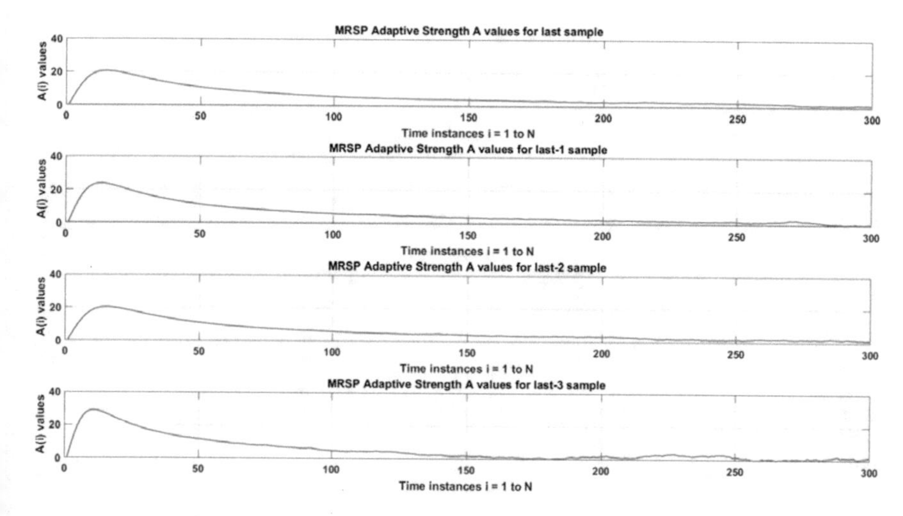

Fig. 10.19 Adaptive MRSP coefficient values (or strength) *A(i)* for i =1, 2, …, 300 increments within the last, last-1, last-2, and last-3 5 min interval out of 21

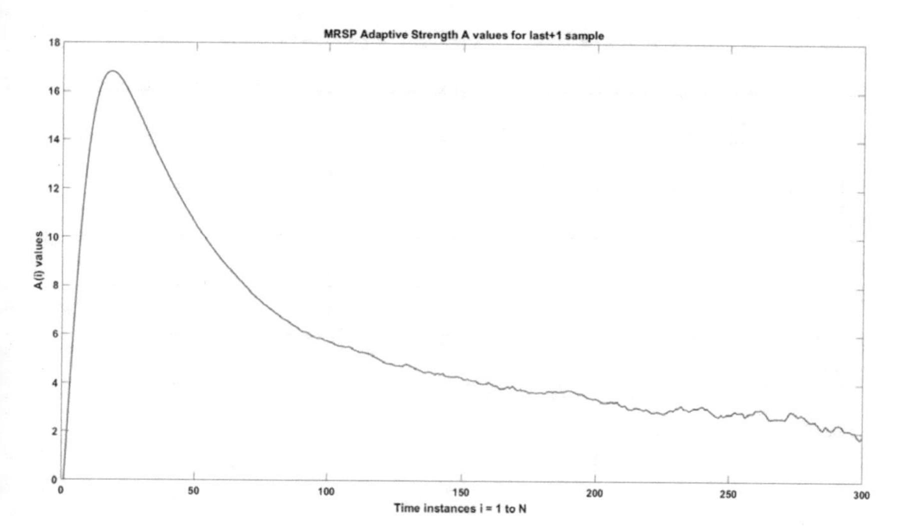

Fig. 10.20 Adaptive MRSP coefficient values (or strength) *A(i)* for i=1, 2, …, 300 increments within the last + 1 (or 22nd) 5 min interval

of the interval starting from a small value increase to a high level reaching a constant value at the end, as designed for the MRSP algorithm for normal 5 min traffic.

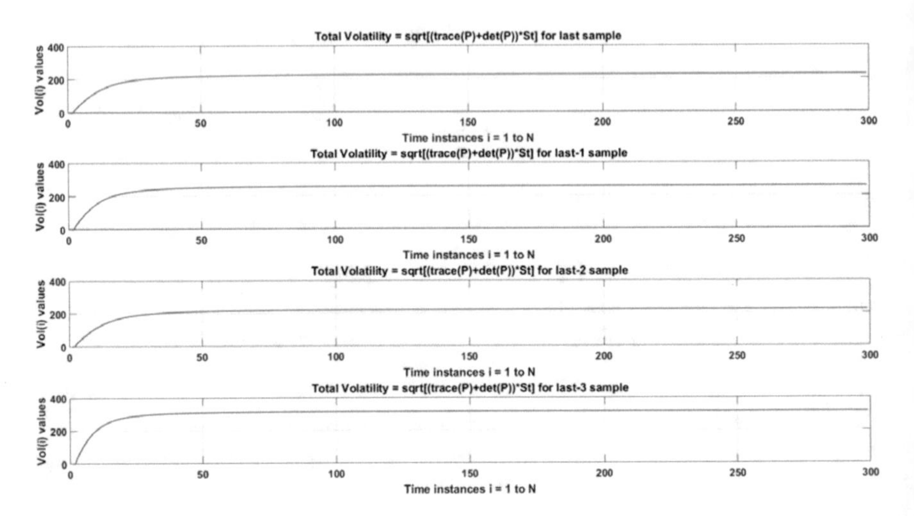

Fig. 10.21 Adaptive MRSP volatility values for $i = 1, 2, \ldots, 300$ increments within the last, last-1, last-2, and last-3 5 min interval out of 21

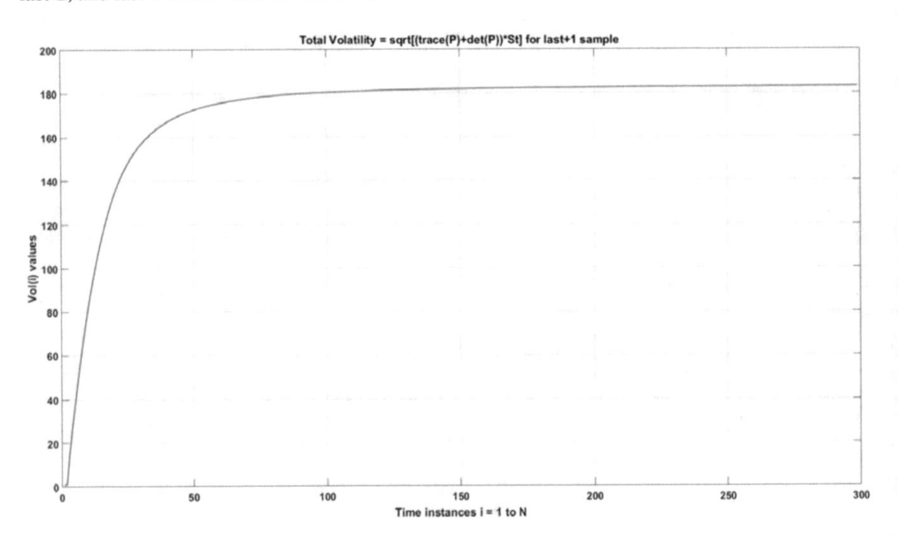

Fig. 10.22 Adaptive MRSP volatility values for $i = 1, 2, \ldots, 300$ increments within the last + 1 (or 22nd) 5 min interval

10.6 Detecting Anomalies

10.6.1 DETECTING a DDoS ATTACK

After the presentation of the algorithm let us now examine how bandwidth-flooding DDoS attacks may be detected. First, let us examine the traffic generated by a typical

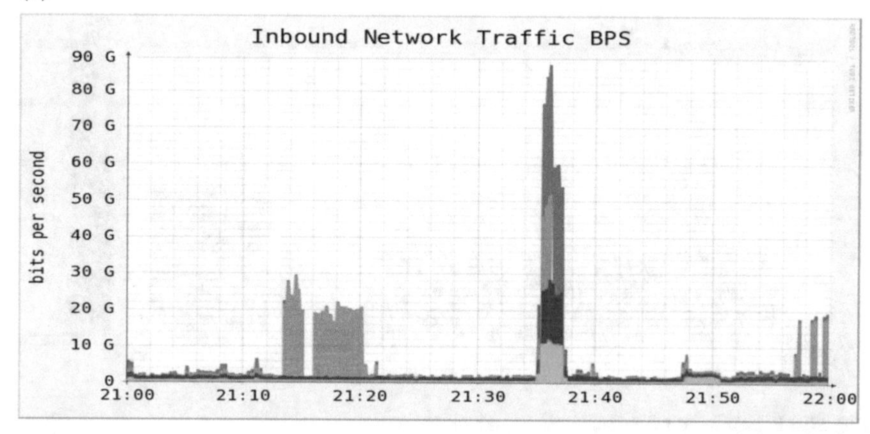

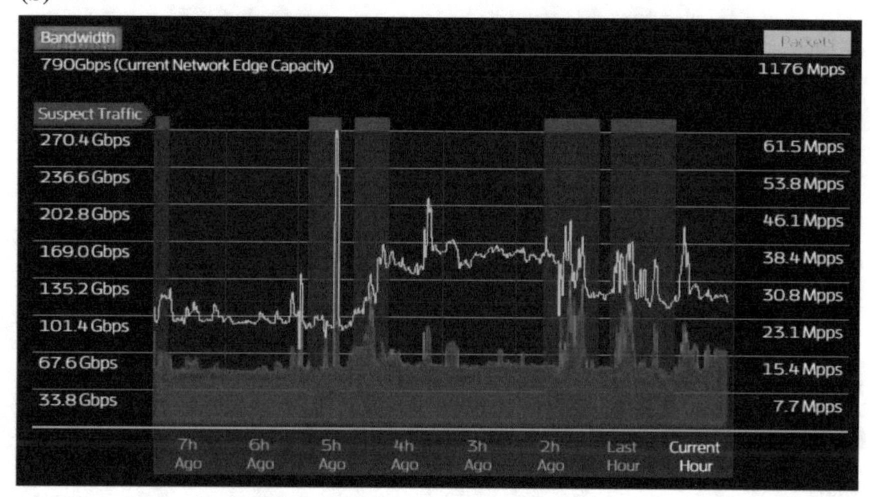

Fig. 10.23 Typical bandwidth-flooding DDoS attack (**a**, **b**)

DDoS attack (Fig. 10.23a, b): the typical characteristic is a peak traffic with limited duration.

For the purposes of this work we have also conducted several DoS attacks to a server of ours (Fig. 10.24) which verified the above results.

The proposed approach produces intra-values within the 5 min intervals and identifies successfully normal, as well as abnormal activities in the incoming traffic. In case of normal traffic, the algorithm behaves appropriately; in case of abnormal sudden peaks in traffic (due to an attack or a misuse), specific parameters of the algorithm take excessive values thus detecting anomaly in the incoming traffic stream. Excessive values appearing in the incoming traffic (exceeding by far the prediction),

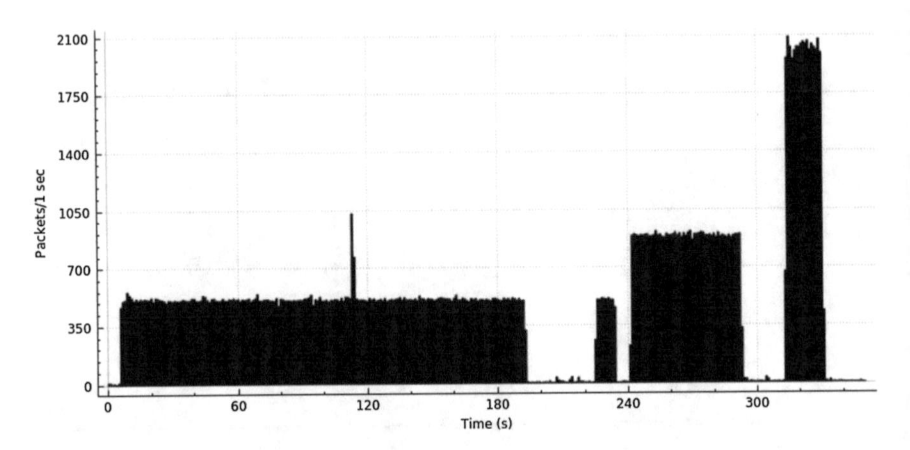

Fig. 10.24 Bandwidth-flooding DoS attacks on our server

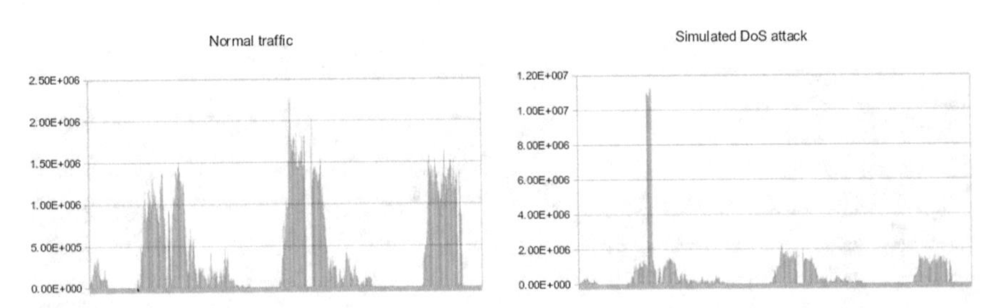

Fig. 10.25 Normal traffic (left) and the same traffic with simulated DoS attack

will indicate a bandwidth-flooding DoS attack, calling the network administrator to take the proper measurements (such as blocking the suspicious IP addresses). In Fig. 10.25b (right), a peak simulating a bandwidth-flooding DoS attack has been added to normal traffic (left).

It is obvious that DoS traffic is much higher in volume in order to set the server out or order. Researchers use this characteristic to detect DoS attacks. In the proposed method:

1. We take the profile of the traffic for workdays as well as weekends;
2. We use an adaptive algorithm predicting the traffic for the next 5 min intervals for each of the above two profiles.
3. We also compare the current incoming traffic with the typical workday or weekend profile (Fig. 10.25).

So when an excessive value appears in the incoming traffic, (exceeding by far prediction), this will indicate a DOS attack, calling the network administrator to take the proper measurements (such as blocking the suspicious IP addresses).

The proposed MRSP algorithm presents abnormal behavior in case of a D/DoS attack. To show this we run the algorithm for the 5 min traffic data from the interval

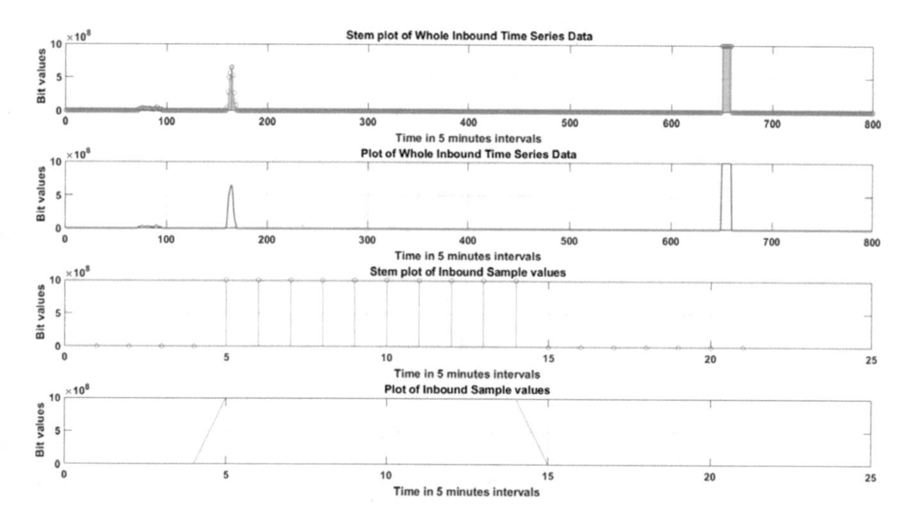

Fig. 10.26 The top first and second graphs show the stem plot and the continuous plot, respectively, of the total inbound time-series real data set consisting of 800 aggregate every 5 min traffic values. The third and fourth graphs show the stem plot and the continuous plot, respectively, of a sample from the total inbound time-series real data set consisting of 21 aggregate every 5 min normal traffic values at times from 646 to 666 which contains the flooding DoS attack values from 650 to 659

646 to 666. In this interval the values from 650 to 659 are simulated values to represent DoS attack values as they are at the router capacity of 1 Gbps (exceeding by far the 50 Mbps threshold value of normal traffic). This time-series real data sample with some normal traffic values before it and some normal traffic values after the DoS attack is presented in Fig. 10.26.

Running the MRSP algorithm on the above sample data set we see the following parameters which are directly affected:

1. [AR(3)-KF] predictions are excessively large at the 5 min interval following the bandwidthflooding DoS attack (see Figs. 10.27 and 10.28).
2. RMS as wells as MAPE errors between predictions and sample inbound data are excessively large, e.g. RMS 10^{15}–10^{16} at the 5 min interval following the DoS attack (see Fig. 10.29) and MAPE 10^{14} at the 5 min interval following the DoS attack (see Fig. 10.30).
3. The variance var(H) has excessively large values, fact which affects the Kalman filter performance (residuals, Adaptive Q and P and state predictions), as depicted in Fig. 10.31.
4. The adaptive MRSP coefficient A(i) takes normal values for normal traffic (top 3 plots in Fig. 10.32) whereas it takes excessively large values at the flooding DoS attack data (last plot in Fig. 10.32).

When the MRSP algorithm processes normal data again after the DoS attack traffic data then the adaptive coefficient A(i) returns back to normal values again as seen in Fig. 10.33.

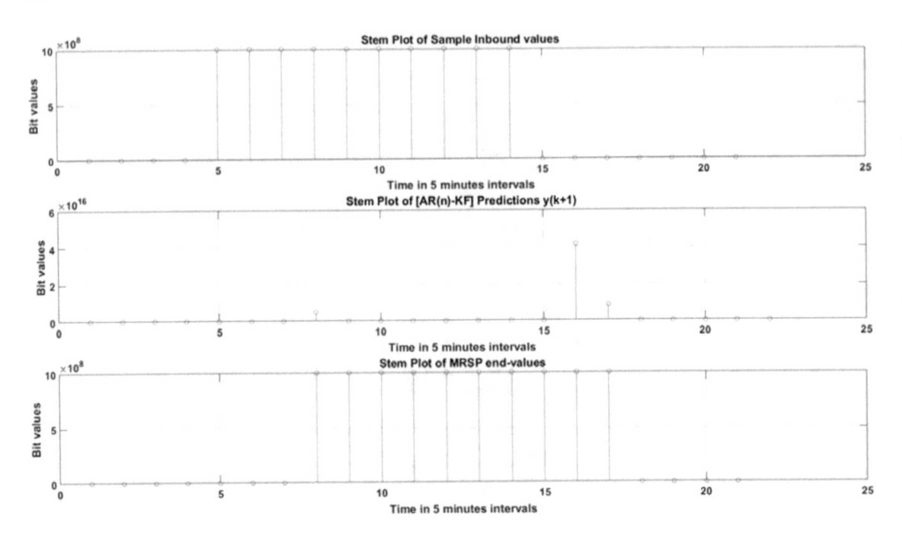

Fig. 10.27 [AR(3)-KF] predictions are excessively large at the 5 min interval following the DoS attack

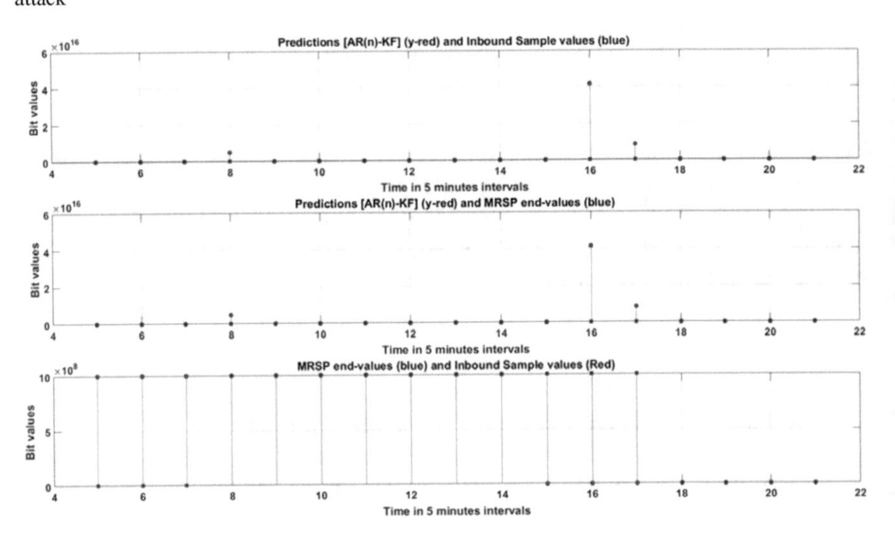

Fig. 10.28 [AR(3)-KF] predictions are excessively large at the 5 min interval following the DoS attack

This is also true for Volatility since it takes normal values for normal traffic (top 3 plots in Fig. 10.35) while it takes excessively large and oscillating values at the DoS attack data (see last plot in Fig. 10.34).

When the MRSP algorithm processes normal data after the DoS attack traffic data then the Volatility returns back to normal values again as seen in Fig. 10.35.

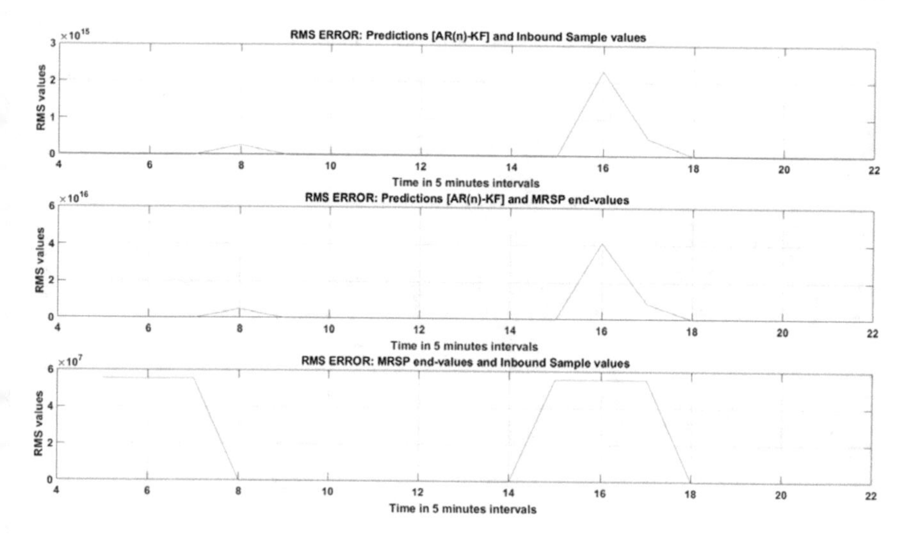

Fig. 10.29 RMS errors for normal traffic data, DoS attack traffic data, and normal again traffic data

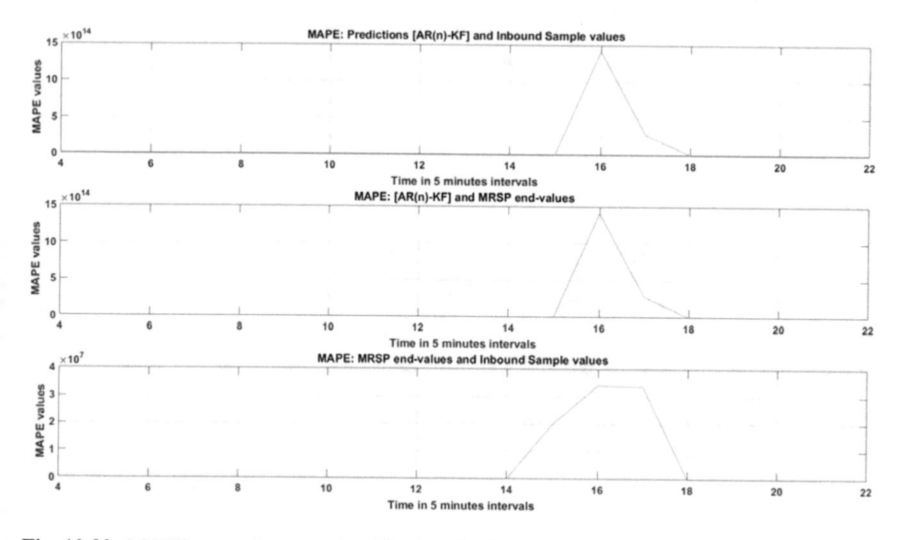

Fig. 10.30 MAPE errors for normal traffic data, DoS attack traffic data, and normal again traffic data

Hence, a sudden change in the incoming data caused by a DoS attack gets immediately detected by the MRSP algorithm because leads the algorithm to abnormal behaviour (instability).

Simulation results for different DoS attacks with different statistical characteristics prove that the MRSP algorithm immediately detects the caused change in behaviour and issues proper warning messages.

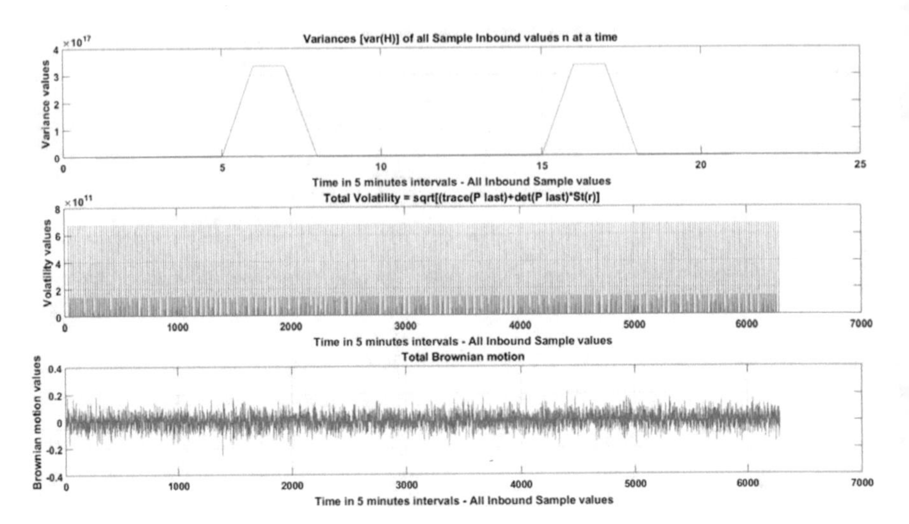

Fig. 10.31 Variance of sample measurements for normal traffic data, flooding DoS attack traffic data, and normal again traffic data

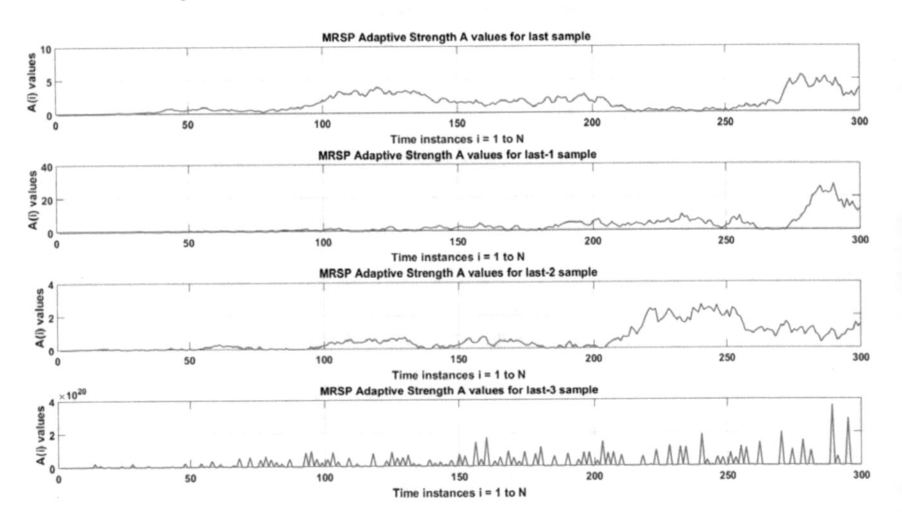

Fig. 10.32 MRSP adaptive coefficient $A(i)$ values for normal traffic data, DoS attack traffic data, and normal again traffic data

10.6.2 Detecting an Anomaly

In Fig. 10.36, the top first and second graphs show the stem plot and the continuous plot, respectively, of the total inbound time-series real data set consisting of 800 aggregate within a 5 min interval traffic values. In the same figure, the values starting slowly at 161 and increasing to the value of 651 Mbps at 165 and then decaying

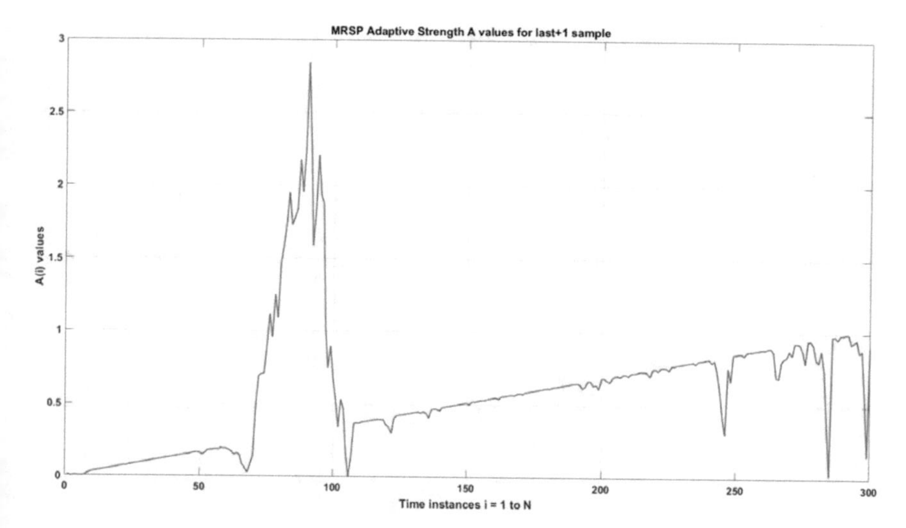

Fig. 10.33 MRSP adaptive coefficient $A(i)$ values for normal traffic data after the previous DoS attack traffic data

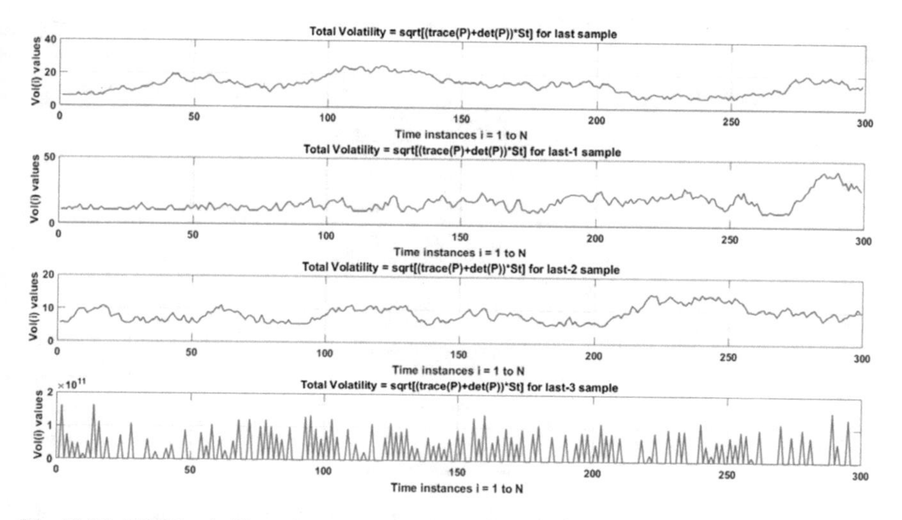

Fig. 10.34 MRSP volatility values for normal traffic data, DoS attack traffic data, and normal again traffic data

slowly back to normal values at 169 are inserted simulated aggregate within a 5 min interval traffic values representing peak attack anomaly values as they exceed the 50 Mbps threshold value. The third and fourth graphs in the same figure show the stem plot and the continuous plot, respectively, of a sample of the total inbound time-series data set consisting of 21 aggregate within a 5 min interval traffic values taken at times from 155 to 175. This range contains the peak attack anomaly values for testing the performance of the MRSP algorithm for some normal traffic values before the peak

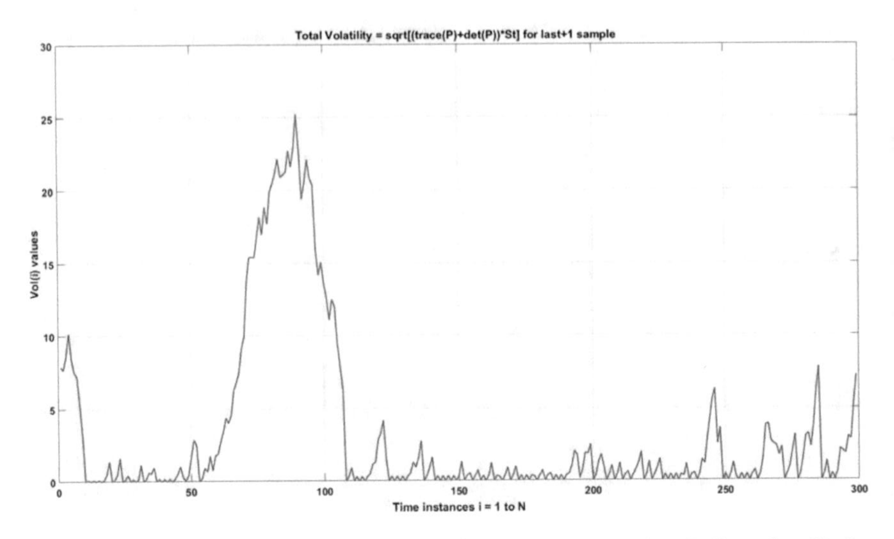

Fig. 10.35 MRSP volatility values for normal traffic data after the previous DoS attack traffic data

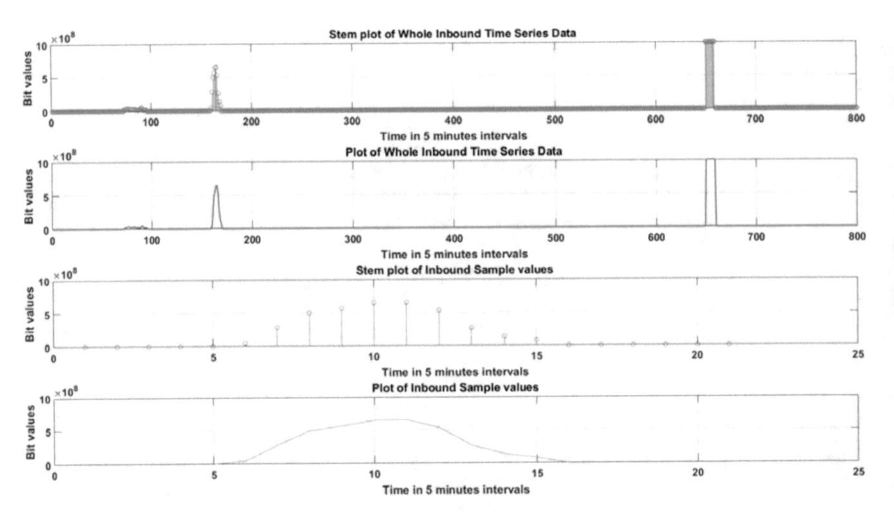

Fig. 10.36 The top first and second graphs show the stem plot and the continuous plot, respectively, of the total inbound time-series real data set consisting of 800 aggregate every 5 min traffic values. The third and fourth graphs show the stem plot and the continuous plot, respectively, of a sample from the total inbound time-series real data set consisting of 21 aggregate every 5 min normal traffic values at times from 155 to 175 which contains the peak attack values from 161 to 169

attack values, during the peak values, and for some normal traffic values after the peak attack values.

The next Fig. 10.37 shows the stem plots for both the 21 sample inbound 5 min traffic values in blue squares and the corresponding [AR(3)-KF] prediction values in red circles. The middle part shows the stem plots for both the [AR(3)-KF] prediction

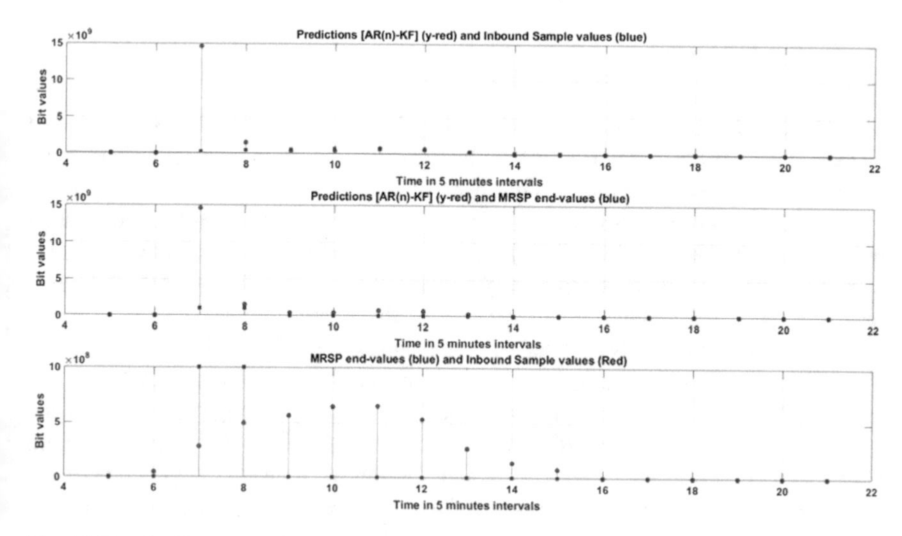

Fig. 10.37 [AR(3)-KF] predictions are excessively large at the 5 min interval following the peak attack values

values in red circles and for the corresponding MRSP end-values in blue squares. The bottom part shows the stem plots for both the MRSP end-values in blue squares and for the corresponding 21 sample inbound 5 min traffic values in red circles. The [AR(3)-KF] predictions are excessively large at the 5 min interval following the peak attack values indicating that the MRSP algorithm immediately can detect the caused change in behaviour and may issue proper warning messages.

10.6.3 Final Remarks

The proposed algorithm produces a warning message when excessive traffic demand is detected. For this purpose a threshold is used, which can be adjusted by the administrator.

The proposed algorithm works the same with outbound data, i.e., it can detect excessive outbound traffic. This is extremely important in cases where firewalls filter only the inbound traffic. However, the proposed method will not detect D/DoS attacks which do not produce excessive traffic exceeding by far the normal traffic.

Finally we note that the proposed algorithm (after slight extension) can be used for predicting the need for extra bandwidth allocation, a function especially useful for ISP providers [10].

10.7 Conclusions

This chapter dealt with network traffic analysis on a time-series input data in rrd format. An algorithm which successfully identifies normal, as well as abnormal activities contained in the datasets has been developed and verified on real traffic.

The algorithm produces intra-values of a time-series network traffic using a mean reverting stochastic process (MRSP) and consists of a deterministic and a stochastic part. The deterministic part consists of an adaptive gain and the [AR(n)-KF] mean reverting term. The stochastic part consists of **three components**: an adaptive diffusion (or volatility) coefficient, based on the square root of the determinant and the trace of the Kalman filter error covariance (which, to the best of our knowledge, has not been used before); the adaptive diffusion is multiplied by the square root of the stochastic process itself according to the Cox-Ingersoll-Ross (CIR) model (which is the second component), and a Brownian motion to capture the dynamics of the evolution from a starting point to reach the AR(n) predictions (which is the third component).

To the best of our knowledge, the specific model has not appeared in the literature so far, is original and is a main contribution of our research work. Experimental results from real traffic data prove that the proposed algorithm works correctly in the sense that:

(a) it can produce estimated intra-values within the 5 min intervals;
(b) it can also predict the value at the end of the next 5 min interval successfully, i.e. close to the real data.

These features are very useful to Network Administrators because they help them detect various anomalies such as excessive capacity demand and bandwidth-flooding D/DoS attacks. It is also helpful to ISPs because it allows them to know if and when the allocated traffic volume will be depleted. Using the prediction we could estimate—within the 5 min interval and under current traffic conditions—when the available capacity of the router will be depleted, after which point the router won't be able to respond. The estimation can be used—as already mentioned—either for asking for extra bandwidth (within the router capabilities) in order to serve the users, or, to early detect a bandwidth-flooding D/DoS attack, which will be detected by the increased rate for bandwidth demand.

References

1. Anderson, B.D.O., Moore, J.B.: Optimal filtering. In: Kailath, T. (ed.) Information and System Sciences Series. Prentice-Hall, Inc., Englewood Cliffs, N.J. (1979)
2. Anjali, T., Scoglio, C., Chen, L.C., Akyildiz, I.F., Uhl, G.: ABEst: an available bandwidth estimator within an autonomous system. In: IEEE Global Telecommunications Conference, Nov 2002
3. Arbor Networks: Worldwide infrastructure security reports series (2005–2012) (2012). http://www.arbornetworks.com/report

4. Bougioukou, A.P., Leros, A.P., Papakonstantinou, V.: Modelling of non-stationary ground motion using the mean reverting stochastic process. Appl. Math. Model. **32**, 1912–1932 (2008)
5. Brockwell, P.J., Davis, R.A.: Introduction time series and forecasting. Springer, New York (2002)
6. Commandeur, J.J.F., Koopman, S.J.: Practical Econometrics: An Introduction to State Space Time Series Analysis. Oxford University Press, New York (2007)
7. Cox, J.C., Ingersoll, Jonathan E., Ross, Stephen A.: A theory of the term structure of interest rates. Econometrica **53**(2), 385–408 (1985)
8. Geva, M., Herzberg, A., Gev, Y.: Bandwidth distributed Denial of service: attacks and defenses. IEEE Secur. Priv. **12**, 54–61 (2013)
9. Giannopoulos, I.K., Leros, A.P., Leros, A.K., Tsaramirsis, G.: A stochastic model with an adaptive proportional controller for the evolution of user-router bandwidth demand for quality of service (QoS) aspects. In: Ad Hoc and Sensor Wireless Networks (2014)
10. Giannopoulos, I.K., Leros, A.P., Leros, A.K.: A model for the evolution of router bandwidth. In: WCE2015, pp. 547–551 (2015)
11. Higham, D.J.: An algorithmic introduction to numerical simulation of stochastic differential equations. SIAM Rev. **43**(3), 525–546 (2001)
12. Kuan Hoong, P., Tan, I.K.T., Yik Keong, C.: Bit torrent network traffic forecasting with ARIMA. IJCNC **4**(4) (2012)
13. Lipschutz, S., Lipson, M.L.: Linear Algebra, 4th edn. In: Schaum's Outline Series. The McGraw-Hill Companies, Inc. (2009)
14. Ludwing, A.: Stochastic Differential Equations: Theory and Applications. Wiley (1973)
15. Mahanta, D., Ahmed, M., Bora, U.J.: A study of bandwidth management in computer networks. Int. J. Innov. Technol. Explor. Eng. 2(2) (2013)
16. Maybeck, P.: Stochastic Models, Estimation and Control, vol. I. Academic Press (1979)
17. Mitrokotsa, A., Douligeris C.: DDoS attacks and defense mechanisms: a classification. In: 3rd IEEE International Symposium on Signal Processing and Information Technology (ISSPIT 2003)
18. Mohamed, A.H., Schwarz, K.P.: Adaptive Kalman filtering for INS/GPS. J. Geodesy **73**(4), 193–203 (1999)
19. Moussas, V.C., Daglis, M., Kolega, E.: Network traffic modeling and prediction using multiplicative seasonal ARIMA models. In: Proceedings of the 1st International Conference on Experiments/Process/System Modeling/Simulation/Optimization, Athens, 6–9 July 2005
20. Moussas, V.C., Pappas, S.S.: Adaptive network anomaly detection using bandwidth utilization data. In: Proceedings of the 1st International Conference on Experiments/Process/System Modeling/Simulation/Optimization, Athens, 6–9 July 2005
21. Moussas, V.C.: Network traffic flow prediction using multi-model partitioning algorithms. In: Tsahalis, D.T. (ed) Proceedings of the 2nd SCCE International Conference "From Scientific Computing to Computational Engineering", Athens, 5–8 July 2006
22. Moussas, V.C.: Adaptive traffic modelling for network anomaly detection (chapter 1). In: Daras, N.J. (ed). Springer (2016)
23. Oetiker, T.: Multi Router Traffic Grapher (MRTG) tool, Software Package and Manuals (2018). http://oss.oetiker.ch/mrtg/
24. Oetiker, T.: MRTG: Multi Router Traffic Grapher (2018). http://people.ee.ethz.ch/oetiker/webtools/mrtg/
25. Oetiker, T.: Round Robin Database Tool (RRD tool), Software Package and Manuals (2018). http://oss.oetiker.ch/rrdtool/
26. P. T. Inc.: Prolexic Attack Report, Q3 2011–Q4 2012 (2011/2012). http://www.prolexic.com/attackreports
27. Shu, Y., Yu, M., Liu, J., Yang, O.W.W.: Wireless traffic modeling and prediction using seasonal ARIMA models. In: IEEE International Conference on Communication, ICC'03, vol. 3, May 2003
28. Thottan, M., Ji, C.: Detection in IP networks. IEEE Trans. Signal Process. **51**(8), 2191–2204 (2003)
29. White Paper: Understanding fiber ethernet bandwidth vs. end user experience. http://fiberinternetcenter.com/WhitePapers-Podcasts/WhitePaperEthervsEndUser.pdf

Chapter 11
Intelligent Data Analysis in Electric Power Engineering Applications

V. P. Androvitsaneas, K. Boulas and G. D. Dounias

Abstract This chapter presents various intelligent approaches for modelling, generalization and knowledge extraction from data, which are applied in different electric power engineering domains of the real world. Specifically, the chapter presents: (1) the application of ANNs, inductive ML, genetic programming and wavelet NNs, in the problem of ground resistance estimation, an important problem for the design of grounding systems in constructions, (2) the application of ANNs, genetic programming and nature inspired techniques such as gravitational search algorithm in the problem of estimating the value of critical flashover voltage of insulators, a well-known difficult topic of electric power systems, (3) the application of specific intelligent techniques (ANNs, fuzzy logic, etc.) in load forecasting problems and in optimization tasks in transmission lines. The presentation refers to previously conducted research related to the application domains and briefly analyzes each domain of application, the data corresponding to the problem under consideration, while are also included a brief presentation of each intelligent technique and presentation and discussion of the results obtained. Intelligent approaches are proved to be handy tools for the specific applications as they succeed to generalize the operation and behavior of specific parts of electric power systems, they manage to induce new, useful knowledge (mathematical relations, rules and rule based systems, etc.) and thus they effectively assist the proper design and operation of complex real world electric power systems.

V. P. Androvitsaneas (✉)
High Voltage Laboratory, School of Electrical and Computer Engineering, National Technical University of Athens, 9 Iroon Politechniou Street, Zografou Campus, 15780 Athens, Greece
e-mail: v.andro@mail.ntua.gr

K. Boulas · G. D. Dounias (✉)
Management & Decision Engineering Lab, Department of Financial and Management Engineering, University of the Aegean, 41 Kountouriotou Street, 82100 Chios, Greece
e-mail: g.dounias@aegean.gr

K. Boulas
e-mail: kboulas@aegean.gr

G. A. Tsihrintzis et al. (eds.), *Machine Learning Paradigms*, Intelligent Systems Reference Library 149, https://doi.org/10.1007/978-3-319-94030-4_11

11.1 Introduction

Nowadays, Artificial Intelligence (AI) and Machine Learning (ML) are the state-of-the-art methodologies for approaching and solving problems, that conventional mathematics cannot do so. Though these techniques seem to be suitable and preferable for complex and insoluble modeling issues in sectors of robotics, medicine and finance, the AI and ML models find a perfect application in practical engineering and, more specifically, in electric power engineering. AI and ML models are widely applied in grounding systems engineering for estimating and forecasting the behavior of these systems in variable weather and soil conditions, as well as in insulators of transmission lines, estimating the value of critical flashover voltage. Similar models have also been developed for load forecasting problems and other optimization tasks in problems related to transmission lines. All these fields are extremely important for the safe, firm and reliable operation and service of electric power systems.

Grounding systems constitute a crucial component for the protection system of electrical installations and facilities, more specifically buildings, high voltage (HV) substations and transmission lines of electric power systems, electric railway, wind farms, etc., against lightning and fault currents. Such currents are usually of great magnitude, resulting in high values of step and touch voltages which, in turn, are usually extremely dangerous for people and equipment. Grounding systems' role is to lead fault and lightning currents into the earth, both in the safest way and in the shortest possible time, limiting the developing over-voltages within the safe limits [1]. One of the featured parameters of grounding systems, concerning their capability and effectiveness to dissipate high fault currents into the earth in a safe way, is ground resistance. The value of this magnitude represents the outcome of the combination of several factors, such as the underground soil structure, the soil electric resistivity ρ (in Ω m), the soil humidity H (%), and the grounding electrode geometry. In general, a well-designed grounding system is characterized by low resistance values of few Ohms. Ground resistance must be estimated as accurately as possible during the design phase, using suitable software for this job, and must be measured after the construction of the grounding system at regular time [2].

However, in most cases, ground resistance measurements become difficult or even impossible, because of dense building infrastructure and/or ground morphology, resulting in the lack of the required space for this measurement. Moreover, in many cases of special electrical installations like HV substations, wind farms, and electric railway, estimation for the behavior of the designed grounding systems over time is very useful and important for electrical engineers. In this way, they are able to take extra technical measures in order to maintain the ground resistance value in low levels, in case of a major seasonal variance. Artificial Neural Networks (ANN), Wavelet neural networks (WN), Inductive Machine Learning (IML), and Genetic Programming (GP) have proved to be precious tools of computational intelligence, giving reliable and accurate estimations for ground resistance value as a function of several unpredictable parameters, e.g. rainfall and soil resistivity; a relation among these variables that conventional mathematical techniques cannot describe and model.

Artificial intelligence also finds considerable applications in the field of insulators of overhead transmission and distribution lines for the estimation of the critical flashover voltage on them. Surface pollution and nonuniform distribution of electric field along the insulator surface cause the appearance of quasi-stable arcs, which can evolve to long discharges along the surface and lead to the final breakdown of air insulation (bridging) when the applied voltage exceeds a critical value. It is obvious that this situation may lead to unexpected malfunction or service interruption of a line, with subsequent disruptions to the electric power system. ANNs are used successfully for the estimation of the critical flashover voltage on polluted insulators considering the pollution level, the salt deposit density, and the dimensions of the insulators. Therefore, via the critical flashover voltage estimation, the insulators' condition can be monitored and evaluated so that engineers plan the proper maintenance in due time.

Another significant application of artificial intelligence in power engineering is, certainly, the short- and mid-term load forecasting in electric power systems. The highest possible accuracy in forecasting the hourly, daily, or monthly, load profiles is of great importance for the power system scheduling in terms of power generation, unit commitment availability of transfer capacity, stability margins, and power control. Conventional methods for estimating the demand in electric power systems are based on many assumptions and approximations, while they have to include parameter relations that they cannot clearly describe. Artificial neural networks have achieved to provide concrete solutions to the above problems, rendering them as the predominant load forecasting technique [3].

This chapter includes an extensive review and practical examples for applied AI and ML models in the fields of grounding and insulators, initiating the reader into the basic principles of computational intelligence and machine learning by presenting practical applications in modern engineering problems.

The rest of the chapter is organized as follows:

In Sect. 11.2 applications of intelligent techniques in grounding systems are described. Section 11.3 contains details regarding the application of AI and ML models and techniques for solving problems related to insulators. In Sect. 11.4 are given applications of AI approaches in load forecasting, and then, Sect. 11.5 refers to Transmission Lines problems handled with the aid of intelligent techniques. Finally, Sect. 11.5 contains conclusions and further research, either on intelligent approaches or on similar electrical engineering applications.

11.2 Intelligent Techniques in Ground Resistance Estimation

Though that artificial intelligence has been applied in many fields of science and technology with great success, the application in grounding systems for estimating and forecasting several parameters and, specifically ground resistance, is in early stage. This fact is mainly attributed to the great difficulty of collection a significant and exploitable number of in situ measurement data from real grounding systems, since the continuous monitoring of ground resistance value is usually laborious and costly.

11.2.1 Grounding Systems

The safe operation of electricity transmission and distribution networks relies on properly designed and effective grounding systems. The role of grounding is particularly essential and extends to any electrical installation, e.g. buildings, HV substations, electric railway etc. It is a crucial structural component of the facilities and serves to quickly and safely dissipate high fault currents into the earth in the shortest possible time. The effectiveness of grounding systems major determines the response time of electric protection devices such as switchgears, fuses, or simple circuit breakers, in order to clear fault currents efficiently and safely for people and devices. Both fault currents (over-currents or short circuit currents) and lightning currents dissipated into the earth generate high ground potential rise, resulting in high values of step and touch voltages in the neighboring area. Hence, an effective grounding system of low resistance value, dissipating these fault currents into the earth in a controlled and safe way, limits these hazardous voltages significantly under maximum permissible limits, recommended by international standardization [1, 2].

Regardless of the fault current type, grounding systems should be appropriately designed to transport the currents into the earth safely [4]. The efficiency of the grounding system is a function of low ground resistance value. The latter should remain at low levels over time and seasonal variation. In most cases, either the cost of constructing a grounding system is too high, or there is a lack of necessary space required for installation. In each case, the type of soil and especially its behavior must be considered. High soil electric resistivity due to underground structure (sand, rock) and strong corrosive environment are crucial factors, being considered for the construction of a grounding system at a certain location.

The soil composition in contact with the surface of grounding electrodes is subject to seasonal variations due to soil moisture, especially for the upper soil layers of 1–2 m. Weather conditions such as rainfall, air temperature and wind speed effect on soil moisture, therefore, altering the percentage of dissolved salts. Thus, soil composition plays the most critical role in ground resistance value [5–8].

Specifically, ground resistance value depends on soil density and composition. There is a variety of soil types and, therefore, a large range of ground resistance values depending on structures like clay soil, sand, rock, wet or dry soil, non-homogeneous soil, etc. The drier and harsher the soil is, the greater its soil resistivity (ρ) is, measured in Ω m. The soil resistivity is affected by the amount of water retained in the soil, the salt deposit and the soil grain size. In other words, as the soil conductivity is mainly electrolytic, it gets higher values through the water retained in the ground. In anisotropic grounds, the resistivity is different to the grounding electrode circumferential and non-linear [9].

In particular, soil moisture has a significant effect on its resistivity. Indicatively, in clay soil with 10% moisture content (% by weight), the resistivity was 30 times greater than the same soil with a moisture content of 20%. Nevertheless, moisture itself does not play a major role in the resistivity value, as it has to be combined with salts and natural ingredients in order to form a conductive electrolyte. Addition of artificial solvents in water, such as sodium chloride (NaCl), calcium chloride ($CaCl_2$), cupric sulphate ($CuSO_4$), or magnesium sulphate ($MgSO_4$), is a practical way of reducing soil resistivity. Seasonal temperature variance also leads to some resistivity variations, especially at areas of ice conditions [9].

In the effort to construct better and safer electrical facilities, researchers and engineers aim to improve grounding techniques. Much work has been done on the ways of studying and designing grounding systems for reducing ground resistance value and some techniques have been developed over the last decades. One method that has been widely applied around the world and is predominant in the field of grounding, nowadays, is the use of ground enhancing compounds [10–14].

Periodic measurement of ground resistance is often a difficult task because of the residence and dense building infrastructure. Moreover, it is usually necessary for engineers to have an estimation of the behavior of grounding systems built or planned regards to time. For this purpose, experiments are being carried out to investigate the phenomenon and obtain useful data to develop tools for estimating and forecasting the earth resistance value of several grounding systems. These experiments last for many years and correlate the values of soil resistivity at the site of interest with the local weather data. Due to the complexity of relationships and the inability to identify and describe the parameters involved in the formation of the ground resistance value, methods of machine learning are judged to be appropriate to address this problem.

11.2.1.1 Field Measurements of Soil Resistivity and Ground Resistance

The data needed for the training of used techniques have been obtained by field measurements performed in the context of an experiment which is still carried out till nowadays, for the evaluation of ground enhancing compounds performance at the National Technical University of Athens Campus, Hellas [14], under the supervision and technical support of High Voltage Laboratory, School of Electrical and Computer Engineering. During the experimental procedure, six ground rods have been tested, five of them encased in various ground enhancing compounds (e.g. G_2 in conductive concrete, G_3 in bentonite, etc. see Fig. 11.9) and one of them in natural soil as a reference electrode (G1). All the measurements have been performed according to [2], since February 2011 till now. Soil resistivity was measured by using the 4-point method (Wenner method), in different depths of 1, 2, 4, 6 and 8 m. According to the Wenner method, four electrodes 0.5 m in length are driven in line, at equal distances each other, and in a depth b. A test current (I_t) is injected at the two terminal probes, and the voltage (U_t) is measured between the two middle probes. The ratio U_t/I_t

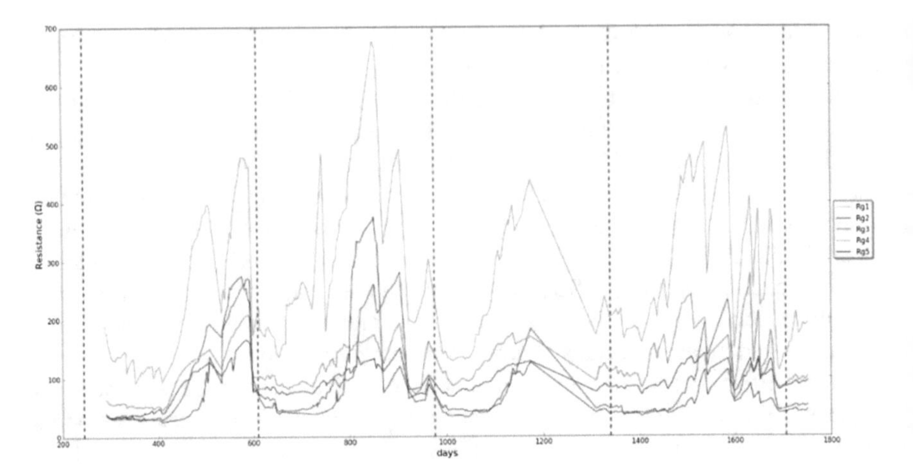

Fig. 11.1 Resistances $Rg_1, \ldots, Rg_5$ versus days of experiment

results the apparent resistance R (in Ω). Consequently, the apparent soil resistivity ρ is given by the Eq. (11.1) [2].

$$\rho = \frac{4\pi\alpha R}{1 + \frac{2a}{\sqrt{a^2+4b^2}} - \frac{2a}{\sqrt{(4a^2+4b^2)}}} = \frac{4\pi\alpha R}{n} \tag{11.1}$$

where variable n depends on the ratio value *b/a* and fluctuates between the values of 1 and 2. In the particular case of this experimental array where $b \ll a$, the Eq. (11.1) is simplified to $\rho = 2\pi\alpha R$. The 3-pole method, also known as the fall of potential method was used to accurately measure the ground resistance of each main rod [14]. In Fig. 11.1 are presented the resistance data for 5 grounding electrodes where season variation is obvious. The 0th day is the 5-2-2010, and the vertical dashed lines represent the first day of the year.

11.2.1.2 Weather Data

The weather data were provided by the Hydrological Observatory of Athens, from the 13th station operated by the National Technical University of Athens. The station is located at the same area where the experiment takes place. The variables are the following (at the end of the name of the variable and inside parenthesis are given the abbreviation and the units of the measurements): wind speed (ws, m/s), wind direction (wd, °),net radiation (nr, W/m^2), diffuse radiation (dr, W/m^2), sunshine duration (sd, min/10 min), barometric pressure (bp, hPa), soil moisture (sm, %), air temperature (at, °C), rainfall (rf, mm), air humidity (ah, %), and evaporation pan (ep, mm).

Weather data consist of a chaotic set of measurements and collected in a two-dimension matrix form, where columns are weather variables or ground parameter measurements, and rows represent time periods specifically days, see [15]. Due to the fact that soil humidity measurements require special and expensive sensors and equipment, rainfall height of the experimental field is used for a collateral estimation of soil humidity.

11.2.2 Application of ANN Methodologies for the Estimation of Ground Resistance

11.2.2.1 Research Review

Though that artificial intelligence has been applied in many fields of science and technology with great success, the application in grounding systems for estimating and forecasting several parameters and, specifically ground resistance, is in early stage. This fact is mainly attributed to the great difficulty of collection a significant and exploitable number of in situ measurement data from real grounding systems, since the continuous monitoring of ground resistance value is usually laborious and costly.

Artificial Neural Networks (ANNs), at this time, are not true copies of the anatomy of cerebral cortex neurons, but simplified models of their function. Since the neural networks first appeared to our ancestors, they have evolved. Their initial operations that ensured the advantage of survival and development have been transformed into complex processes such as pattern recognition, memory and understanding.

ANNs were designed to automate complex pattern recognition tasks. The mathematical structure of these tools allows quantification of patterns and parameter estimation. Comparisons made on a wide variety of problems have shown that a properly designed ANN gives better results or in the worse case, their results are as well as conventional methods.

The first attempt of estimating electric soil resistivity (ρ) and, indirectly ground resistance (R_g) of vertical grounding rods with no use of ANNs, was done by Blattner in 1980 [16]. He formulated an empirical logarithmic formula which substantially approached the soil resistivity curves, drawn using experimental data, with good accuracy. This relationship has linked the known values of soil resistivity to the respective ones of ground resistance for the first few meters of depth. Solving the constants of the empirical formula, Blattner managed to extract a general formula which could forecast the values of soil resistivity in deeper layers of the ground beyond those with available measurements (extrapolation). In this way he was able to estimate the respective values of ground resistance for each tested grounding electrode.

Several years later, Salam et al. published the results of their model, based on artificial neural networks, for modeling and forecasting the relationship between ground resistance and electrode length. They performed measurements of ground resistance on vertical grounding rod, which was driven into the ground gradually with a step of 6 m [17]. They also trained a feed-forward neural network of three layers, where the input layer included the electrode length and the month of the measurement and the output layer provided the ground resistance estimator. The R^2 coefficient for the training set reached the value of 0.995 and for the test set 0.925. In another research work [18] artificial neural networks have been developed for grounding system designing consisted of vertical rods.

In the next few years, research focused on estimating the seasonal variation of ground resistance of vertical grounding rods buried in natural ground [19, 20]. The soil resistivity in the field of interest and the rainfall height were used as input variables for the neural network [19, 20], while the ground resistance of the tested electrode was the output variable. The researchers have developed a feed-forward multilayer perceptron for this application. The network was composed of three layers, its training was performed according to the back-propagation algorithm (BP) and the dataset was divided into three subsets, the training set (53 cases), the evaluation set (14 cases) and the test set (10 cases). The training of the network was performed by the error back-propagation algorithm and nearly all its variations. In this application, the conjugate gradient algorithm combined with Fletcher-Reeves equation and the use of three stopping criteria provided the best results and, finally, was selected for the training of the developed ANN. The convergence results for the training set, the test set, and the confidence intervals were quite satisfactory and encouraging.

The use of new techniques and materials for constructing grounding systems motivated researchers and engineers to continue the development of AI models, adapted to the new techniques, for estimating and forecasting the power frequency resistance of grounding systems. More particularly, the use of ground enhancing compounds—materials that are poured around the grounding electrodes for decreasing the ground resistance value—has brought new processing data for training and evaluating ANNs. Therefore, a typical application of ANNs is the study of ground enhancing compounds behavior over time and the estimation of the ground resistance on grounding electrodes encased in such materials.

11.2.2.2 ANN Architecture Paradigm in Grounding Systems

In a research work of 2012 [21], researchers have studied six vertical grounding rods of 1.5 m in length, encased in various enhancing materials and in natural soil under field conditions. The regular measurements performed at the experimental field for one year (February 2011 to February 2012) during this study, were: (a) soil resistivity (ρ) in the depth of 1, 2, 4, 6, and 8 m, (b) the ground resistance (R_g) of the five tested rods, and (c) rainfall height. The developed network was a feed-forward neural network of three layers the input, the hidden, and the output layer. The input vector for the first layer comprised: (a) the average value of the soil resistivity measurements for each individual depth for the last seven days before the day of ground resistance measurement (ρ_{1w}, ρ_{2w}, ρ_{4w}, ρ_{6w}, ρ_{8w}), (b) the average value of the

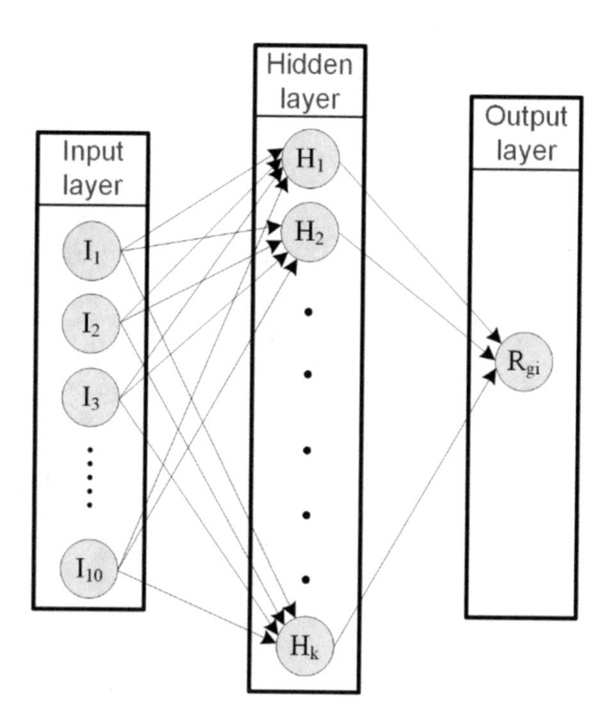

Fig. 11.2 Architecture of the ANN

soil resistivity measurements for the depths of 1 and 2 m for the last thirty days before the day of ground resistance measurement (ρ_{1m}, ρ_{2m}), (c) the cumulative total rainfall height in the day of ground resistance measurement, (d) the cumulative total rainfall height of the last seven days before the day of ground resistance measurement, and (e) the cumulative total rainfall height of the last thirty days before the day of ground resistance measurement. The output of the network was the ground resistance value for each of the six tested grounding rods (R_{gi}). The structure of the developed ANN is illustrated in Fig. 11.2.

The number of neurons in both the input and the output layers are equal to the size of the input and output data vector respectively, while the number of neurons in the hidden layer (or layers) has to be determined. According to Kolmogorov's theorem if the number of neurons in the hidden layer is properly selected, then a single hidden layer is enough [21].

The ground resistance of each vertical rod is estimated by applying the methodology presented in Fig. 11.3. Prior to the training stage, the input and output values were normalized in order for the training process to avoid saturation problems, caused in the case of using nonlinear activation functions. The normalized value $\hat{x}$ (for the variable x) is given by Eq. (11.2):

$$\hat{x} = \alpha + \frac{b - \alpha}{x_{\max} - x_{\min}}(x - x_{\min}) \tag{11.2}$$

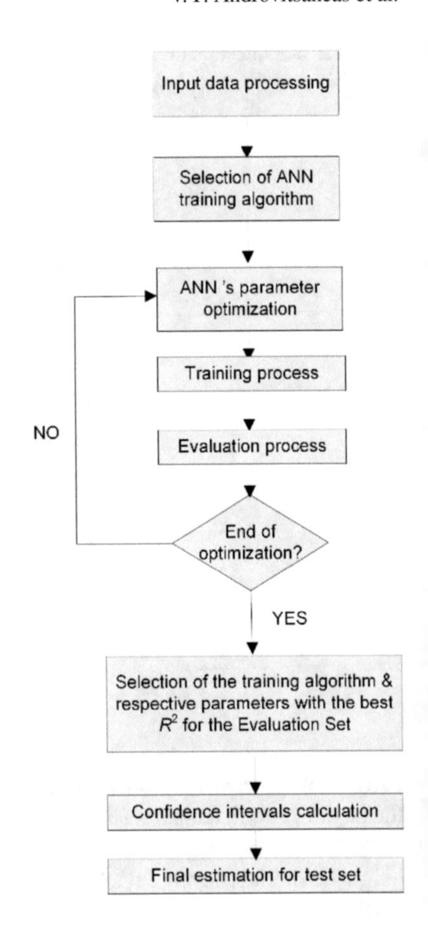

Fig. 11.3 Flowchart of the ANN methodology for the estimation of ground resistance [21]

where $\hat{x}$ is the normalized value for variable x, $x_{\min}$ and $x_{\max}$ are the lower and the upper values of variable x, a and b are the respective values of the normalized variable.

From the field measurements performed in one year, as aforementioned, 126 measurements have been chosen to constitute the dataset for the training of the ANN, i.e. 126 input–output patterns, which was divided randomly into three sets:

- The training set (102 cases) is used until the network has learned the relationship between the inputs and the output.
- The evaluation set (26 cases) is used for the optimal selection of the ANN parameters (i.e. the number of the neurons in the hidden layer, the type and the parameters of the activation functions, the learning rate, and the momentum term).
- The test set (24 cases) verifies the generalization ability of the ANN by using an independent dataset.

The ANN has been trained with the use of Stochastic training with learning rate and momentum term (decreasing exponential functions). The purpose of the training process was to minimize the average error function between the estimated and the actual value, by adjusting the free parameters (weights) of the network. The adjustment of the weights was performed as follows: each input pattern was randomly presented; the adjustment of the weights was performed after the completion of the random presentation of all the input patterns, in order for the average error function between the estimated and the actual value to be minimized. The average error function for all the N patterns is given by Eq. (11.3):

$$G_{av} = \frac{1}{2N}\sum_{n=1}^{N}\sum_{j\in C}(d_j(n) - y_j(n))^2 \tag{11.3}$$

where C is the set of neurons, $d_j(n)$ the target value and $y_j(n)$ the estimated value of the j-neuron.

The weights of the ANN have been adjusted until one of the stopping criteria was fulfilled. The three stopping criteria were: the weights' stabilization criterion, the error function's minimization criterion and the maximum number of epochs' criterion, which are respectively described by the following expressions:

$$\left|w_{kv}^{(l)}(ep) - w_{kv}^{(l)}(ep-1)\right| < \text{limit}_1, \forall k, v, l \tag{11.4}$$

$$|RMSE\,(ep) - RMSE\,(ep-1)| < \text{limit}_2 \tag{11.5}$$

$$ep \le \max\_epochs \tag{11.6}$$

where $w_{kv}^{(l)}$ is the weight between l-layer's k-neuron and (l-1)-layer's v-neuron, $RMSE = \sqrt{\frac{1}{m_2 \cdot q_{out}}\sum_{m=1}^{m_2}\sum_{k=1}^{q_{out}} e_k^2(m)}$ is the root mean square error of the evaluation set with m_2 members and q_{out} neurons of the output layer (in this case $q_{out}=1$), *max_epochs* is the maximum number of the epochs.

The parameters have been selected in a way that the minimum G_{av} for the evaluation set was achieved. At first, the optimal number of neurons N_n is determined. All the other parameters of the network are given fixed values while the number of neurons varies. The maximum number of epochs was set to 7000. The optimal N_n is selected as the one with the smallest average error function (G_{av}) for the evaluation set. The G_{av} for the evaluation set is presented in Fig. 11.4, with the neurons varying from 2 to 25. Finally, the number of neurons was chosen to be $N_n = 21$. The number of neurons is subsequently kept constant (equal to 21) while the algorithm parameters were varied in a proper interval. The time parameter T_α and the initial value of the momentum term α_0, were varied as illustrated in Fig. 11.5, where the values T_α = 1500 and $\alpha_0 = 0.4$ have been selected as the optimal for these parameters. Figure 11.6 shows the variance of G_{av} for the evaluation set as a function of the time parameter T_n and the initial value of the learning rate η_0, where the values $T_n = 2400$ and η_0 = 0.8 have been selected as the optimal ones. Then the type of the activation functions

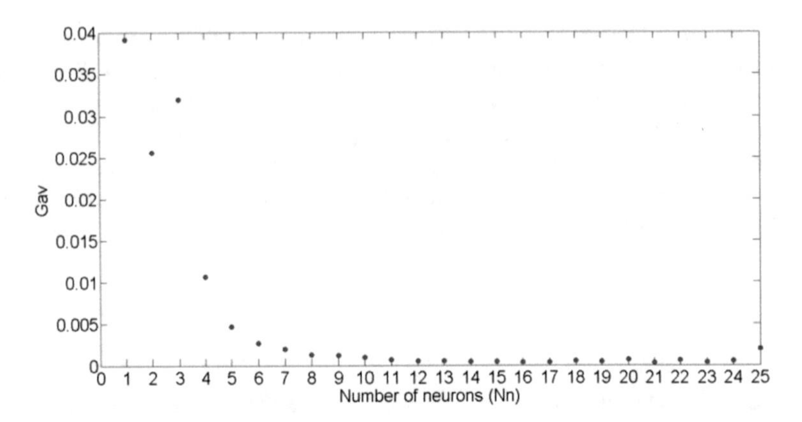

Fig. 11.4 G_{av} for the evaluation set as a function of neurons number [21]

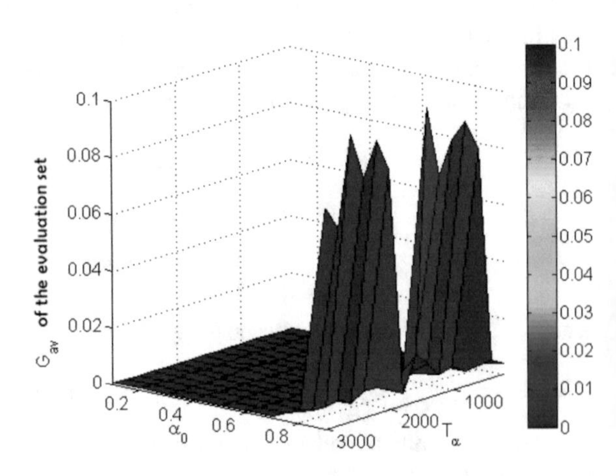

Fig. 11.5 G_{av} for the evaluation set as a function of the momentum term [21]

for the neurons of the hidden and the output layers was determined. The following activation functions have been examined:

Logistic:

$$f(x) = 1/\left(1 + e^{-ax}\right) \tag{11.7}$$

Hyperbolic tangent:

$$f(x) = tanh(ax + b) \tag{11.8}$$

Linear:

$$f(x) = ax + b \tag{11.9}$$

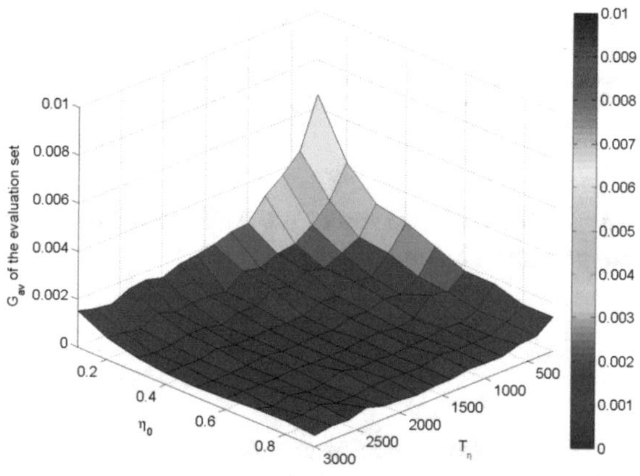

Fig. 11.6 G_{av} for the evaluation set as a function of the learning rate [21]

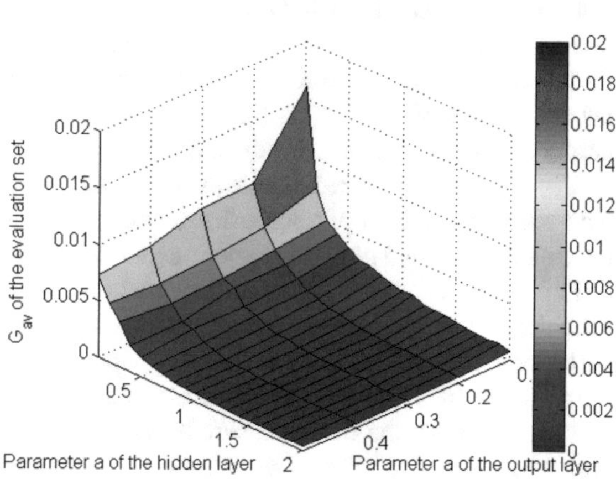

Fig. 11.7 G_{av} for the evaluation set with hyperbolic tangent activation function for the hidden layer and logistic activation function for the output layer [21]

Testing all the possible combinations for the activation functions of the hidden and the output layers and changing the values of the parameters a and b, the most suitable functions for each method were selected. Specifically, $f_1(x) = tanh(2x)$ was selected for the hidden layer and $f_2(x) = 1/(1 + e^{-0.5x})$ was selected for the output layer. Hence, for this particular combination, the graph of the G_{av} as a function of the parameter a is illustrated in Fig. 11.7.

The developed ANN has shown remarkable ability in learning the relationship among the input variables (soil resistivity and rainfall in particular time windows) and the output variable (ground resistance of each grounding rod) estimating the output variable with very good accuracy. Moreover, the model has demonstrated quite good flexibility and adaptability to the special conditions each ground enhancing compound induced in the forming of ground resistance value. The network managed to learn and generalize the relationship among the input and output variables in every case of enhancing material. That means that the network is adaptable to several different soil conditions around the electrodes. The R^2 coefficient is given by

Table 11.1 R^2 coefficient for the test set

	G1	G2	G3	G4	G5	G6
R^2	0.932	0.990	0.924	0.958	0.851	0.979

Eq. (11.10) while the values the ANN model has achieved for the data of the test set are presented in Table 11.1.

$$R^2 = r_{y-\hat{y}}^2 = \frac{\left(\sum_{i=1}^{n}\left((y_i - \bar{y}_{real}) \cdot (\hat{y}_i - \bar{y}_{est})\right)\right)^2}{\sum_{i=1}^{n}(y_i - \bar{y}_{real})^2 \cdot \sum_{i=1}^{n}\left(\hat{y}_i - \bar{y}_{est}\right)^2} \quad (11.10)$$

The field experiment was carried out for the following months, thus, the dataset has been enriched with new field measurements and the researchers had the chance to train the aforementioned ANN with new and more data, using several variations of the BP algorithm at the same time. Aim of the new training was the study of the ANN behavior, trained by different algorithms, and the highlighting of that providing the best convergence outcome for the particular dataset [22].

The ANN architecture remained similar to that of Fig. 11.2 and the training of the network was based on two scenarios for collecting the dataset. In the first scenario, the dataset consisted of purely experimental data (field measurements from February 2011 to October 2012), thus, the data comprised 185 input–output patterns which were divided into three subsets i.e. training set, evaluation set, and test set. The input variables were exactly the same as the previous ANN model. In the second scenario, the underground structure at the experimental field has been modeled by a two-layer soil model, using the field measurements of soil resistivity at the aforementioned various depths. Therefore, the dataset consisted of simulation results for ground resistance, using a proper simulation package for this job, providing 190 input–output patterns. The dataset was similarly divided into three subsets, while the input vector comprised seven input variables: (a) the average soil resistivity of the upper layer of the two-layer soil model for the last seven days before the day of ground resistance measurement ($\rho_{\text{upper, w}}$), (b) the average soil resistivity of the lower layer of the two-layer soil model for the last seven days before the day of ground resistance measurement ($\rho_{\text{lower, w}}$), (c) the average soil resistivity of the upper layer of the two-layer soil model for the last thirty days before the day of ground resistance measurement ($\rho_{\text{upper, m}}$), (d) the average soil resistivity of the lower layer of the two-layer soil model for the last thirty days before the day of ground resistance measurement ($\rho_{\text{lower, m}}$), and (e) the three time windows for rainfall height as before.

The algorithms used for the ANN training are demonstrated in Table 11.2. The network parameters are optimized through an optimization process, similar to the previous ANN model of [21], which was applied repetitively for each training algorithm of Table 11.2 and for each scenario. The steps followed for the training and validation process are concisely illustrated in the flowchart of Fig. 11.8. Graphs of the optimization process for the ANN parameters are indicatively illustrated in Figs. 11.9, 11.10

Table 11.2 ANN training algorithms

No	BP training algorithms
1	Stochastic training, constant learning rate
2	Batch mode, constant learning rate
3	Batch mode with momentum term and use of adaptive rules for the learning rate
4	Batch mode, quasi-Newton algorithm
5	Batch mode, Levenberg-Marquardt algorithm

and 11.11 for the first scenario. The stopping criteria and the activation functions used for this application are given by Eqs. (11.4)–(11.6) and (11.7)–(11.9) respectively.

The estimation results for ground resistance of electrode 6, given by the two training scenarios and by two training algorithms, are illustrated in Fig. 11.12, so that the reader may have a clear view for the performance of the applied ANN model. Grounding rod 6 was selected for the illustration of the ANN estimation results, as it presented the largest and steepest variance in ground resistance values, so it was the most difficult case of input–output function that ANN had to approach from the six tested grounding electrodes. Moreover, for evaluating the performance of each training algorithm, Figs. 11.13 and 11.14 show the convergence results of each algorithm with a polygon shape for both scenarios, based on the results of R^2 coefficient. Studying the graphs of Figs. 11.13 and 11.14, the Levenberg-Marquardt algorithm seems to be the more effective for ANN training concerning this particular application in grounding systems. Its predominance against the others is quite clear for the case of the first training scenario and this is obvious from the almost total convergence of the ANN estimation results to the actual values according to Levenberg-Marquardt algorithm, as it is shown in Fig. 11.12. However, in the case of the second scenario the Levenberg-Marquardt algorithm seems to provide slightly better results but a clear advantage is not obvious.

Regarding the above results, ANN technique seems to be a powerful and flexible tool for estimating and forecasting the ground resistance value in variable soil and weather conditions that definitely influence the behavior of grounding systems. In such technical matters the condition monitoring of grounding systems is possible only by field in situ measurements, which, in most cases, are difficult, costly and sometimes impossible. Thus, a well-designed and well-trained ANN is a credible solution for an accurate estimation of grounding systems condition.

As a next step to the effort for modeling and forecasting ground resistance variation and based on the previous work with ANN models, the researchers aimed to develop and train a neural network by using only local rainfall data at the experimental field uncoupling completely the input vector from soil resistivity variables [23]. This effort

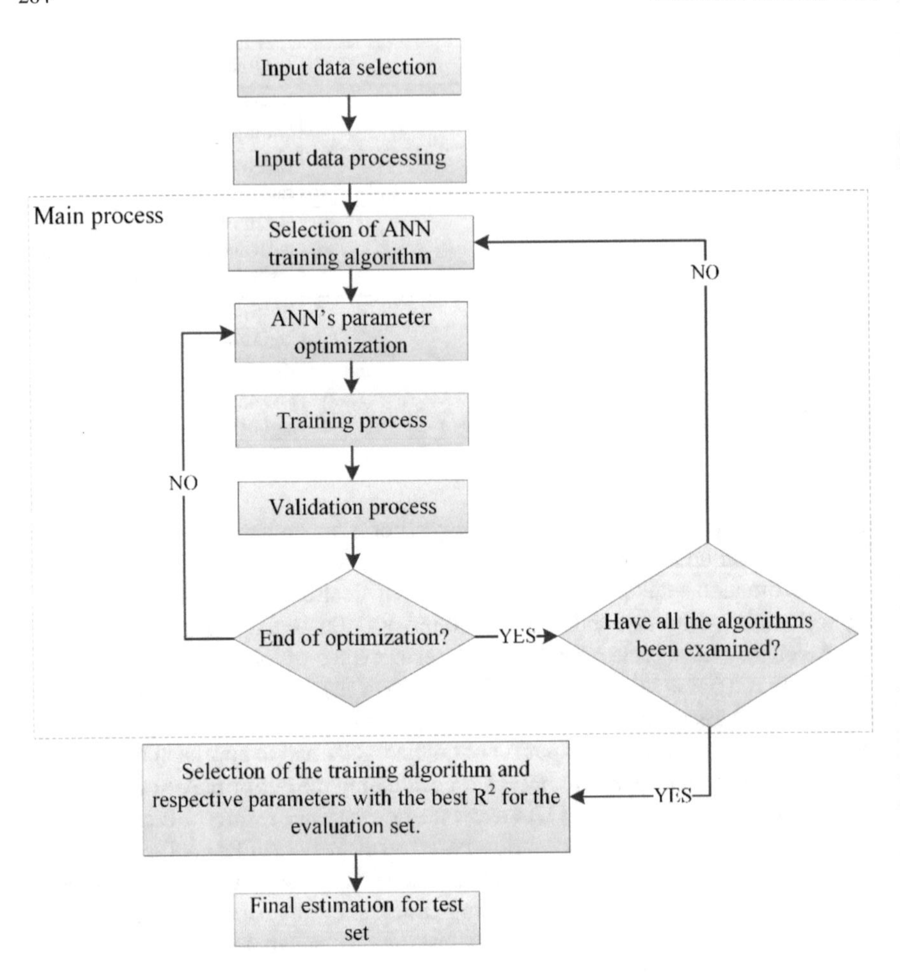

Fig. 11.8 Flowchart of the ANN methodology [22]

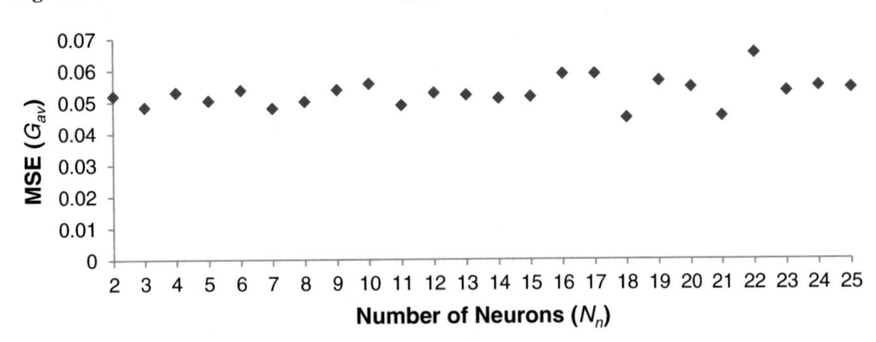

Fig. 11.9 G_{av} for the evaluation set as a function of neurons number [22]

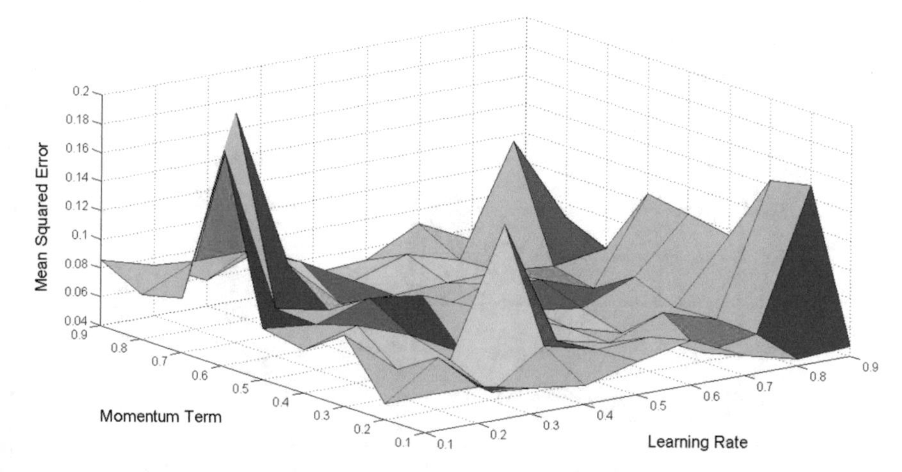

Fig. 11.10 G_{av} for the evaluation set for $\alpha_0 \in [0.1, 0.9]$ and $\eta_0 \in [0.1, 0.9]$ [22]

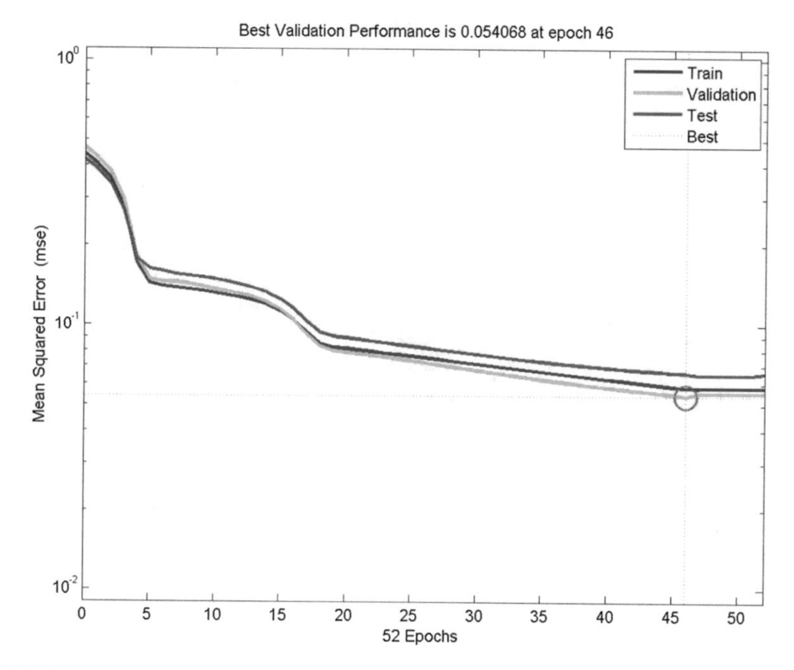

Fig. 11.11 G_{av} for the training, the evaluation and the test sets during the execution of algorithm 3 in the first scenario [22]

makes sense because the use of soil resistivity values in such a model demands a large series of measurements at the field of interest, which is a costly and time-consuming procedure. On the other hand, detailed local rainfall data are simply acquired at every place of interest, just addressing the national meteorological services.

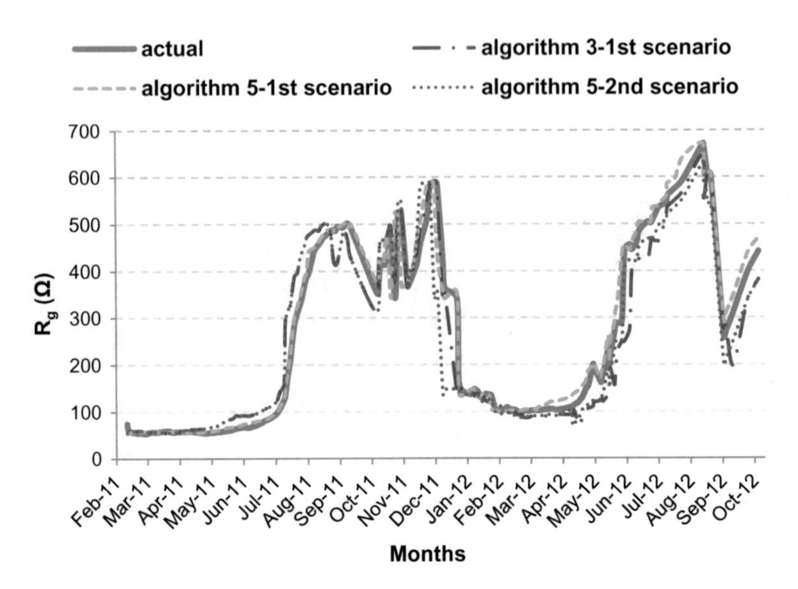

Fig. 11.12 Actual values and estimation results for ground resistance of electrode 6 versus time [22]

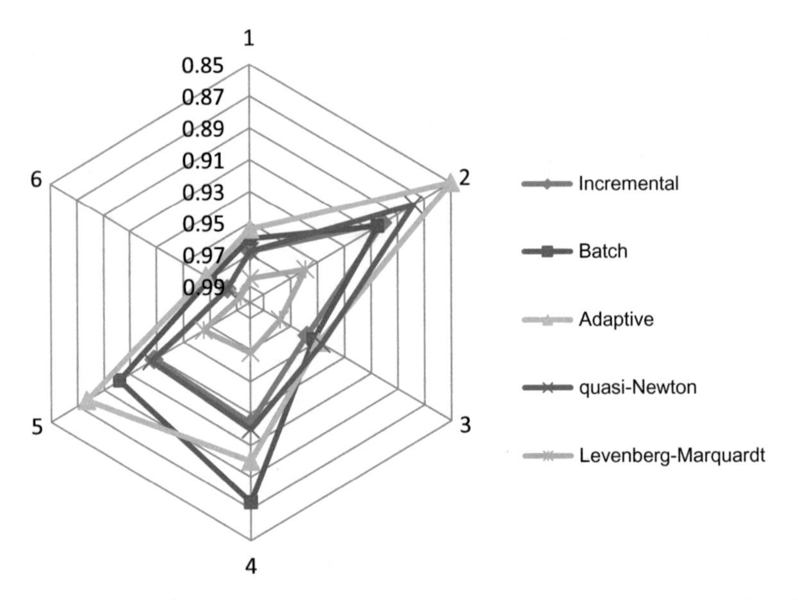

Fig. 11.13 Graph of R^2 coefficient for the 6 grounding systems including all the training algorithms (1st scenario) [22]

Thus, in the initial process only one ground rod has been investigated for the ANN modeling and the ANN architecture was similar to that of Fig. 11.2. The input layer can include: (a) the cumulative total rainfall height in the day of ground resistance

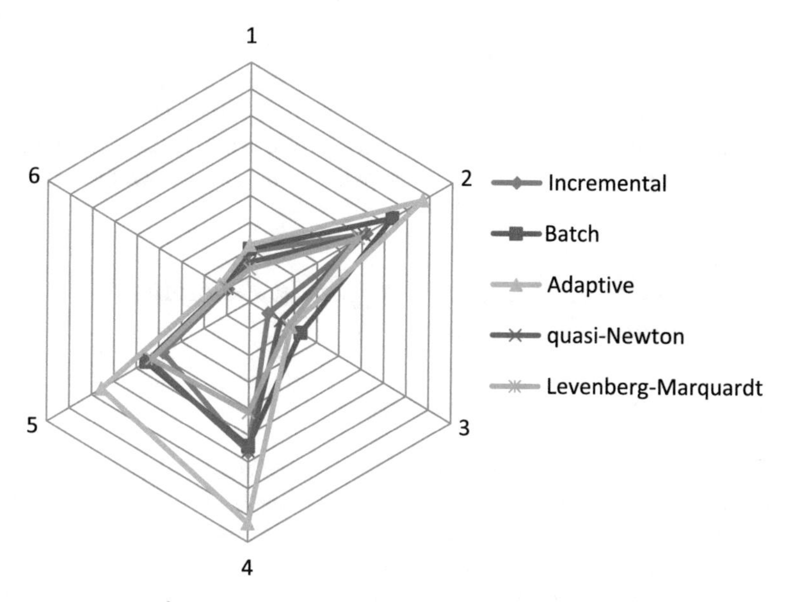

Fig. 11.14 Graph of R^2 coefficient for the 6 grounding systems including all the training algorithms (2nd scenario) [22]

measurement ($r_d(t)$), (b) the cumulative total rainfall height of the last seven days before the day of ground resistance measurement ($r_w(t)$), (c) the cumulative total rainfall height of the last thirty days before the day of ground resistance measurement ($r_m(t)$) and, finally, (d) the sinusoidal functions with period of one year, that describe the periodicity of the phenomenon and are given by $\cos(2\pi t/T)$ and $\sin(2\pi t/T)$, with T the days population per year (365 for the regular year, 366 for the bisect year). The output variable (output layer) of the ANN is the ground resistance of the grounding rod [23]. Several scenarios for the training of the network have been examined, aiming to assess the significance and relevance of each input independent variable to the output of the network. In real problems it is important to determine correctly the independent variables. Including unnecessary input variables in the model may cause the increase of model complexity and training time, while the predictive capability of the model is reduced.

The important role of the time periodicity on the final estimation of ground resistance, as a supplementary factor for the improvement of the generalization ability of the network, is one of the noteworthy conclusions from this work. Besides, the use of the sinusoidal functions as the only input variables to the network results in a correlation of almost 0.9 for the test sample. As for the individual use of rainfall variables $r_d(t)$ and $r_m(t)$, their relevance to the network output is too small with the R^2 coefficient lying in very low levels, even if the respective results of $r_m(t)$ for validation sample are quite satisfactory.

Concluding, the model is effective in estimating and forecasting ground resistance values quite accurately. Further work, on developing an algorithm that correctly identifies the insignificant variables, is of major importance. Aiming to this direction, more classifications of the rainfall data could be assessed for their significance in the accurate estimation of ground resistance. Therefore, the suitable selection of the input variables, with the aid of such algorithms, is expected to lead to more rapid and effective training of the network and increase its performance to the highest levels.

11.2.3 Wavelet Networks Modeling for the Estimation of Ground Resistance

11.2.3.1 Introduction to Wavelet Analysis

Wavelet analysis has been highlighted as a valuable mathematical tool used for the analysis of a large number of time series and has been applied, so far, in the fields of image processing, signal denoising, density estimation, signal and image compression, and signal decomposition in time domain with great success. It is often considered as a microscope in mathematics [24] and a powerful tool for representing nonlinearities [25]. For finding an alternative tool, instead of classic neural networks, which would be more flexible and less time-consuming in its training, research has turned to the development of more flexible networks, named Wavelet Neural Networks (WNNs) or simply Wavelet Networks (WN) as referred to in literature. These networks have kept enough characteristics from the classic neural networks, using the wavelet functions as activation functions instead.

WNs have been used in a large number of technical applications such as short-term load forecasting, time series forecasting, signal classification and compression, signal denoising, and nonlinear modeling. WNs are a generalization of radial basis function networks (RBFN). They are networks of a single hidden layer, using a wavelet as activation function instead of the classic sigmoid function [26]. The nodes of the wavelet networks are the wavelet coefficients of the expanded function which have a significant value. These networks usually consist of three layers. The first layer is the input layer, the middle is the hidden layer and the last layer is the output layer. The explanatory variables of the problem are introduced to WN through the input layer. The hidden layer consists of the hidden neurons, or else, *hidden units* (HUs). The hidden units are usually referred to as *wavelons* and they are similar to the neurons of hidden layer in classic sigmoid ANNs. In the hidden layer, the input variables are transformed to a dilated and translated version of the mother wavelet. Finally, the output layer gives the estimated value of the target variable [26].

The output of a feed-forward single-hidden-layer WN is given by the following expression:

$$g_\lambda(\mathbf{x};\mathbf{w}) = \hat{y}(\mathbf{x}) = w_{\lambda+1}^{[2]} + \sum_{j=1}^{\lambda} w_j^{[2]} \cdot \Psi_j(\mathbf{x}) + \sum_{i=1}^{m} w_i^{[0]} \cdot x_i \tag{11.11}$$

In the above expression, $\Psi_j(\mathbf{x})$ is a multidimensional wavelet which is constructed by the product of m scalar wavelets, $\mathbf{x}$ is the input vector, m is the number of network inputs, λ is the number of hidden units (HUs) and w stands for a network weight. The multidimensional wavelets are computed as follows:

$$\Psi_j(x) = \prod_{i=1}^{m} \psi(z_{ij}) \tag{11.12}$$

where ψ is the mother wavelet and

$$z_{ij} = \frac{x_i - w^{[1]}_{(\xi)ij}}{w^{[1]}_{(\zeta)ij}} \tag{11.13}$$

In the above expression, $i=1, \ldots m, j=1, \ldots \lambda+1$ and the weights w correspond to the translation ($w^{[1]}_{(\xi)ij}$) and the dilation ($w^{[1]}_{(\zeta)ij}$) factors. The complete vector of the network parameters comprises $w = \left(w_i^{[0]}, w_j^{[2]}, w_{\lambda+1}^{[2]}, w^{[1]}_{(\xi)ij}, w^{[1]}_{(\zeta)ij}\right)$. These parameters are adjusted during the training phase.

11.2.3.2 WN Paradigm in Grounding Systems

Though WNs have found significant applications in many technical and economic problems, their application in grounding field is trivial. Androvitsaneas et al. [27] developed a feed-forward single-hidden-layer for forecasting the ground resistance of five vertical grounding rods encased in ground enhancing compounds. The dataset used for the training and the validation of the constructed network was an enriched version of the dataset used in ANN modeling. More particularly, the experimental dataset of 337 input–output patterns was randomly divided into two sets:

- The training set (or in-sample set) consisted of 237 patterns and was used until the network has learned the relationship between the inputs and the output.
- The validation set (or out-of-sample set) consisted of 100 patterns and was used for the initialization of the WN parameters, for the model and variable selection, as well as for the evaluation of the learning and generalization ability.

The input variables were exactly the same as the ANN model of Fig. 11.2 and the output variable was the ground resistance of the five grounding rods, since one of tested rods has been withdrawn from the experiment. The Backward Elimination (BE) method [26, 28] was used for the initialization of the network parameters, starting the regression by selecting all the available wavelets from the wavelet library. This method provided effective initialization of the network parameters, resulting in less iterations during the training stage and, certainly, avoidance of local minima for the loss function. Then, the wavelet that contributed the least in the fitting of the training data was repeatedly eliminated. The WN has been trained using the Batch mode of BP algorithm with constant learning rate $\eta = 0.1$ and zero momentum term. The BP algorithm may be less faster compared to other training algorithms of higher order, but is less prone to sensitivity, regarding the initial conditions, at the same time. Thus, the network weights were constantly adjusted, so that the network obtain the final vector of the parameters $w = \hat{\mathbf{w}}_n$ minimizing the loss function, given by Eq. (11.14):

$$L_n = \frac{1}{n}\sum_{p=1}^{n} E_p = \frac{1}{2n}\sum_{p=1}^{n} e_p^2 = \frac{1}{2n}\sum_{p=1}^{n} \left(y_p - \hat{y}_p\right)^2 \tag{11.14}$$

where y_p is the target value, $\hat{y}_p$ the network output and n the number of the patterns in the training set.

The activation function used for the wavelons of WN was the second derivative of the Gaussian, the so-called "Mexican Hat" wavelet, due to the shape of its curve, which is given by the following expression:

$$\psi\left(z_{ij}\right) = \left(1 - z_{ij}^2\right)e^{-\frac{1}{2}z_{ij}^2} \tag{11.15}$$

More precisely, the activation function of the constructed wavelet network in [27] was slightly different from the expression (11.14) and was:

$$\psi_{a,b}(t) = \frac{2}{\sqrt{3}\pi^{1/4}}e^{-\frac{z^2}{2}}\left(1 - z^2\right), \quad z = \frac{t-b}{a} \tag{11.16}$$

The stopping criteria used for the termination of the training phase were the following [27]:

$$|L_n(ep) - L_n(ep-1)| \le \text{limit}_1 \tag{11.17}$$

$$\left|\frac{\partial L_n(ep)}{\partial w_t} - \frac{\partial L_n(ep-1)}{\partial w_t}\right| \le \text{limit}_2 \tag{11.18}$$

The optimal WN architecture should comprise the least necessary number of HUs in order to be able to describe the variability of the training data efficiently. Therefore, researchers have chosen the Minimum Prediction Risk (MPR) criterion as the most appropriate measure for the WN generalization ability [26, 27]. According to this principle, the prediction risk is the prospective performance of the network on new and unseen data, i.e. unknown data to the network during the training phase. Its mathematical expression is:

$$P_\lambda = E\left[\frac{1}{n}\sum_{p=1}^{n}\left(y_p^* - \hat{y}_p^*\right)^2\right] \tag{11.19}$$

For estimating the prediction risk and resulting in a network with the optimal forecasting ability, the Bayesian Information Criterion (BIC) was chosen for the WN construction, as BIC is considered to be the most suitable among other information criteria [27]. Its main advantage is the small computational burden and the precision on estimation results. The BIC is calculated based on the expression (11.20) [26]:

$$J_{BIC} = \frac{1}{n}\sum_{p=1}^{n}(y_p - \hat{y}_p)^2 + \frac{k\hat{\sigma}^2 \ln(n)}{n} \tag{11.20}$$

In the above expression (11.20), k is the number of the parameters of the network, n the number of the training patterns and σ^2 the noise variance estimator.

Eventually, for determining the most significant explanatory variables for the input vector—the limitation of the input vector would reduce the computational burden, and the estimation results for the output variable would be more accurate and targeted, as unnecessary explanatory variables reduce the model's forecasting power—Sensitivity Based Pruning (SBP) method [26] was chosen for variable selection, among several sensitivity and model fitness criteria. This method investigates the relevance of each explanatory variable to the model and quantifies this relationship by measuring the variation in the empirical loss function (L_n) that brings the replacement of the investigated variable $\mathbf{x}$ by its mean value. The SPB method is described by the following expressions [26, 27]:

$$\mathrm{SBP}(x_j) = L_n(\mathbf{x}; \hat{\mathbf{w}}_n) - L_n(\bar{\mathbf{x}}^{(j)}; \hat{\mathbf{w}}_n) \tag{11.21}$$

$$\bar{\mathbf{x}}^{(j)} = (x_{1,t}, x_{2,t}, \ldots, \bar{x}_j, \ldots, x_{m,t}) \tag{11.22}$$

$$\bar{x}_j = \frac{1}{n}\sum_{t=1}^{n} x_{j,t} \tag{11.23}$$

An indicative sample about the results of the developed WN model for forecasting and modeling the seasonal variation of ground resistance, affected by weather and soil conditions, is given in Fig. 11.15. The graph illustrates the convergence between forecasting and target values, resulted by the model for the ground rod 3. The constructed WN model for this technical application presented remarkable convergence results, regarding the R^2 and adjusted R^2 coefficients, reaching the values of 0.9898 and 0.9795 respectively for out-of-sample set. Furthermore, the model noted low values for SMAPE in out-of-sample set, where the lowest value reached the 2.79%. Therefore, the developed WN architecture noted quite high performance in assessing and modeling the relationship among heavily variable and unpredictable parameters as soil resistivity, rainfall height and ground resistance.

11.2.4 Inductive Machine Learning

Entropy information based inductive learning techniques were applied for the estimation of the ground resistance of grounding systems, used for the safe operation of electrical installations, substations and power transmission lines. Experimentation

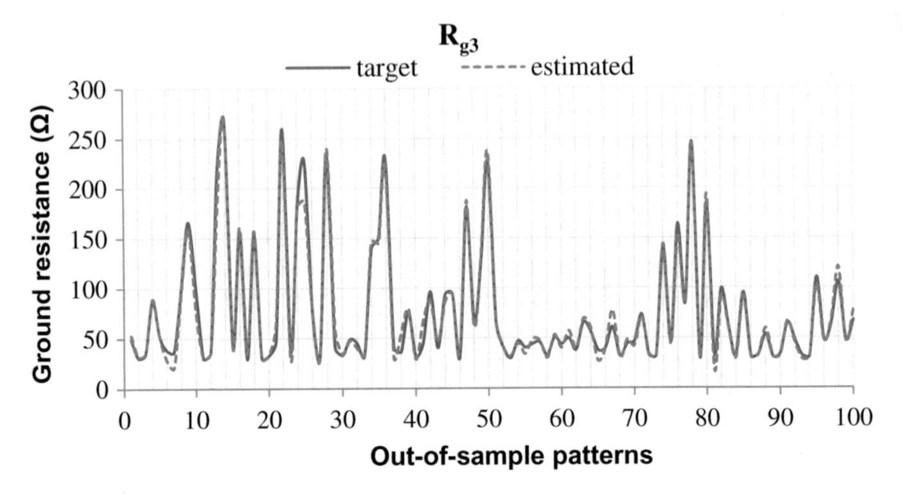

Fig. 11.15 Convergence results between forecasting and target values for out-of-sample set [27]

took place with measurements of soil resistivity in various ground depths and of rainfall height, which have been carried out over a period of four years or so, at a particular field. At the same time, the ground resistance values of few grounding rods, encased in ground enhancing compounds, have been recorded as a function of time. The applied computational method generalized over numerical data corresponding to these ground resistance measurements. For the modeling of the data, classes were represented by discrete intervals of measurements. Decision trees were constructed in [29] for approximating the discrete valued target function of ground resistance and, then, they were represented by production rules in order to improve the model comprehensibility. The error rates and the performance of the model on unseen cases were determined by a v-fold cross validation approach. Results proved promising for further development of the method. Inductive machine learning was used not primarily as a classifier, aiming at obtaining high accuracy, but more as a knowledge discovery tool, finding interesting rules and decision patterns of high quality, to be checked further with statistical techniques.

Decision tree learning has been one of the most widely used and practical methods for inductive inference. It is a method for approximating discrete-valued target functions that is robust to noisy data, in which the learned function is represented by a decision tree and capable of learning disjunctive expressions [30].

For the particular problem of ground resistance estimation a decision tree has been constructed, classifying the available ground resistance values of the training set, resulted from the field measurements, in predefined classes [31]. The algorithm C5.0, a newer version of C4.5, has been used for the decision trees construction and their produced rules [32].

The thirteen attributes (parameters) that determine the classified values of ground resistance, used in the particular developed tree, are the daily value of soil resistivity at the depth of 1, 2, 4, 6 and 8 m on the day of measurement (ρ_{id}), the mean weekly

Table 11.3 Results of the IML methodology for the grounding rods G_2 and G_3, see [29]

	R_{G2a}	R_{G2b}	R_{G2c}	R_{G3a}	R_{G3b}	R_{G3c}
Nodes	51	67	69	45	39	80
Trees error (%)	1.4	0.8	0.5	0.3	0.3	0.8
Rules	42	55	59	43	29	68
Rules error (%)	2.2	1.4	0.8	0.3	0.5	1.1
CV error (%)	19.5	26.0	24.6	15.1	14.0	27.4
Extended CV error (%)	3.6	5.2	5.8	5.5	3.8	9.0

value of soil resistivity at the same depths (ρ_{iw}), the mean monthly value of soil resistivity at depths of 1 and 2 m (ρ_{im}) and the total rainfall height during the last seven days before the measurement day (r_w). It is noted that $i = 1, 2, 4, 6, 8$ m in depth.

The variable to be classified was the ground resistance of each tested grounding rod. The training set consisted of 365 cases (covering a 4-years period) for the training and the validation of the decision tree. The categories (classes) to which cases were to be assigned, had been established beforehand. Three different scenarios were established for the class discrimination assigned to each rod, corresponding to different value intervals, considering both the uniformity of intervals and the uniform distribution of the cases among the classes.

A decision tree was finally constructed for each individual hypothesis, considering all the 13 attributes, each time. Afterwards, a 10-fold cross validation (cv) run was performed for each individual tree, as it is a more robust estimation of accuracy on unseen cases. The results of the developed IML model for each grounding rod are tabulated in Table 11.3. The "extended cv error" was the cv error considering each time an extended class, the default with the neighboring classes.

The rules extracted using the IML methodology and proposed in [29] are simple, understandable and cover a large number of data. The inductive decision tree is comprehensible, straightforward and is a near-perfect classifier in the training set, presenting a 0.3% error. A number of 29 rules described the available sample of data giving a small error rate in the cross-validation data [33]. This error varies depending on how the classes are classified and depending on the amount and variety of available data also. In Table 11.4 five indicative rules are presented, obtained using the methodology presented above. In the first rule of the table, for example, the number of cover cases shows that it is sturdy and has a possibility of 99.5% of correctly classifying new data.

The results for the rod G_1 show that, despite the remarkably low misclassification errors of decision trees and production rules, the cross validation error on unseen cases is quite high in all the scenarios. Moreover, the cross validation error increases as the range (in Ohms) of the class becomes smaller, e.g. in scenarios (b) and (c) with 38.3% and 37.8% respectively, against the scenario (a) with 35.4%. These errors are higher in the case of the rod G_1 than the respective errors for the rods G_2 and G_3, due to the large variance of values, the ground resistance of G_1 presents, as the Fig. 11.4

Table 11.4 Five indicative rules derived from the proposed IML methodology

Rule id	Cover cases	Extracted rule	Probability of correct new data classification expressed in [0, 1]
1	195	$\rho_{1m} \leq 203.89 \rightarrow$ class "25–60"	0.995
2	120	$\rho_{2m} \leq 148.74 \rightarrow$ class "25–60"	0.992
3	123	$\rho_{4d} > 117.12$ AND $\rho_{2w} \leq 229.96 \rightarrow$ class "25–60"	0.992
4	119	$\rho_{1d} \leq 325.47$ AND $\rho_{8d} \leq 189.601$ AND $\rho_{6w} > 143.533 \rightarrow$ class "25–60"	0.967
5	78	$\rho_{6d} > 154.566$ AND $\rho_{8d} \leq 210.11$ AND $\rho_{2w} \leq 199.72 \rightarrow$ class "25–60"	0.950

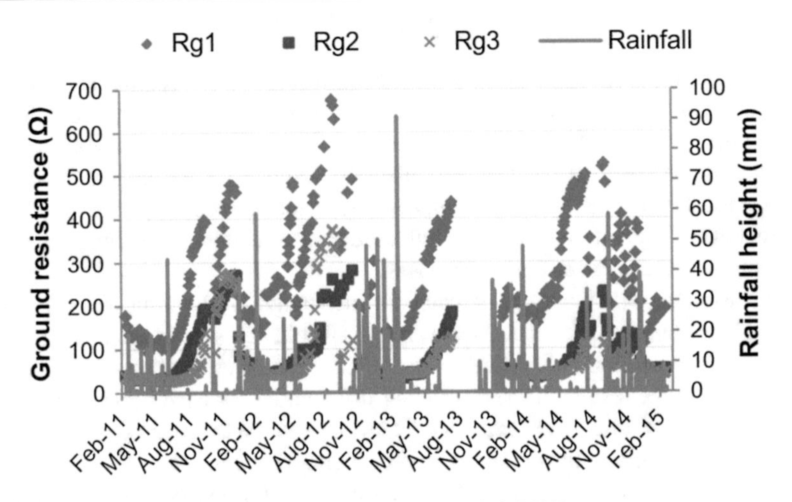

Fig. 11.16 Ground resistance as a function of time and rainfall [29]

does point out. On the contrary, considering the default class and the class next to it, as an extended class, the cv error significantly decreases to approximately 1/3 of the original. The rods G_2 and G_3, on the other hand, present some proposed scenarios, as R_{G2a}, R_{G3a} and R_{G3b}, with satisfactory results. The error in the scenario (a) of G_2 reaches the value of 19.5% while the corresponding values for scenarios (a) and (b) of G_3 are just 15.1% and 14%, respectively. It seems that the soil alleviation, the ground enhancing compounds achieve and the consequent consistence in the ground resistance values (Fig. 11.16), result in a classification task with compensative results on unseen data. This means that the ground resistance forecasting, given soil resistivity and rainfall data, is a most promising task. The main cause of the great difference in cross validation errors, among the various scenarios of each individual rod, is probably the nonuniform range of the classes in each scenario. This could be a field for further investigation on the best class establishment.

Overall, the results are encouraging enough to keep on the work on constructing similar models for the estimation of grounding systems performance. The great difference, in cross validation errors, points out the need for better and more focused establishment of the necessary classes. Furthermore, a probable application of a suitable attribute selection algorithm could result in much lower errors on classified unseen cases. In conclusion, the results confirm the successful application of learning decision trees in the field of grounding and, after the necessary modifications of the algorithm, it is expected to be a powerful and reliable tool.

11.2.5 *Genetic and Gene Expression Programming Versus Linear Regression Models*

Linear Regression and Gene Expression Programming (GEP) [34] were used in [15] to develop models for describing how ground resistance acts as a function of weather conditions. A series of 378 measurements of resistance (Rg_1, …, Rg_5) and 1753 cases of weather data were used. In order to obtain the desired models—one for each ground enhancing compound—the appropriate time interval should be estimated, starting one day before resistance measurement i.e. day zero and is extended to the past for some days. The letter i denotes the number of these days. The number of days for the desired interval is calculated using the multiple correlation coefficients of all the independent variables to the measured, dependent resistance value. The vector of the independent variable in the range i has a dimension of $N = 13$ and is presented in Eq. (11.24), where exponents are indices, α denotes variable average in the base of the interval i and s indicates summation accordingly.

$$p^i = \left(ws^a, wd^a, nr^a, dr^a, sd^s, bp^a, sm^a, at^a, rf^s, ah^a, ep^s, tr^a, fr^s\right) \quad (11.24)$$

Using the vector $\mathbf{c} = (r_{x_1y}, r_{x_2y}, \ldots, r_{x_Ny})^\top$ of correlations and the correlation matrix R_{xx}, which is represented in Eq. (11.25), of inter-correlations among the predictor variables, the squared coefficient of multiple correlations can be computed as $R^2 = \mathbf{c}^\top R_{xx}^{-1}\mathbf{c}$. R in this case (without index) is the Pearson Correlation Coefficient. For each resistance R_j, the smaller value of i is looked for, where $dR^2/di = 0$. Thus, the value of i is derived from the plot $R^2 = f(i)$.

$$R_{xx} = \begin{pmatrix} r_{x_1x_1} & r_{x_1x_2} & \cdots & r_{x_1x_N} \\ r_{x_2x_1} & \ddots & & \vdots \\ \vdots & & \ddots & \\ r_{x_Nx_1} & \cdots & & r_{x_Nx_N} \end{pmatrix}. \quad (11.25)$$

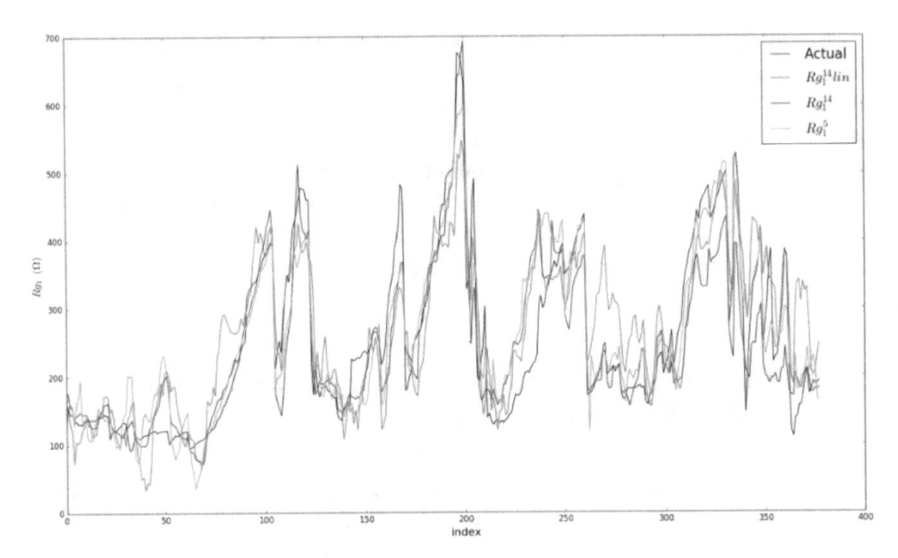

Fig. 11.17 Plot of the actual versus predicted values for Rg_1 produced by Linear Regression & GEP

At this point, for each dependent variable R_j of grounding electrodes the value of a time window, that the vector of the weather data has the highest correlation coefficient, is obtained. Using this value of the time window a modular process was initiated for the development of models starting from the linear models, continuing with models using GEP. In order to derive some improved models, the value of interval i was estimated using the GEP instead R^2 directly.

For each i, $i \in \{2, 5, 7, 10, 15, 20, 30, 50, 80, 100, 150, 180\}$, a quick GEP execution was performed with population of 50 individuals. The number of maximum generations was set to 2000 using the entire data set and derived the corresponding models computing the R^2 between actual and predicted values, for each of them. For the highest value of R^2, the corresponding i is chosen and the same procedure is followed for the values of i in the interval $[i - a,\ i + a]$ and $a \in \{5,\ 10\}$, aiming to achieve a model with the highest value of R^2.

GEP implementation was also performed for the corresponding value of i. A training set of 70% randomly chosen measurements from the data set and the other 30% as a testing set were used. GEP experiments were configured using a population of 2,000 individuals, 10 genes per chromosome, 15 gene head length, the maximum number of generation 10,000 and the Function set was $\{+, -, *, /, a^{1/2}\}$. There were not used homeotic genes and the linking function was the addition. An example of a derived model with this methodology is represented in Eq. (11.26). The model is referred to the electrode G_1 and the time interval for whom weather data elaborated is 5 days. Estimated variable is Rg_j^i where i, j are time window and grounding electrode respectively (here 1 and 5 days). Figure 11.17 presents the model of Eq. (11.26) against experimental values and linear and the other GEP model.

$$Rg_1^5GEP = \sqrt{at} + at\sqrt{at \cdot yr} + \frac{2bp}{sm} + 2sd + sm - tr + \frac{tr}{nr - rf - 13.002} - 9.275 + \frac{2592.705}{sm - 1} \quad (11.26)$$

The results showed good performance of actual versus predicted values of ground resistance for five different ground enhancing compounds. GEP model fits the actual data better than linear models and they can give an intuition of the phenomenon. GEP models have given better values of R^2 index and MAE and proved that they could be used as a feature selection method, for future research in order to achieve better approximations of the ground resistance. The time of CPU required to train the models of GEP is about 10^3–10^5 longer with respect to the time taken to calculate the Linear Regression models. The results also showed that genetic programming variants like GEP could provide formulas capable of describing the phenomenon and, indeed, with similar accuracy to ANNs but without forming a black box. The research team proceeded to implement genetic programming approaches with improved accuracy, although the resulting models are more complex. Running experiments will produce a sufficient number of training data for machine learning methods to adequately address the problem. GP based approaches seem to be promising for generating generalized formulas as models for expressing ground resistance estimation.

11.3 Estimation of Critical Flashover Voltage of Insulators

One of the most critical problems with high-voltage insulators when operating under pollution conditions is the occurrence of the flashover phenomenon. The knowledge of the parameters the flashover depends on, as well as their critical values, is particularly useful. The way each of these parameters contributes to the phenomenon is unknown; therefore, researchers investigate models to approximate the flashover on polluted insulators. Intelligent techniques are proved to be handy tools for estimating and predicting the critical strain in overcoming an insulator under polluted conditions.

11.3.1 Problem Description

The phenomenon of flashover on insulators due to pollution refers to the fact that an electric arc bridging, which runs through the air of the gap, is created between the point of attachment of the line conductor to the insulator and the grounded mounting or suspension of the insulator. The apparent paradox of overtaking insulators due to pollution is that destructive electrical discharges, expanding in air meters, are generated by electrical forces which, under normal conditions, could be intercepted by air gaps of a few centimeters in length.

In some ways, the presence of inconspicuous conductive particles deposited on a surface that would otherwise be strongly insulating reduces its actual electric resistance by a factor of not less than 100. The reasons for this are two: (a) the localized dehydration of an electrode layer increases the discontinuities in the conductive layer—also known as dry bands—along which stresses are produced sufficient to ionise the air and (b) the arcs in a gas, which create they can easily be expanded without impairment. For much of its life, an insulator will operate with dry bands on its surface which occasionally penetrate electrical discharges. These discharges are harmless, except for the problems of causing interference and surface damage for which they are responsible. Very rarely the combination of conduction and electric voltage will be so strong that it will allow the development of an arc of such intension that it is self-preserving and propagated; then it is caused a split.

The technical problem is that the surface conduction that causes the breakdown remains, even when the arc has been eliminated by the function of the protection, which allows for other decays to follow [35]. Pollution can be caused by various sources, such as fly ash, sea salt, dust from industries etc. The deposition of soil contaminants creates, when wet, a conductive layer on its surface. The inert components, on the other hand, are the percentage of the solid material that does not dissolve but forms a mechanical sheath in which the conductive layer is incorporated [36]. Humidity can be produced under mist or frost during the morning hours. Also, drizzle and rain can have the same effect.

The combination of all these dirt and moisture deposits creates an electrolyte conductive layer across the insulator. From this, a surface current pass that heats the electrolyte layer, causing the water to evaporate locally. The solution becomes supersaturated with salt and creates dry zones with higher ohmic resistance. Between the boundaries of the dry zones, the most significant potential difference is observed due to the voltage applied to the ends of the insulator, and then the occurrence of some discharges is found. In case that these discharges are extended to the rest of the wet surface of the insulator, then the flashover phenomenon occurs. If the arc created arises can expand and cover a critical length; the split can no longer be avoided. Because of the above phenomenon, the electricity transmission networks are stressed daily, and they often go out of order, with a more typical case of glacier fog has caused some of the most severe incidents, as in 1962 the multiple problems and a temporary interruption in England's transition network [35].

Addressing the problem has led to the design and execution of laboratory experiments to analyze the phenomenon. These experiments are carried out on artificially contaminated insulators of various types but require a lot of time for their execution. From these experiments, numerous experimental data have been obtained, like minimum flashover voltage against pollution, leakage current etc. Using the results and mathematical procedures, the arc constants of the insulator can be estimated according to its type. In Fig. 11.18 the basic parameters of Cap and Pin insulator type are depicted.

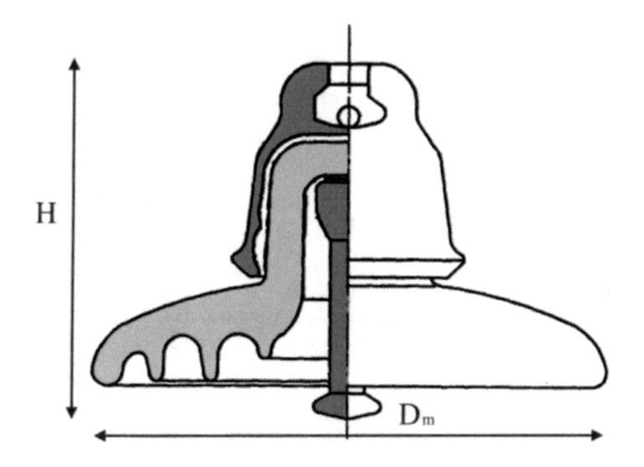

Fig. 11.18 Basic design parameters of Cap and Pin Insulator

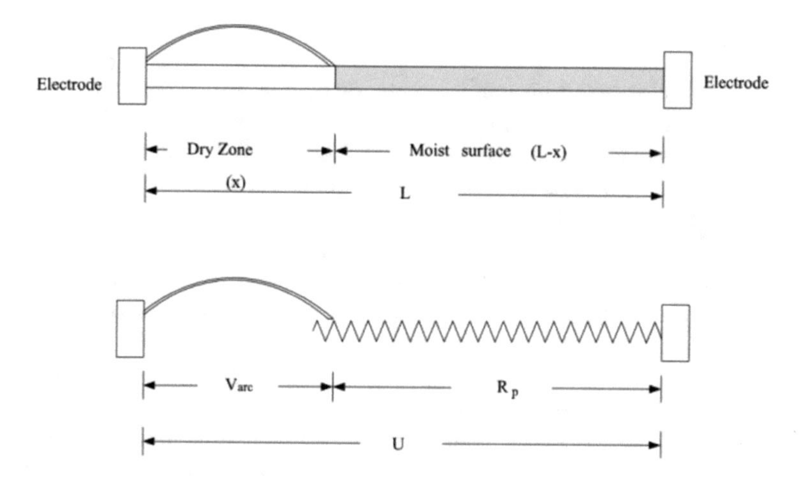

Fig. 11.19 Equivalent circuit for the evaluation of the flashover voltage

11.3.1.1 Mathematical Model and Experimental Data

A number of researchers have investigated the flashover mechanism on polluted insulators. Obenaus has developed the simplest model [37] and for more details see [38]. The equivalent circuit for the evaluation of the critical flashover voltage consists of a partial arc spanning over a dry zone in series with the resistance of the polluted wet zone as shown in Fig. 11.19, where V_{arc} is the arcing voltage, R_p the resistance of the pollution layer and U the voltage between the two ends of the insulator. The critical flashover voltage U_c according to that model is given by the Eqs. (11.26–11.30) in Table 11.5.

In Eq. (11.27), A and n are the arc constants, L is the leakage length, D_m is the maximum diameter of the insulator disc, F is the form factor a constructional characteristic and is determined from its dimensions using the integral of Eq. (11.28) where $D(l)$ is the diameter of the insulator, varying across its leakage length L.

Table 11.5 Equations of the model in order to figure out the critical flash over voltage U_c

$U_c = \frac{A}{n+1} \cdot (L + \pi \cdot n \cdot D_m \cdot F \cdot K) \cdot (\pi \cdot A \cdot D_m \cdot \sigma_s)^{-(n/(n+1))}$	(11.27)
$F = \int_0^L \frac{dl}{\pi \cdot D(l)}$	(11.28)
$K = 1 + \frac{n+1}{2 \cdot \pi \cdot F \cdot n} \ln(\frac{L}{2 \cdot \pi \cdot R \cdot F})$	(11.29)
$R = 0.469 \cdot (\pi \cdot D_m \cdot \sigma_s)^{1/(2(n+1))}$	(11.30)
$\sigma_s = (369.05 \cdot C + 0.42) \cdot 10^{-6}$	(11.31)

The coefficient of the resistance of the pollution layer K depends on the type of the insulator. In case of cap and pin insulators is figured out using Eq. (11.29), where R is the radius of the arc foot (in cm) and is given by Eq. (11.30). The surface conductivity σ_s (in Ω^{-1}) is figured out using Eq. (11.31), where C is the equivalent salt deposit density expressed in mg/cm^2.

In order to be able to apply the above mathematical model to calculate the critical flashover voltage, the arc constants need to be determined. The last requires a series of arduous experiments examining a set of different types of insulators used in the power distribution lines. The specimens are examined in an insulator test station on two chambers according to the solid layer-cool fog method to simulate the industrial pollution [38]. The overall process was in line with IEC norm [39].

The result of the process is the maximum withstand voltage depending on the geometric characteristics of the insulator and the degree of contamination. Details about the performed experiments one can find in [38, 40]. The set of the experimental data that will be used as training set to computational intelligent models completes a series of similar experiments [41, 42].

11.3.2 Genetic Algorithms

Genetic Algorithms (GA) belong to the heuristic methods of optimisation and the capability of searching for the solution is based on the mechanism of natural selection and survival of the fittest [43, 44]. The function of the genetic algorithm is based on the appropriate encoding of the solution and is a genotype-phenotype system [45]. GA can find near-optimal solutions, where other conventional methods do not do just as well. In [46] GA have been used successfully to find the constants of the arc A, n of Eq. (11.27).

The estimation of the arc constants is a difficult task and is one of the significant difficulties and challenges faced by researchers in trying to develop reliable mathematical models that describe the dielectric behaviour of contaminated insulators.

The use of different experimental methodology and the complexity of the followed mathematical approaches resulted that the values of A and n have been presented in the literature with several deviations ($A = 131.5,\ n = 0.374$ [38]; $A = 63.0, n = 0.76$ [47], $A = 270 - 461, n = 0.42 - 0.66$ [48]). Since σ_s is a function of C and F and K are a function of known L, Dm which are depended by insulator type the critical voltage is a function of a form of Eq. (11.32).

$$Uc = f(A, n, L, Dm, C) \tag{11.32}$$

Three types of insulators were selected from the bibliography for which experiments were conducted to determine the Uc against σ_s. Since the geometrical characteristics of the insulators are known, the calculation of the arc constants becomes a minimization problem of F_g of Eq. (11.33)

$$F_g = \sum_{i=1}^{284} |Uc_i - f_i(A, n)| \tag{11.33}$$

The number of experimental data that forming the training set is 284 vectors. The measured value of Uc_i is the experimental result, and the f_i is the evaluation of the model using data for the ith experiment. Taking into account the values of A and n from the literature, the following intervals, $A \in (0, 500)$ and $n \in (0, 1)$ were selected for investigation from the GA and were mapped to a 16-bit length binary strings each. Thus, each chromosome consists of 32 bits and represents both variables.

The population of each generation is limited to 20 individuals and the algorithm stops after 500 generations. From the population, a pair of two parental chromosomes are selected and each of them is divided into six parts, that interchange their genetic material between two parents. Four children are created from each pair of parents. The children then undergo a mutation with probability 1%. The mutation operator selects bits randomly from the progeny chromosome and inverts its value, i.e. zero becomes one, and one becomes zero. The process results in the population increases because the population of the parents is added to a double number of children. In order to conserve the population number steady only the better 20 individuals remain to the population for the next generation.

The results of the genetic algorithm quickly converge to an $A = 124.8$ and $n = 0.409$ after about 20 generations. The model that emerged using the genetic algorithm gave an accurate approximation of the experimental data [38, 41] for both fog-type and cap and pin type measurements. The model was tested in a critical voltage range of 27–7 kV against surface conductivity range of (0.1–2) × 10^{-4} Ω^{-1} respectively and passed through the experimental data.

11.3.3 Application of ANNs

Increasing air pollution with pollutants affects electricity distribution networks. An electrolyte layer spreads to the surface of the insulators and when the voltage exceeds a critical value, discharges strain the power lines. Thus, the network operator should monitor the status of the insulators and maintain maintenance at the appropriate time.

That is the reason why many researchers have addressed the problem and, among other things, they have developed new technologies to control the condition of insulators like ANNs. Neural Network algorithms have been successfully applied to estimate the equivalent salt deposit density, by using information about weather data to schedule maintenance tasks.

Another ANN [49] has been applied in order to estimate the critical flashover voltage on polluted insulators using as input variables the geometric characteristic of an insulator and the equivalent salt deposit density using experimental data and the mathematical model to form the training set. In [50], an ANN has been used in order to estimate the time-to-flashover when the applied voltage, the creepage length and the resistance per unit length are given. The experimental data were obtained by studies performed on a flat plate model for a polluted insulator under a power frequency voltage. In [51] an ANN was trained to deduce whether or not a breakdown is imminent using data collected from two pollution-related monitoring devices and the predictions were validated with a test of flashover experiments. Several other ANNs have been developed in order to analyze the insulator surface tracking on solid insulators. The system can protect the insulator from immoderate damage warning the user [52]. Another study [53] presents a multilayer ANN model that classifies the development of the arc gradient into three stages. ANNs used to estimate the partial discharge inception voltage [54] and the leakage current on silicone rubber insulators [55, 56]. In (Asimakopoulou et al. 2009), a methodology was developed to form an ANN to be used in the calculation of the critical flashover voltage on polluted insulators. The process that is being developed aims to select the appropriate ANN training algorithm and optimize the values of the parameters that regulate its behavior in order to achieve the best possible accuracy.

The training, validation and test of the designed ANNs were performed using two datasets. (a) Experimental data from previous works [40–42] and (b) the mathematical model of [38] to enrich the available data. The quantities used had common units of measurement and the independent variables for the training are those of the Eq. (11.32) except layer conductivity σ_s (in mS) against equivalent salt deposit density C.

A check was made to eliminate gross errors in order to deduct noise. The values were normalized to avoid saturation problems due to the use of sigmoid functions outside the interval $[-1, 1]$.

The neural network developed has three layers. In the input layer, the independent variables are assigned. Their values are obtained from the training, validation and test set.

The ANN training is performed with a series of variants of the Back Propagation (BP) algorithm. The appropriate algorithm selection is made with the R^2 index correlating the actual values of the critical flashover voltage with those calculated by the ANN for the validation set. Index R^2 is used for what is an undiminished number, describing the approximation estimated with actual values regardless of the units of measurement and the absolute values of the quantities. The algorithm that presented the highest value of R^2 is more appropriate for the task of critical flashover voltage estimation.

Confidence intervals are calculated for the validation set using the resampling method [57]. At the overall end of the process, the estimation of value of the critical flashover voltage is done using the data of the test set. The trained ANN using the BP algorithm showing the larger R^2 index and the others corresponding design parameters as derived from the process followed is used for the estimation of critical flashover voltage.

Depending on the way of adaptation of the weights, there are two ways of training: (a) the case of stochastic training where weights are adjusted after the presentation of each training model on the network, (b) the case of batch mode where weights are updated after the presentation in the network of all the set of patterns of the epoch. The mean value of square errors is determined by Eq. (11.3). The adjustment of all weights takes place once at the end of each epoch.

The stopping criteria are the same as Eqs. (11.4–11.6). In each ANN, two approaches were followed. In the first (from now on referred to as a), all three criteria mentioned above were used, while in the second (hereafter referred to as b) only the first and third were used.

There are several parameters to be selected, depending on the variation of the BP algorithm that is being used each time in order to train the ANN. The parameters that are common in all methods are: the number of neurons N_n, the type and the parameters (a, b) of the activation functions, Eqs. (11.7–11.9) and the maximum number of epochs (max_epochs). The uncommon parameters were selected based on each algorithm, see [58].

The development of the most appropriate ANN requires the investigation of every possible combination of parameters for each input variable. In order to reduce this combination of parameters, a sequential stages process was followed in order to initially determine the optimal parameters of the stage being examined and then keeping them fixed to examine the remaining parameters of BP algorithm. Thus, the appropriate number of hidden layer neurons (i.e., the one providing the smallest error) was first determined. This was achieved for each BP algorithm by examining the number of neurons in the interval 2 … 25 with step 1. The optimal choice is that minimizing the value of G_{av} (Eq. 11.3). Holding the corresponding optimal number of neurons for each BP algorithm, the sequential determination of the other parameters (e.g. activation function parameters, momentum term etc.) follows, in asimilar way. More details and an exemplify paradigm can be found in [58]. The overall optimization methodology is the same as Fig. 11.8, except for one more stage before the final estimation for test set, where the confidence intervals calculation takes place.

The most accurate ANN is that constructed using 3 neurons in the hidden layer, the scaled conjugate gradient algorithm with three stopping criteria in estimating the critical flashover voltage. The respective correlation between the actual and the estimated values R^2 for the test set is 0.9853 and for the evaluation set is 0.9972 and the respective Mean Absolute Percentage Error (MAPE) is 3.84%. The ANN is superior than the mathematical model [38] which respective correlation R^2 is 0.9801 and MAPE is 4.59%. The comparison of mathematical model and proposed ANN methodology for 24 experimental vectors of the test set confirmed that ANN methodology gave better results.

11.3.4 Multilayer Perceptron ANNs

A multilayer perceptron approach was followed in [59], where an attempt has been made to model the flashover phenomenon estimating the critical flashover voltage. For this purpose a set of 140 cases from the model of have been used and a set of 28 experimental observations [38]. The variables are the same as Eq. (11.6).

The main database was divided into two parts for all methods of work. The first part was used for the development of the predictive models of critical flashover voltage and the second part was used to test the final predictive model. The Training-Validation Set is derived from 150 data cases (130 cases of mathematical data and 20 cases of actual data) and the Test Set is derived from 18 data cases (10 cases of actual data and 8 cases of mathematical data). Data were normalized in the interval of $[-1, 1]$ to avoid saturation problems.

The ANN model belongs to the category of the feed-forward artificial neural networks (Multilayer Perceptron—MLPs. On finding the optimal MLP for the estimation of critical flashover voltage, 351 MLPs were developed and compared with each other. All MLPs had one hidden layer and the activation function of hidden nodes was the logistic function [60].

The training of each MLP was supervised into batch mode and was performed using the Backpropagation Algorithm (BP) [61]. Moreover, the validation of each MLP was held together with the process of education, using the 10-fold cross validation method [62] based on the same dataset (Training-Validation Set).

The experimental procedure followed at this point of the work consists of 4 parts comparisons of different MLP architectures. The goal in each test part was to minimize the MAE, RMSE, RAE and RRSE and to maximize the R^2 between the Uc ($_{actual}$) and Uc ($_{predicted}$). The aim of the first test part of the experimental procedure was to find the best combinations between the learning-rate (T_η) and the momentum term (T_a) for specific numbers of epochs (E). The searching, which took place at the second test part of the experimental procedure, is concerned the finding of the best combinations between the number of epochs and the number of nodes to hidden layer of MLPs, for the combinations $T_\eta - T_a$, which were resulted from the first test part. In the third test part and having into account the best combinations $E - T_\eta - T_a - N_n$ of the second test part of experimental procedure, was observed the behavior

Table 11.6 Performances of ANN [59]

Model: ANN2	R^2	MAE (kV)	RMSE (kV)	RAE (%)	RRSE (%)
Training	0.9991	0.146	0.2522	2.8568	4.3264
Test	0.9904	0.731	0.8256	–	–

Table 11.7 Performance of ANN model [63]

	Training-validation set	Test set
R^2	0.9994	0.9842
MAE (kV)	0.1370	1.3249
RMSE (kV)	0.1886	1.6957
RAE (%)	2.9136	–
RRSE (%)	3.3936	–

of MLPs by altering the number of nodes to hidden layer (N_n) and keeping fixed the combinations $E- T_\eta - T_a$. Finally, in the fourth test part, using the combinations of third test part was observed the behavior of MLPs by altering the number of epochs. The performance of the model is presented in Table 11.6. The number of nodes in hidden layer is 20.

Previous work [59] was expanded in [63] where, using the same data, 1895 architectures of MLPs have been developed and compared to each other in order to estimate U_c. The training process follows 4 stages accordingly. The ANN architecture has 3 layers where hidden layer has 21 neurons. For its training the BP algorithm adjust to learning and momentum term 0.1 and 0.9 respectively. The correlation between actual and predicted values of the model for the train-validation and test set shows the good performance of proposed ANN, Table 11.7. The comparison with similar ANN models from related literature shows that the model generalizes well on the estimation of critical flashover voltage on polluted cap and pin porcelain insulators.

11.3.5 Genetic Programming

Genetic programming (GP) is a nature-inspired optimization algorithm which follows the model of biological evolution. It was invented by Koza [64] and has since gained popularity due to its robustness and performance so many applications of GP have been reported. GP approaches the solution of a given problem, performing an evolutionary search within the solution space i.e. the space of the possible program syntaxes, for the expression that best solves the problem. GP is used to evolve solutions to different types of problems with a large variety of the applications. Symbolic Regression is an application of GP and aims to find the appropriate expression that best fits a given set of training data. These expressions are usually being reenacted as syntax trees where the variables and constants are the leaves of the tree and functions are internal nodes. Without constraints in size these trees along to evolution process grow in size and produce solutions difficult to read and results are often no general.

A simple method to control the expression growth is the use of limits on the depth or the size of the tree that represents the candidate solution. This point needs carefully estimation because tightly constrained leads to weakness to express good solutions [65]. To accomplish that task some trial and error maybe required in order to establish a good setup of the algorithm to obtain accurate results, sufficiently generalized, readable and more comprehensible.

In [59] a GP was developed (GP-1). The first set of 150 cases was partitioned into a k = 10 subset of equal size. Of the k subsets, a single subset was retained as the testing data and the remaining k − 1 subsets were used as training data. The cross-validation process was repeated k times (k-fold Cross-Validation), with each of the k subsets used exactly once as testing data. For each fold 30 intendedly runs executed with the same modulation. All runs trained their solution with k-1 subsets and all compared in the rest testing data. So, we had 10 candidate models which compared in the second set of 18 cases, and the better model in terms of correlation coefficient had been chosen.

Each run was modulated with a number of 50 generations and the population number was 1000 individuals. The maximum tree depth was 12 and the maximum tree length was 150 nodes. Person correlation coefficient used to evaluate the obtained solutions which produced by primitives from the Function set: {+, −, *, /, EXP, LOG, POW, SQRT} and the Terminal set was constants and problem variables. The internal crossover probability was 0.9 and the mutation probability was 0.15. Details of the model, which is rather complex, someone can find in [59]. In Table 11.8 performances of the model are presented under the name GP-1.

11.3.6 Gravitational Search Algorithm Technique

Aiming to a more comprehensible method, a Gravitational Search Algorithm (GSA) [66] was used in [59] for building a model that fits on data. GSA is an Evolutionary Algorithm and it is known for using vectors for every agent in every dimension of the problem and, thus, the vectors of the solutions are relocated in a better solution in solution space, based on the best solution that attracts all agents in every dimension. However, the main disadvantage of Nature Inspired Algorithms, such as GSA, in this kind of problems is that they do not check different mathematical function models, while GP and ANN do. A Nature Inspired Technique is based on the model that the user sets and tries different values of the weights of independent variables.

This has led to the thought that Evolutionary Algorithms could optimize the proposed model from other algorithms like the proposed model of the Genetic Programming, named GP-2. As a result, GSA provided a more accurate model.

For the initialization of population, Harmony Search (HS) algorithm [67] was used. Harmony Search alternates the vector of weights slightly and produces a population of solutions very close to the proposed one by GP. Then GSA was implemented

Table 11.8 Performances of GP-1, GP-2 and GSA-1 models [59]

	r	MAE (kV)	RMSE (kV)	RAE (%)
Model: GP-1				
Training	0.9979685	0.169555	0.2499	1.2901
Validation	0.999247	0.171671	0.2177	1.2332
Test	0.993614	0.276758	0.4538	3.5585
Model: GP-2				
Training	0.99506	0.32137	0.41889	2.5808
Test	0.99564	0.26586	0.29275	2.2182
Validation	0.98985	0.50008	0.62641	3.4887
Model: GSA-1				
Training	0.99728	0.33031	0.19303	–
Test	0.99958	0.03388	0.01541	–
Validation	0.99027	0.46951	0.38056	–

for 1000 iterations, given a population of 100 agents. The initial value of Gravity G_0 was 1000, the bandwidth of HS was 0.01, the Harmony Memory Consideration Rate was 0.8 and the Pitch Adjustment Rate was 0.3. The fold of data that produced the best model was used also as input here. In Eq. (11.34), the simpler model by GP can be seen, while in Eq. (11.35) is the optimized model by GSA-HS scheme. In Table 11.8 the performances of the models are presented under GP-2 and GSA-1 respectively.

$$U_c = (log(0.054012C))^2(1.298L + 1.4616D_m) \cdot 0.0053419 + 1.9647 \quad (11.34)$$

$$U_c = (log(0.0544783402089791C))^2 \cdot (1.29462312655721L + 1.46957461951861D_m) \cdot 0.0053419 + 1.9579433609812 \quad (11.35)$$

Three models based on GP were developed. These models show a high correlation between estimated and actual values for all sets, training, validation and test. The most accurate of all is the GP-1 model where r for the test set was 0.999247. However, the model is highly complex. Simpler Models have been developed with satisfactory precision such as that of GSA-1 with a corresponding r in the test set is equal to 0.99958 with the model being more straightforward than the mathematical model of Eq. (11.27).

11.4 Other Applications of Electric Power Systems

In this section a brief description is provided, of advances of AI techniques applied in two other important electric power engineering applications, namely load forecasting and optimization in transmission lines.

11.4.1 Load Forecasting

The short- and mid-term load forecasting in electric power systems is another important application of artificial intelligence in electric power engineering. As already mentioned before, the highest possible accuracy in forecasting the hourly, daily, or monthly, load profiles is of great importance for the power system scheduling in terms of power generation, unit commitment, availability of transfer capability, stability margins, and power control. Conventional methods for estimating the demand in electric power systems are based on many assumptions and approximations, while they have to include parameter relations that they cannot clearly describe. Artificial neural networks have achieved to provide concrete solutions to the above problems, rendering them as the predominant load forecasting technique [3].

In international literature of electric power systems a serious number of methods have been proposed for short-term load forecasting with various rates of approach to real data, such as ARMAX models [68], regression models [69] artificial neural networks (ANNs) [70], fuzzy modeling [71] etc. Some of the first studies for load forecasting in power system of Greece were that of Bakirtzis et al. [72, 73] and of Tsekouras et al. [74–76]. Researchers have proved that the use of classical multilayer ANNs trained with error back propagation algorithm (BP) results in small mean absolute percentage errors of the order of 1.5–2.5% [3], rendering ANNs the most effective load-forecasting method.

11.4.2 Lightning Performance Evaluation in Transmission Lines

Furthermore, ANNs have been used with great success for evaluating the lightning performance of overhead high-voltage transmission lines, which constitute a very important part of electric power systems. Noteworthy ANN models have been developed and trained with real experimental data collected from high-voltage transmission lines of 150 and 400 kV under service in the Hellenic electric power system [77–79]. More particularly, researchers developed feed-forward (FF) and radial basis function (RBF) neural networks for the estimation of total lightning failure N_T, shielding failure N_{SF}, and back flashover failure N_{BF} (variables for the output vector) on overhead transmission lines, using data for total footing resistance R, peak lightning current I_{peak}, lightning current derivative *di/dt*, and keraunic level T from Hellenic transmission lines as variables for the input vector. The proposed ANN models proved to be very efficient and accurate in estimation results of the target variables, providing compact distributed representations for complex datasets.

11.5 Conclusions and Further Research

In this chapter different intelligent approaches for modelling, generalization and knowledge extraction from data were presented, as they were applied in (a) ground resistance estimation, (b) estimation of critical flashover voltage on insulators, (c) load forecasting and (d) optimization tasks in transmission lines.

The intelligent techniques perform very well and with the aid of proper data carefully collected, can be used for the efficient modelling and construction of real world systems, related to grounding, insulators, load forecasting and power transmission lines. Neural networks (ANNs) prove a powerful tool for generating models from numerical data, inductive machine learning can generate highly comprehensible knowledge in the form of simple IF/THEN rules, and finally, genetic algorithms and genetic programming can utilize the evolution principles properly for obtaining high quality solutions to hard optimization problems in the abovementioned domains of application for electric power engineering.

The authors direct their research towards the implementation and application of other promising intelligent techniques such as support vector machines, fuzzy rule based systems (FRBS), as well as hybrid and adaptive intelligent algorithmic approaches for knowledge extraction, generalization and solution of hard optimization and forecasting problems. They also examine and study other versions of the problems under analysis, through the collection of additional and more specialized data, like (a) data for different types of ground in grounding systems modelling, (b) aggregation off of weather data and other similar parameters for analysis in grounding systems, (c) data collection for different types of insulators in collaboration to authorities that cope with the problem of flashover voltage in real world conditions, (d) specialized load forecasting optimization problems, etc.

In addition, other important electric power engineering applications of AI techniques are considered (sometimes on the interface and with mechanical engineering areas in power systems), such as:

- Fault diagnosis and reliability (network faults, imbalances, frequency deviation, network restoration, stability analysis and enhancement, etc.)
- Control of power systems (stability control, power flow control, etc.)
- Energy management systems
- Alarm analysis and protection systems
- Distribution systems applications
- Demand side management and demand response
- Reactive power planning
- Behavior prediction of power systems
- Power system operation and monitoring (unit commitment, economic dispatch, hydro thermal coordination, maintenance scheduling, congestion management, etc.)
- Electric markets (bidding, market analysis, etc.)

Finally, another interesting point for researchers would be to observe the degree to which proposed methods and research findings are adopted in practice and become routine in the abovementioned domains of application.

References

1. Std 80–2013: IEEE guide for safety in AC substation grounding. IEEE Std 80-2013 Cor 1-2015 (2015)
2. S. 81–2012: Std A.E.E.E.: IEEE guide for measuring earth resistivity, ground impedance, and earth surface potentials of a grounding system. IEEE Std 81-2012 (2012)
3. Tsekouras, G.J., Kanellos, F.D., Mastorakis, N.: Short term load forecasting in electric power systems with artificial neural networks. In: Mastorakis, N., Bulucea, A., Tsekouras, G. (eds.) Computational Problems in Science and Engineering, vol. 343, pp. 19–58, Cham, Springer International Publishing (2015)
4. Androvitsaneas, V.P., Gonos, I.F., Stathopulos, I.A.: Experimental study on transient impedance of grounding rods encased in ground enhancing compounds. Electr. Power Syst. Res. **139**, 109–115 (2016)
5. Banton, O., Cimon, M.-A., Seguin, M.-K.: Mapping field-scale physical properties of soil with electrical resistivity. Soil Sci. Soc. Am. J. **61**(4), 1010–1017 (1997)
6. Gonos, I.F., Stathopulos, I.A.: Estimation of multilayer soil parameters using genetic algorithms. IEEE Trans. Power Deliv. **20**(1), 100–106 (2005)
7. Gonos, I.F., Moronis, A.X., Stathopulos, I.A.: Variation of soil resistivity and ground resistance during the year. In: Presented at the Proceedings of the 28th International Conference on Lightning Protection (ICLP 2006), pp. 740–744. Kanazawa, Japan (2006)
8. Sudha, K., Israil, M., Mittal, S., Rai, J.: Soil characterization using electrical resistivity tomography and geotechnical investigations. J. Appl. Geophys. **67**(1), 74–79 (2009)
9. Tagg, G.F.: Earth resistances. George Newnes Limited, London (1964)
10. Ahmad, W.F.W., Rahman, M.S.A., Jasni, J., Ab Kadir, M.Z.A., Hizam, H.: Chemical enhancement materials for grounding purposes. In: Proceedings of ICLP, pp. 1233–1241 (2010)
11. Galván, A.D., Pretelin, G.G., Gaona, E.E.: Practical evaluation of ground enhancing compounds for high soil resistivities. In: Presented at the Proceedings of ICLP, pp. 1233–1241 (2010)
12. Jasni, J., Siow, L.K., Ab Kadir, M.A., Ahmad, W.W.: Natural Materials as Grounding Filler For Lightning Protection System, pp. 1101–1111 (2010)
13. Gomes, C., Lalitha, C., Priyadarshanee, C.: Improvement of earthing systems with backfill materials. In: 2010 30th International Conference Lightning Protection (ICLP). Cagliary, Italy (2010)
14. Androvitsaneas, V.P., Gonos, I.F., Stathopulos, I.A.: Performance of ground enhancing compounds during the year, in 2012 Int. Conf, Lightning Protection (ICLP) (2012)
15. Boulas, K., Androvitsaneas, V.P., Gonos, I.F., Dounias, G., Stathopulos, I.A.: Ground resistance estimation using genetic programming. In: 5th International Symposium and 27th National Conference Operation Research, pp. 66–71. Aigaleo, Athens (2016)
16. Blattner, C.J.: Prediction of soil resistivity and ground rod resistance for deep ground electrodes. IEEE Trans. Power Appar. Syst. **PAS-99**(5): 1758–1763 (1980)
17. Salam, M.A., Al-Alawi, S.M., Maqrashi, A.A.: An artificial neural networks approach to model and predict the relationship between the grounding resistance and length of buried electrode in the soil. J. Electrost. **64**(5), 338–342 (2006)
18. Gouda, O., Amer, M., El Saied, M.: Optimum design of grounding systems in uniform and non-uniform soils using ANN. Int. J. Soft Comput. **1**(3), 175–180 (2006)
19. Asimakopoulou, F.E., Kourni, E.A., Kontargyri, V.T., Tsekouras, G.J., Stathopulos, I.A.: Artificial neural network methodology for the estimation of ground resistance. In: Presented at the

15th WSEAS International Conference on Systems 2011, pp. 453–458. Corfu Island, Greece (2011)
20. Asimakopoulou, F.E., Tsekouras, G.J., Gonos, I.F., Stathopulos, I.A.: Estimation of seasonal variation of ground resistance using artificial neural networks. Electr. Power Syst. Res. **94**, 113–121 (2013)
21. Androvitsaneas, V.P., Asimakopoulou, F.E., Gonos, I.F., Stathopulos, I.A.: Estimation of ground enhancing compound performance using artificial neural network. In: Paper Presented at the 2012 International Conference on High Voltage Engineering and Application (ICHVE), pp. 145–149. Shanghai, China (2012)
22. Androvitsaneas, V.P., Gonos, I.F., Stathopulos, I.A.: Artificial neural network methodology for the estimation of ground enhancing compounds resistance. IET Sci. Meas. Technol. **8**(6), 552–570 (2014)
23. Androvitsaneas, V.P., Tsekouras, G.J., Gonos, I.F., Stathopulos, I.A.: Design of an artificial neural network for ground resistance forecasting. In: Proceedings of 9th Mediterranean Conference on Power Generation, Transmission, Distribution and Energy Conversion (Med Power 2014), Athens, Greece (2014)
24. Cao, L., Hong, Y., Fang, H., He, G.: Predicting chaotic time series with wavelet networks. Phys. Nonlinear Phenom. **85**(1), 225–238 (1995)
25. Fang, Y., Chow, T.W.S.: Wavelets based neural network for function approximation. In: Advances in Neural Networks—ISNN 2006, vol. 3971, pp. 80–85 (2006)
26. Alexandridis, A.K., Zapranis, A.D.: Wavelet neural networks: with applications in financial engineering, chaos, and classification. Wiley, New Jersey (2014)
27. Androvitsaneas, V.P., Alexandridis, A.K., Gonos, I.F., Dounias, G.D., Stathopulos, I.A.: Wavelet neural network methodology for ground resistance forecasting. Electr. Power Syst. Res. **140**, 288–295 (2016)
28. Zhang, Q.: Using wavelet network in nonparametric estimation. IEEE Trans. Neural Netw. **8**(2), 227–236 (1997)
29. Androvitsaneas, V., Gonos, I., Dounias, G., Stathopulos, I.: Ground resistance estimation using inductive machine learning. In: Presented at the the 19th International Symposium on High Voltage Engineering, Pilsen Czech Republic (2015)
30. Mitchell, T.M.: Machine Learning, vol. 1, McGraw-Hill (1997)
31. Quinlan, J.R.: Induction of decision trees. Mach. Learn. **1**(1), 81–106 (1986)
32. Quinlan, R.J.: C4.5: Programs for Machine Learning. San Mateo, California, USA: Morgan Kaufmann (1993)
33. Androvitsaneas, V.P.: Contribution to Behavioral Study of Grounding Systems Encased in Ground Enhancing Compounds, PhD Thesis, NTUA, Athens, Greece, (in Greek) (2016)
34. Ferreira, C.: Gene Expression Programming: Mathematical Modeling by an Artificial Intelligence. Springer, Berlin, Heidelberg (2006)
35. Looms, J.: Insulators for high voltages. IET (1988)
36. Mackevich, J., Shah, M.: Polymer outdoor insulating materials. Part I: Comparison of porcelain and polymer electrical insulation. IEEE Electr. Insul. Mag. **13**(3), 5–12 (1997)
37. Obenaus, F.: Fremdschichtueberschlag und kriechweglaenge. Dtsch. Elektrotechnik **4**, 135–136 (1958)
38. Topalis, F.V., Gonos, I.F., Stathopulos, I.A.: Dielectric behaviour of polluted porcelain insulators. IEE Proc. Gener. Transm. Distrib. **148**(4), 269–274(5) (2001)
39. International Electrotechnical Commission: Artificial pollution tests on high voltage insulators to be used on AC systems. Int. Stand. IEC **507** (1991)
40. Ikonomou, K., Katsibokis, G., Panos, G., Stathopoulos: Cool fog tests on artificially polluted insulators. In: Presented at the 5th International Symposium on High Voltage Engineering, Braunschweig, vol. 2, p. paper 52.13 (1987)
41. Guan, Z., Zhang, R.: Calculation of DC and AC flashover voltage of polluted insulators. IEEE Trans. Electr. Insul. **25**(4), 723–729 (1990)
42. Sundararajan, R., Sadhureddy, N.R., Gorur, R.S.: Computer-aided design of porcelain insulators under polluted conditions. IEEE Trans. Dielectr. Electr. Insul. **2**(1), 121–127 (1995)

43. Goldberg, D.E.: Genetic Algorithms in Search Optimization and Machine Learning (1989)
44. Holland, J.H.: Adaptation in Natural and Artificial Systems: An Introductory Analysis with Applications to Biology, Control, and Artificial Intelligence, MIT Press (1992)
45. Kramer, O.: Genetic Algorithm Essentials, vol. 679. Springer (2017)
46. Gonos, I.F., Topalis, F.V., Stathopolos, I.A.: Genetic algorithm approach to the modelling of polluted insulators. IEE Proc. Gener. Transm. Distrib. **149**(3), 373–376(3) (2002)
47. Rizk, F.A.: Mathematical models for pollution flashover. Electra **78**(5), 71–103 (1981)
48. Ghosh, P., Chatterjee, N.: Polluted insulator flashover model for AC voltage. IEEE Trans. Dielectr. Electr. Insul. **2**(1), 128–136 (1995)
49. Kontargyri, V.T., Gialketsi, A.A., Tsekouras, G.J., Gonos, I.F., Stathopulos, I.A.: Design of an artificial neural network for the estimation of the flashover voltage on insulators. Electr. Power Syst. Res. **77**(12), 1532–1540 (2007)
50. Ghosh, P., Chakravorti, S., Chatterjee, N.: Estimation of time-to-flashover characteristics of contaminated electrolytic surfaces using a neural network. IEEE Trans. Dielectr. Electr. Insul. **2**(6), 1064–1074 (1995)
51. Cline, P., Lannes, W., Richards, G.: Use of pollution monitors with a neural network to predict insulator flashover. Electr. Power Syst. Res. **42**(1), 27–33 (1997)
52. Ugur, M., Auckland, D.W., Varlow, B.R., Emin, Z.: Neural networks to analyze surface tracking on solid insulators. IEEE Trans. Dielectr. Electr. Insul. **4**(6), 763–766 (1997)
53. Dixit, P., Gopal, H.G.: ANN based three stage classification of arc gradient of contaminated porcelain insulators. In: Proceedings of the 2004 IEEE International Conference on Solid Dielectrics, 2004. ICSD 2004, vol. 1, pp. 427–430. Toulouse, France (2004)
54. Ghosh, S., Kishore, N.: Modeling PD inception voltage of epoxy resin post insulators using an adaptive neural network. IEEE Trans. Dielectr. Electr. Insul. **6**(1), 131–134 (1999)
55. Jahromi, A.N., El-Hag, A.H., Cherney, E.A., Jayaram, S.H., Sanaye-Pasand, M., Mohseni, H.: Prediction of leakage current of composite insulators in salt fog test using neural network. In: CEIDP'05. 2005 Annual Report Conference on Electrical Insulation and Dielectric Phenomena, pp. 309–312 (2005)
56. Jahromi, A.N., El-Hag, A.H., Jayaram, S.H., Cherney, E.A., Sanaye-Pasand, M., Mohseni, H.: A neural network based method for leakage current prediction of polymeric insulators. IEEE Trans. Power Deliv. **21**(1), 506–507 (2006)
57. da Silva, A.P.A., Moulin, L.S.: Confidence intervals for neural network based short-term load forecasting. IEEE Trans. Power Syst. **15**(4), 1191–1196 (2000)
58. Asimakopoulou, G., Kontargyri, V., Tsekouras, G., Asimakopoulou, F., Gonos, I., Stathopulos, I.: Artificial neural network optimisation methodology for the estimation of the critical flashover voltage on insulators. IET Sci. Meas. Technol. **3**(1), 90–104 (2009)
59. Karampotsis, E. et al.: Computational intelligence techniques for modelling the critical flashover voltage of insulators: from accuracy to comprehensibility. In: Advances in Artificial Intelligence: From Theory to Practice: 30th International Conference on Industrial Engineering and Other Applications of Applied Intelligent Systems, IEA/AIE 2017, Arras, France, June 27–30, In: Benferhat, S., Tabia, K., Ali, M. (eds.) Proceedings, Part I, pp. 295–301 Cham, Springer International Publishing (2017)
60. Popescu, M.-C., Balas, V.E., Perescu-Popescu, L., Mastorakis, N.: Multilayer perceptron and neural networks. WSEAS Trans. Circuits Syst. **8**(7), 579–588 (2009)
61. Ganatra, A., Kosta, Y., Panchal, G., Gajjar, C.: Initial classification through back propagation in a neural network following optimization through GA to evaluate the fitness of an algorithm. Int. J. Comput. Sci. Inf. Technol. **3**(1), 98–116 (2011)
62. Refaeilzadeh, P., Tang, L., Liu, H.: Cross-validation. In: LIU, L., ÖZSU, M.T. (eds.) Encyclopedia of database systems, pp. 532–538. Boston, MA, Springer, US (2009)
63. Androvitsaneas, V.P., Karampotsis, E., Gonos, I.F., Dounias, G., Stathopolos, I.A.: Critical Flashover Voltage on Polluted Insulators Estimated Using Conventional and Intelligent Techniques. In: Presented at the 20th International Symposium on High-Voltage Technology (ISH). Buenos Aires, Argentina (2017)

64. Koza, J.R.: Genetic Programming: on the Programming of Computers by Means of Natural Selection, vol. 1. MIT Press (1992)
65. Crane, E.F., McPhee, N.F.: The effects of size and depth limits on tree based genetic programming. In: Yu, T., Riolo, R., Worzel, B. (eds.) Genetic Programming Theory and Practice III, pp. 223–240. Boston, MA, Springer, US (2006)
66. Rashedi, E., Nezamabadi-Pour, H., Saryazdi, S.: GSA: a gravitational search algorithm. Inf. Sci. **179**(13), 2232–2248 (2009)
67. Geem, Z.W., Kim, J.H., Loganathan, G.V.: A new heuristic optimization algorithm: harmony search. Simulation **76**(2), 60–68 (2001)
68. Yang, H.-T., Huang, C.-M., Huang, C.-L.: Identification of ARMAX model for short term load forecasting: an evolutionary programming approach. IEEE Trans. Power Syst. **11**(1), 403–408 (1996)
69. Haida, T., Muto, S.: Regression based peak load forecasting using a transformation technique. IEEE Trans. Power Syst. **9**(4), 1788–1794 (1994)
70. Hippert, H.S., Pedreira, C.E., Souza, R.C.: Neural networks for short-term load forecasting: a review and evaluation. IEEE Trans. Power Syst. **16**(1), 44–55 (2001)
71. Mastorocostas, P.A., Theocharis, J.B., Bakirtzis, A.G.: Fuzzy modeling for short term load forecasting using the orthogonal least squares method. IEEE Trans. Power Syst. **14**(1), 29–36 (1999)
72. Bakirtzis, A.G., Petridis, V., Kiartzis, S.J., Alexiadis, M.C., Maissis, A.H.: A neural network short term load forecasting model for the Greek power system. IEEE Trans. Power Syst. **11**(2), 858–863 (1996)
73. Kiartzis, S.J., Zoumas, C.E., Theocharis, J.B., Bakirtzis, A.G., Petridis, V.: Short-term load forecasting in an autonomous power system using artificial neural networks. IEEE Trans. Power Syst. **12**(4), 1591–1596 (1997)
74. Elias, C.N., Tsekouras, G., Kavatza, S., Contaxis, G.: A midterm energy forecasting method using fuzzy logic. WSEAS Trans. Syst. **3**(5), 2128–2135 (2004)
75. Tsekouras, G.J., Hatziargyriou, N.D., Dialynas, E.N.: An optimized adaptive neural network for annual midterm energy forecasting. IEEE Trans. Power Syst. **21**(1), 385–391 (2006)
76. Tsekouras, G. et al.: A comparison of artificial neural networks algorithms for short term load forecasting in Greek intercontinental power system. In: Presented at the WSEAS International Conference on Circuits, Systems, Electronics, Control & Signal Processing, Canary Islands, Spain, pp. 15–17 (2008)
77. Ekonomou, L., Iracleous, D., Gonos, I., Stathopulos, I.: Lightning performance identification of high voltage transmission lines using artificial neural networks. Eng. Intell. Syst. Electr. Eng. Commun. **13**(3), 219–223 (2005)
78. Ekonomou, L., Liatsis, P., Gonos, I.F., Stathopulos, I.A.: Artificial neural network-based software tool for calculating the lightning performance of high-voltage transmission lines. IEE Proc. Sci. Meas. Technol. **153**(5), 188–193(5) (2006)
79. Ekonomou, L., Gonos, I.F., Iracleous, D.P., Stathopulos, I.A.: Application of artificial neural network methods for the lightning performance evaluation of Hellenic high voltage transmission lines. Electr. Power Syst. Res. **77**(1):55–63 (2007)

Part IV
Data Analytics for Digital Forensics

Chapter 12
Combining Genetic Algorithms and Neural Networks for File Forgery Detection

Konstantinos Karampidis, Ioannis Deligiannis and Giorgos Papadourakis

Abstract Today's electronic devices are so ubiquitous that the collection and use of digital evidence has become a standard part of many criminal and civil investigations. The uncovering and examination of those shreds of evidence is a relatively new and important process to provide crucial information in a court of law. Suspects routinely have their laptops and cell phones examined for corroborating evidence. However, digital forensic investigators are facing several challenges such as file obfuscation, encryption, alteration and a massive amount of evidence. These challenges often lead to incomplete analysis and inadequate conclusions. Consequently, a digital forensic examiner uses specialized forensic software to accurately identify the file types to determine which of them may contain potential evidence.

12.1 Introduction

The 21st century has seen a dramatic increase in new and ever-evolving technologies available to consumers and industry alike. Generally, the consumer—level user base is now more adept and knowledgeable about what technologies they employ in their day-to-day lives. The field of digital forensics has grown increasingly over the last few years as both the computer and the cellular market has grown. This inevitable technological progress is followed up by numerous cyber-attacks carried

---

K. Karampidis (✉)
Department of Information & Communication Systems Engineering,
University of the Aegean, 83200 Karlovasi, Samos, Greece
e-mail: karampidis@aegean.gr

I. Deligiannis
Department of Cultural Heritage Management and New Technologies,
University of Patras, G. Seferi 2, 30100 Agrinio, Greece
e-mail: i_-_-_s@outlook.com

G. Papadourakis
Department of Informatics Engineering, Technological Educational Institute of Crete,
71500 Heraklion, Crete, Greece
e-mail: papadour@cs.teicrete.gr

G. A. Tsihrintzis et al. (eds.), *Machine Learning Paradigms*, Intelligent Systems Reference Library 149, https://doi.org/10.1007/978-3-319-94030-4_12

out by criminals who have the appropriate technical background. For that reason, computer forensics has become an essential part of the litigation process, and electronic evidence plays an increasingly vital role in the Court's to prove or disprove an individual's actions to secure a conviction. Nevertheless, most people immediately think that the only sources for digital evidence are computers, cell phones, and the Internet. The truth behind that is that almost any piece of technology that processes information can be used criminally. For example, hand-held games can carry messages between criminals and even newer household appliances, such as a refrigerator with a built-in TV, could be used to store, view and share illegal images. The important thing to know is that responders need to be able to recognize and properly seize potential digital evidence.

However, the identification and preservation of evidence in digital forensic investigations in emerging environments have always presented a challenge. To develop an understanding of these problems and their place in the digital forensic process we could consider two predominant models for digital forensic investigations proposed by McKemmish [1] and NIST [2].

12.1.1 McKemmish Predominant Model

- Identification: in this stage the location and format of evidence is identified to enable an appropriate mechanism to be determined for the purpose of recovering evidence. Digital evidence can be found in a myriad of places; computers, mobile phones, smart cards, set top boxes etc.
- Preservation: it is imperative that evidence is preserved as in many cases it will be the subject of judicial scrutiny. In some circumstances changes to data are unavoidable. In these cases, change should be minimized and the process causing the change documented along with an explanation/justification of why the change was required.
- Analysis: consists of the extraction, processing and interpretation of digital evidence. It forms the main element of forensic computing. Following extraction, processing is often required to make data human readable. Processing of extracted data may be part of the extraction stage or a separate stage in its own right.
- Presentation: the final stage of the process involves a presentation of both the evidence and the process by which the evidence was gathered along with the presenter's qualifications.

12.1.2 Kent Predominant Model

- Collection: encompasses identification, preservation and acquisition of relevant evidence.
- Examination: uses automated and manual tools to extract data of interest.

- Analysis: the derivation of useful information from the results of the examination stage.
- Reporting: is concerned with the preparation and presentation of the evidence and forensic analysis process.

12.1.3 Digital Evidences

Both the previously mentioned models result to digital evidences which could be defined as valuable information in binary form that could be received or transmitted by an electronic device and may be relied on in court. These evidences can be acquired when electronic devices are seized and secured for examination and may have one or more of the following properties:

- Can only be seen, understood, analyzed, and presented with and through tools.
- Sometimes exists for very short time periods.
- Easily destroyed or modified.
- Easily mishandled.
- Patterns of information combine to provide substance.
- Easily misinterpreted.
- Often misleading or patently false.

Information that is stored electronically is said to be 'digital' because it is literally stored in a form of digits; binary units of ones (1) and zeros (0), that are saved and retrieved using a set of instructions called software or code. Any information, photographs, words, spreadsheets can be created and saved using these types of instructions. Finding and exploiting evidence saved in this way is a growing area of forensics and constantly changes as the technology evolves. Digital evidence may come into play in any serious criminal investigation such as murder, rape, stalking, carjacking, burglary, child abuse or exploitation, counterfeiting, extortion, gambling, piracy, property crimes and terrorism. Pre and post-crime information are most relevant, for example, if a criminal was using an online program like Google Maps™ or street view to case a property before a crime; or posting stolen items for sale on Craigslist or eBay; or communicating via text message with accomplices to plan a crime or threaten a person. Some crimes can be committed entirely through digital means, such as computer hacking, economic fraud or identity theft. In any of these situations, an electronic trail of information is left behind for a savvy investigation team to recognize, seize and exploit. As with any evidence gathering, following proper procedures is crucial and will yield the most valuable data. Not following proper procedures can result in lost or damaged evidence or rendering it inadmissible in court.

Computer documents, emails, text and instant messages, transactions, images and Internet history are examples of information that can be gathered from electronic devices and used efficiently as evidence. For example, mobile devices use online based backup systems, also known as the 'cloud', that provide forensic investigators with access to text messages and pictures taken from a particular phone. These systems keep an average 2000 or more of the last text messages received from that phone. Moreover, many mobile devices store information about the location history of the device. To gain this knowledge, investigators can access almost the last 200 cell locations accessed by the mobile device. Satellite navigation systems and satellite radios in cars can provide similar information. Even photos posted to social media such as Facebook may contain location information. Photos taken with a Global Positioning System (GPS) enabled on the device, contain file metadata that shows when and exactly where a photo was taken. By gaining a subpoena for a mobile device account, investigators can retrieve a great deal of history related to a device and the person using it.

There are a few common misperceptions about the retrieval and usefulness of digital evidence, including:

- Anything on a hard drive or other electronic media can always be retrieved. This is incorrect as overwritten or damaged files, or physical damage to the media can render it unreadable. Highly specialized laboratories with clean rooms may be able to examine hard drive components and reconstruct data, but this process is very laborious and extremely expensive.
- Decrypting a password is quick and easy, with the right software. With the increasing complexity of passwords including capitals, numbers, symbols and password length, there are billions of potential passwords. Decryption can take a great deal of time, up to a year in some cases, using system resources and holding up investigations. Gathering passwords from those involved in a case is much more efficient and should be done whenever possible.
- Any digital image can be refined to high definition quality. Images can be very useful for investigations, but a low-resolution image is made by capturing fewer bits of data (pixels) than higher resolution photos. Pixels that are not there in the first place cannot be refined. Investigators can look at digital evidence at the crime scene or any time.
- Just looking at a file list does not damage the evidence. It is crucial to note that opening, viewing or clicking on files can severely damage forensic information because it can change the last access date of a file or a piece of hardware. This changes the profile and can be considered tampering with evidence or even render it completely inadmissible. Only investigators with the proper tools and training should be viewing and retrieving evidence.
- First responder training lags advancements in electronics. Without regular updates to their training, responders may not be aware of what new digital devices might be in use and subject to collection. For example, there should be an awareness that thumb drives and SD cards can be easily removed and discarded by a suspect during an encounter with law enforcement.

12.1.4 File Type Identification

If evidence collection and analysis is conducted properly, investigators can secure information that can assist criminal activity claims through dialog or message exchange, images and documents. The examiner will provide all the supporting documentation, by highlighting significant information and delivering a report describing what was done to extract the data. As with evidence of other types, a chain of custody and proper collection and extraction techniques are critical to the credibility of evidence and must be thoroughly documented. However, there are several limitations which could lead to an unsuccessful investigation. For example, encryption and proprietary systems that require decoding before data can even be accessed as well as the usually enormous amount of data that must be examined. To solve those issues, there are several techniques that could automate the process such as password cracking tools and some more sophisticated software which are able to identify suspicious files.

There is always the likelihood that files may be altered in order to hide evidence. Files must also have checked whether are 'genuine' or not. A file format is the blueprint of a file. It tells the processing device (e.g. a computer) how data within a file are organized and specifies the way the information is encoded in a digital storage medium. File formats may be either proprietary e.g. .dwg for an Autocad file, free which is not burdened by any copyrights, patents or other restrictions, or open which anyone can read and study but it may be burdened by restrictions on use. One popular method used by many operating systems, including Windows—the most popular operating system among computer end users—is to determine the format of a file based on the end of its name, the letters following the final period. This is known as the filename extension. For example, text documents are identified by names that end with .doc (or .docx), and PNG images by .png. In the original FAT filesystem, file names were limited to an eight-character identifier and a three-character extension, known as an 8.3 filename (also called a short filename or SFN). Many formats still use three-character extensions even though modern operating systems and applications no longer have this constraint. Some file formats are designed for very particular types of data e.g. doc or docx stands for document files, jpg declares a compressed picture etc., while png extension relates to images using lossless data compression. Nevertheless, other file formats are intended for storage of several different types of data: the flash video (flv, f4v) format can act as a container for video and audio from Adobe Systems. There are thousands of file formats and the list is getting bigger day by day. Since there is no standard list of extensions and given the fact that more than one format can use the same extension, this could lead to confuse both the operating system and end users. From a user's perspective this confusion might be just ignorance or could hide deceit.

Criminals usually attempt to hide their activity/traces by altering file's extension, file's signature (magic bytes) or a combination of both. The first method i.e. altering file extension can be easily detected by specialized forensic software. The second method of file type identification is based on the magic bytes. Magic bytes are

Table 12.1 A list of some widely used file types and their file signatures

File type	Signature
DOC	D0 CF 11 E0 A1 B1 1A E1
FLV	46 4C 56 01
PDF	25 50 44 46
JFIF, JPE, JPEG, JPG	FF D8 FF E0 xx xx 4A 46 49 46 00
MP3 audio file	49 44 33
PNG	89 50 4E 47 0D 0A 1A 0A
RAR (v5) compressed archive file	52 61 72 21 1A 07 01 00
MS Windows/DOS executable file (EXE)	4D 5A
GIF87a	47 49 46 38 37 61
GIF89a	47 49 46 38 39 61

predefined signatures found on file's header. A list of some widely used file types and their file signatures can be found in Table 12.1. A file header is the first portion of a computer file that contains metadata. Metadata may enclose information about the content, quality and condition of the file. The file header also contains necessary information for the corresponding application to recognize and understand the file. Magic bytes may also include some extra information regarding the tool and the tool's version that is used to produce the file. Kessler [3] started in 2002 to record file signatures and right now over 5000 known file types are identified. Checking the magic bytes of a file is indeed much slower method than just checking its extension since the file should be opened—usually in a standalone or in build hex editor—and its magic bytes should be read and compared with the predefined ones. Magic bytes method is adopted by many UNIX based operating systems and file type can be easily found by typing in a terminal the 'file' command. However, this method of identifying a file type has also weaknesses as the extension-based method:

- The magic bytes are not used in all file types.
- They only work on the binary files and are not an enforced or regulated aspect of the file types.
- They vary in length for different file types and do not always give a very specific answer.

There are several thousands of file types for which magic bytes are defined and there are multiple lists of magic bytes that are not completely consistent. Since there is not any standard for what a file may contain, the creators of a new file type usually include something to uniquely identify their file type. It is common that some programs or their developers may never put any magic bytes at the beginning of the file header. To identify if a file has altered signature, forensic software compares file's extension and its signature. If file's signature is altered the software highlights the file as a mismatch between extension and signature.

The most difficult case a forensic investigator may meet is the likelihood that a file has altered signature and extension at the same time. The only way probably to tell if the file is forged is to examine file's content. It can reveal the malicious file types that their contents do not match with their claimed types. The contents of a file are a sequence of bytes and a byte has 256 unique permutations (0–255). Thus, counting the occurrence of byte patterns that is often referred as byte frequency distribution, gives distinguishable patterns to identify file types. There are many content-based file type identification schemes that use byte frequency distribution to build the representative models for file type and use any statistical and data mining techniques to identify file types.

McDaniel and Heydari [4, 5], were the first who actually suggested a way for content-based file type detection. They proposed three different algorithms for the content-based file type detection. The accuracy varied from 23 to 96% depending upon the algorithm used. Li et al. [6] made a few changes on McDaniel's and Heydari's method, in order to improve its accuracy. They proposed to compute a set of centroid models and use clustering to find a minimal set of centroids with good performance while the use of more pattern data is necessary. This approach resulted to 82% accuracy (one centroid), 89.5% accuracy (multi-centroid) and 93.8% accuracy (more exemplar files). Dunham et al. [7] used neural networks for classification and achieved 91.3% accuracy. Amirani et al. [8] used the Principal Component Analysis and unsupervised neural networks for the automatic feature extraction. The classifier they used was a neural network, achieving an accuracy of 98.33% which was the best so far. Cao et al. [9] used Gram Frequency Distribution and vector space model with results of 90.34% accuracy. Ahmed et al. [10] proposed two very interesting methods. Primary they used the cosine distance as a similarity metric when comparing the file content. Subsequent they decomposed the identification procedure into two steps. They used 2000 files of 10 file types as a dataset and achieved an accuracy of 90.19%. Ahmed et al. [11] also proposed two new techniques to reduce the classification time. The first method was a feature selection technique and the K-nearest neighbor (KNN) classifier was used. The second method was the content sampling technique, which used a small portion of a file to obtain its byte-frequency distribution. Amirani et al. [12] then proposed an improved version of their first approach by using a Support Vector Machine classifier and finally succeeded in raising the accuracy of the method to 99.16%. Finally, Evensen et al. [13] used an n-gram analysis with naïve Bayes classifier to a large dataset of 60,000 files (6 file types) with very good results achieving 99.61% topmost. The earlier methods showed by poor to excellent results in file type identification, but the actual problem during a forensic analysis relies on the alteration of file's signature and its extension at the same time. When this occurs the majority—if not all—of the forensic software cannot recognize correctly the file type.

12.2 Methodology of the Proposed Method

The evaluation of the earlier mentioned methodologies showed poor to excellent results in file type identification, but the actual problem during a forensic analysis relies on the alteration of file's signature and its extension at the same time. When this occurs the majority—if not all—of the forensic software cannot recognize correctly the file type. Initially all files from the dataset are loaded and the features are extracted. Afterwards, feature selection is accomplished using a genetic algorithm and finally a neural network performs the classification. Byte Frequency Distribution (BFD) is used as a feature extraction method. In order to create the BFD, the number of occurrences of each byte value in an input file is counted and an array with elements from 0 to 255 is created. Then each element of the array is normalized by dividing with the maximum occurrence. The final result is a file containing 256 features for each instance. The next stage is feature selection, in order to decrease the number of features. Feature selection is the procedure of finding and selecting the minimum number of the most informative relevant features (Fig. 12.1).

As a search method a genetic algorithm was used. The idea of using a genetic algorithm, for feature extraction is not new [14–16], since they can provide candidate solutions. Each candidate solution (chromosome) is represented by a binary feature vector of dimension 256, where zero (0) indicates that the respective feature is not selected, and one (1) indicates that the feature is selected. The score of each candidate solution is evaluated by a fitness function. As a fitness function the Correlation based Feature Selection (CFS) [17] algorithm is utilized. This algorithm evaluates the candidate solutions from the genetic algorithm and choses those which include features highly associated to the file type category and low correlated with each other, by calculating each candidate's solution merit. Let S be a candidate solution consisting of k features. The merit of each candidate solution is calculated by Eq. 12.1.

$$Merits_k = \frac{k\overline{r_{cf}}}{\sqrt{k + k(k-1)\overline{r_{ff}}}} \tag{12.1}$$

where:

$\overline{r_{cf}}$ is the average value of all feature-classification correlations and
$\overline{r_{ff}}$ is the average value of all feature-feature correlations.

CFS stops when five consecutive fully expanded candidate solutions show no improvement [17]. The utilization of the genetic algorithm as a search method and CFS as an evaluator led to the reduction of the 256 extracted features to 44. The third and final stage is classification, performed with a one hidden layer neural network using the backpropagation algorithm. A neural network with one hidden layer was also used by Harris [18] in order to identify file types. Initially, the data are separated into a training set (70%) and a test set (30%).

Furthermore, in order to estimate the accuracy of classification during the training phase a stratified 10 fold cross validation is used [19]. Subsequently, unseen instances from all categories are presented to the model for evaluation.

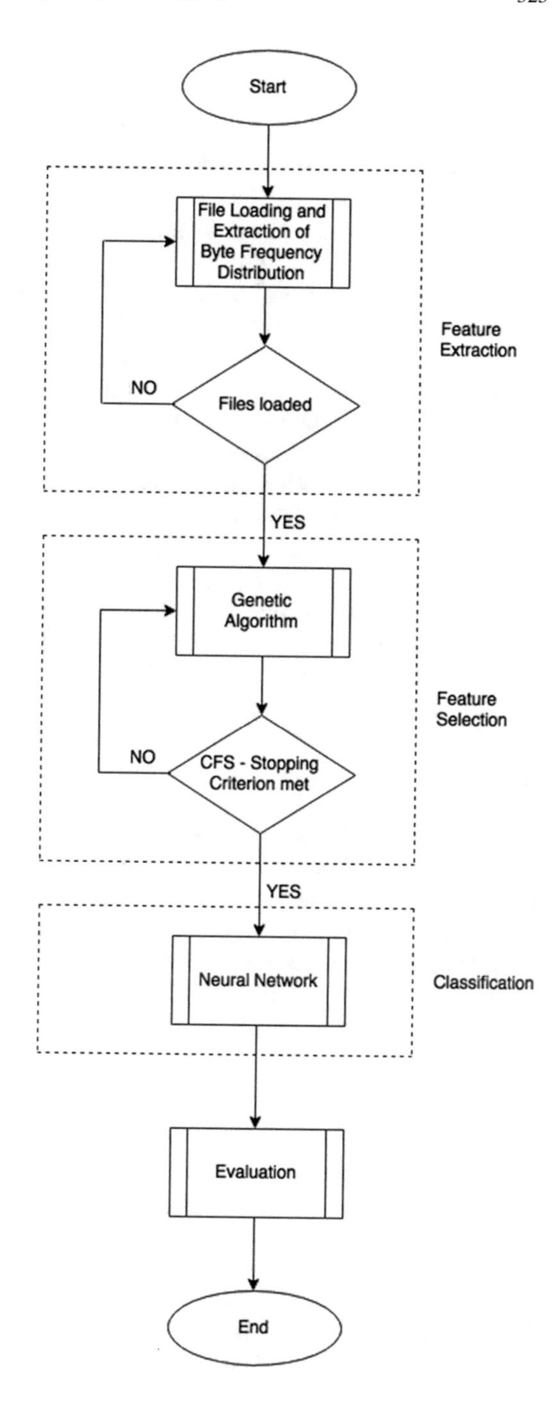

Fig. 12.1 Flowchart of the proposed method

12.3 Experimental Setup and Results

Due to thousands of known file types, this research has focused only in images and portable documents, because of their significance to Digital Forensics. In particular, this research only included jpeg png, gif (not animated), tiff and pdf files. Furthermore, only whole files and not fragments of files were examined. Caltech 101 [20] was used as dataset. It is a dataset made by Caltech University containing 9144 images in jpeg format from 101 categories. These images are in 101 subfolders each one representing one category, therefore the images were extracted to a single folder. From this jpeg dataset, 5519 images were utilized. Afterwards, these images were renamed (image 0001 to image 9144) and one third of them were converted to png format and a similar number to gif format. Total Image Converter [21] (free trial version) was used, in order to convert those files. There were no alterations to converted files regarding size, rotation, crop, further compression, filtering, transparency or watermark embedded. Table 12.2 shows the conversion parameters used.

The dataset was divided into a training set (70%) and a test set (30%) and the exact numbers of any file type used from Caltech dataset in both sets, are shown in Table 12.3. Additionally, 1840 pdf files were added, which were open access undergraduate theses found online from the library of the Technological Educational Institute of Crete [22]. The final dataset is uniformly distributed, and its exact numbers are indicated in Table 12.4. In order to examine if the proposed methodology identifies the correct file type if the file is altered, one third of the testing pdf files (168) were replaced by image files of which their extension and signature was changed to pdf.

More specifically, the extension of 168 jpeg images was changed from .jpg to .pdf. Also, with a hex editor the signature of each jpeg image was also changed. The same procedure was performed to png and gif images and therefore three new test sets were created. The first contained 168 altered files of jpeg format, the second contained 168 files of png format and the third contained 168 files of gif format. Table 12.5 shows the changes made to the 168 files in each new test set. A script written in MATLAB® [23] was implemented to create the BFD containing 256 features. Waikato Environment

Table 12.2 Conversion parameters

From jpg	To png	To gif
Image size changed	No	No
Rotation of the image	No	No
The image was cropped	No	No
Image was compressed	No	–
Any filter applied	No	–
Transparency of source file used	Yes	Yes
Watermark embedded	No	No

Table 12.3 Images utilized from Caltech dataset

Type	Training Set		Test Set	
	Number of images	Image number	Number of images	Image number
jpg	1288	0001–1288	552	1289–1840
png	1288	1841–3128	552	3129–3680
gif	1287	3681–4967	552	4968–5519
Total	3863		1656	

Table 12.4 The final dataset

Dataset			
Total files		Training	Testing
jpeg	1840	1288	552
png	1840	1288	552
gif	1839	1287	552
pdf	1840	1288	552
Total	7359	5151	2208

Table 12.5 Changes made to image files

From		To	
Extension	Signature	Extension	Signature
jpg	FF D8 FF E0 xx xx 4A 46 49 46 00	pdf	25 50 44 46
png	89 50 4E 47 0D 0A 1A 0A	pdf	25 50 44 46
gif	47 49 46 38 37 61	pdf	25 50 44 46

for Knowledge Analysis (Weka) [24], a popular machine learning software developed at the University of Waikato, New Zealand was used for all the experiments. Weka uses Goldberg's Genetic Algorithm [25]. The population size was 256, the number of generations 100, crossover was set to 0.8 and mutation probability to 0.033. CFS was the fitness function, roulette wheel selection was used to probabilistically select individuals and the single-point crossover operator was selected. The use of CFS as a filter selection evaluator and the genetic algorithm as a search strategy resulted to the selection of 44 features (82.81% reduction).

A multilayer neural network using the backpropagation algorithm was implemented as a classifier in Weka. The neural network consisted of one hidden layer with 3 nodes. The number of inputs was the 44 selected features and the number of outputs the 4 possible categories namely jpeg, png, gif and pdf. The learning rate was set to 0.3 and in order to avoid local minimum and to accelerate the learning process, the momentum parameter was set to 0.2. The training time (epochs) after experi-

Table 12.6 Confusion matrix—identifying forged jpg images

Test set	Classified as			
Image type	jpg	pdf	png	gif
jpg	552	0	0	0
pdf (168 jpg)	171	377	2	2
png	0	3	548	1
gif	0	1	7	544

Table 12.7 Confusion matrix—identifying forged png images

Test set	Classified as			
Image type	jpg	pdf	png	gif
jpg	552	0	0	0
pdf (168 png)	3	379	168	2
png	0	3	548	1
gif	0	1	7	544

Table 12.8 Confusion matrix—identifying forged gif images

Test set	Classified as			
Image type	jpg	pdf	png	gif
jpg	552	0	0	0
pdf (168 png)	3	377	2	170
png	0	3	548	1
gif	0	1	7	544

mentation was set to 500. When the training of the neural network was completed the three test sets described previously were evaluated and the results are shown in Tables 12.6, 12.7 and 12.8.

Table 12.6 shows the confusion matrix when the neural network tried to identify forged jpg images (168). When the output of the neural network was compared to the testing dataset, the "misclassified" files were the altered jpg images. The accuracy of the proposed method to altered jpg images was 100%. Table 12.7 shows the confusion matrix when the neural network tried to identify forged png images (168) and 166 out of 168 images were detected. Two png images were wrongly identified as pdf files. In the two misclassified png images there were large areas of a specific color or small variations of a color. Small variations of a color can be found also on pdf files, which led to misclassification of the images. Therefore 2 out of 168 png altered files were not predicted correctly. The accuracy of the proposed method to altered png images was 98.81%. Table 12.8 shows the confusion matrix when the neural network tried to identify forged gif images (168). The "misclassified" files were the altered gif images, thus the accuracy of the proposed method in this case was 100%.

Table 12.9 Final confusion matrix of the proposed method

168 forged files actual type	Classified as			
	jpg	pdf	png	gif
jpg	168	0	0	0
png	0	2	166	0
gif	0	0	0	168

Table 12.10 Results of the k-means algorithm

Altered type (168 altered files)	Clustered as			
	jpg	pdf	png	gif
jpg	1	99	0	68
png	4	84	72	8
gif	30	3	1	134

The accuracy results for the altered images (jpg, png, gif) of the proposed method are summarized in Table 12.9.

The above results showed that a very simple neural network achieved excellent results therefore, other traditional classification methods were implemented as well such as:

- the k-means algorithm,
- a decision tree,
- a Support Vector Machine (SVM),
- Logistic Regression (LR) and
- the k-Nearest Neighbor (k-NN).

The k-means algorithm was implemented in Weka and the three testing sets were clustered into four categories. The algorithm first computed randomly the initial centers of the four clusters, then assigned every instance of the testing file to the cluster whose center was the closest to that instance, by calculating the Euclidean distance. This was repeated until the assignment of the instances has not been changed during one iteration. The output of the clustering algorithm was compared to altered files of the three testing sets and the predictions of this method are summarized on Table 12.10.

The accuracy of the k-means algorithm to altered jpg images was 0.006%, to altered png images 42.85% and to gif images 79.76%. The clustering method failed to identify correctly the exact type of the altered files.

The algorithm selected for decision tree building was C4.5, developed by Quinlan [26]. More specifically an open source implementation of the C4.5 algorithm in Weka known as J48 was utilized. The algorithm has a top down approach. It is a recursive divide and conquer algorithm. The training data are classified instances, while each one of these instances consists of features along with the class the specific instance belongs. One feature is selected as root node and the algorithm creates a branch

Table 12.11 Parameters of the J48 learning algorithm

Parameter	Default value	Chosen value
Minimum number of instances per leaf	2	1
Use of unpruned trees	False	False
Confidence factor used for pruning	0.25	0.25
Consider subtree raising operation when pruning	True	True
Use of binary splits on nominal attributes	False	False

Table 12.12 The default parameters of the algorithm

Parameters in Weka	
Maximum iterations (MaxIts)	−1
Ridge value (ridge)	1.0E−8

for each possible feature value. That splits the instances into subsets, one for each branch that extends from the root node. The splitting criterion the algorithm uses is the normalized information gain. The feature with the highest normalized information gain is chosen to make the decision. Then the procedure is repeated recursively for each branch, selecting a feature at each node and only instances that reach that node are used to make the selection. This machine learning algorithm can be fine-tuned by setting up a lot of parameters. The parameters which were optimized in the experiment are shown on Table 12.11.

A Support Vector Machine (SVM) is a machine learning method based on statistic learning theory. SVM try to find the maximum margin hyperplane that separates two classes. An adaptation of the LIBSVM [27] implementation was used in the following. Four types of kernel function linear, polynomial, radial basis function, and sigmoid are provided by LIBSVM. A Support Vector Classification (C-SVC) was used with Radial Basis Function (RBF) kernel. After various conducted experiments, it was found that the optimal value of gamma (G) parameter of the RBF kernel was 2.

The idea of Logistic Regression (LR) is to make linear regression produce probabilities. Instead of predicting classes, it predicts class probabilities. These class probabilities are estimated directly using the logit transform. In Weka the Logistic algorithm was utilized with the default parameter setup as shown on Table 12.12.

k-Nearest Neighbor (k-NN) is a simple algorithm used for classification. The purpose of the k-NN algorithm is to use a training set—in which each one of instances is already classified—in order to predict the classification of a new unknown instance in a test set. It is a lazy algorithm as it does not use the instances in training set to do any generalization. When a new instance is presented from a given test set, the algorithm searches the entire training set for the k most similar instances (the neighbors). To determine which of the k instances in the training set are most similar to a new input, a distance measure is used. The distance measure utilized in this implementation was the Euclidean distance. The output then can be calculated as the class with the highest frequency from the k-most similar instances. Each instance votes for their class and the class with the most votes is taken as the prediction. In order to find

Table 12.13 Confusion matrix—decision tree (J48)

Forged file's actual type	Classified as			
	jpg	pdf	png	gif
168 jpg	167	1	0	0
168 png	8	3	157	0
168 gif	0	0	0	168

Table 12.14 Confusion matrix—SVM

Forged file's actual type	Classified as			
	jpg	pdf	png	gif
168 jpg	168	0	0	0
168 png	3	8	155	2
168 gif	0	1	0	167

Table 12.15 Confusion matrix—logistic regression (lr)

Forged file's actual type	Classified as			
	jpg	pdf	png	gif
168 jpg	168	0	0	0
168 png	7	6	150	5
168 gif	0	0	0	168

Table 12.16 Confusion matrix—k-nearest neighbor (KNN)

Forged file's actual type	Classified as			
	jpg	pdf	png	gif
168 jpg	167	1	0	0
168 png	8	3	157	0
168 gif	0	0	0	168

the optimum number of k, different implementations were done in Weka and it was found that the optimal value of k is 10.

All the above-mentioned classifiers were trained with the same dataset (as in Table 12.3) and the three test sets were presented to each classification model. The results for each one are presented in Tables 12.13, 12.14, 12.15 and 12.16.

The combined confusion matrix for every classifier utilized in the experiments is shown on Table 12.17. The italic cells indicate the maximum accuracy achieved.

It is obvious that the artificial neural network outperformed all other classification methods. Finally, two more experiments were conducted. The first one concerned the possibility that the proposed model learned to detect the software used (Total Image Converter) when converting jpg images to gif and png format. We repeated the experiment using this time Pixillion [28] (free version) another popular converter.

Table 12.17 Combined confusion matrix for the five classifiers

Forged file types	Prediction accuracy (%)					
	J48	SVM	NN	LR	kNN	k-means
jpg	99.40	*100.00*	*100.00*	*100.00*	99.40	0.006
png	93.45	92.26	*98.81*	89.28	93.45	42.86
gif	*100.00*	99.40	*100.00*	*100.00*	*100.00*	79.76

Table 12.18 Confusion matrix of the proposed method

168 forged files actual type	Classified as			
	jpg	pdf	png	gif
jpg	168	0	0	0
png	0	8	160	0
gif	0	0	0	168

Table 12.19 Confusion matrix—identifying forged tiff images

Test set	Classified as			
Image type	jpg	pdf	png	tiff
jpg	552	0	0	0
pdf (168 tiff)	3	367	9	173
png	0	3	545	4
tiff	4	4	4	540

Again, no alterations made to the converted images regarding size, crop, filtering or embedding watermark. The same procedure was repeated to create the final dataset, a new classification model (ANN) was trained again and the resulted confusion matrix when the unseen instances from the three test sets with the forged files was presented to the model is shown in Table 12.18.

Comparing the two confusion matrices i.e. Tables 12.9 and 12.18 it is obvious that the proposed model shows extremely high accuracy, regardless the software used to convert jpg images to png and gif format.

The second experiment examined if the proposed method worked as well as for uncompressed tiff images and whether the proposed model depends on file compression, although tiff images are not widely used and by a digital forensics viewpoint are not frequently met. For this, a new dataset was created replacing the gif images with uncompressed tiff images and the proposed methodology was applied. Once more, 30% of the pdf files in the testing set were tiff images with altered extension and signature (in a digital forensics viewpoint). Table 12.19 shows the confusion matrix when the neural network tried to identify forged tiff images (168).

Only three altered tiff images were misclassified. Two of them were classified as png images and one as pdf file. Images which misclassified as png, had high color depth and this led to misclassification. The image which misclassified as pdf, had large areas of a specific color, something also found on pdf files. Thus, the accuracy of the proposed method in uncompressed tiff images was 98.21%.

12.4 Conclusions

In this research a new methodology was proposed—in a digital forensics perspective—to identify altered file types with high accuracy by employing computational intelligence techniques. The proposed methodology was applied to the three most common image file types (jpg, png and gif) as well as to uncompressed tiff images. A three-stage process involving feature extraction (BFD), feature selection (genetic algorithm) and classification (neural network) was proposed. Experimental results were conducted having files altered in a digital forensics perspective. The accuracy of the proposed method to altered jpg images and to gif images was 100%, to altered png images was 98.81% and to altered tiff images was 98.21%. This proposed method outperformed other examined classifiers such as a traditional k-means clustering algorithm, a SVM, a decision tree, Logistic Regression and kNN. Experiments were also conducted regarding the scenario the proposed model learned to detect a specific converter and the results were promising again. Finally, although tiff images are not frequently met, the proposed model identified forged tiff images with extremely high accuracy.

References

1. McKemmish, R.: What is forensic computing? Trends Issues Crime Crim. Justice **118**(118), 1–6 (1999)
2. Kent, K., Chevalier, S., Grance, T., Dang, H.: Guide to integrating forensic techniques into incident response (2006)
3. Kessler, G.: File signatures (2015). http://www.garykessler.net/library/file_sigs.html. Accessed 26 Oct 2015
4. McDaniel, M.: Automatic File Type Detection Algorithm. James Madison University (2001)
5. McDaniel, M., Heydari, M.H.: Content based file type detection algorithms. In:. Proceedings of 36th Annual Hawaii International Conference on System Sciences (2003)
6. Li, W.J., Wang, K., Stolfo, S.J., Herzog, B.: Fileprints: identifying file types by n-gram analysis. In: Proceedings from 6th Annual IEEE Systems Man and Cybernetics (SMC) Information Assurance Workshop 2005, vol. 2005, pp. 64–71, June 2005
7. Dunham, J., Sun, M., Tseng, J.: Classifying file type of stream ciphers in depth using neural networks. In: The 3rd ACS/IEEE International Conference on Computer Systems and Applications (2005)
8. Amirani, M.C., Toorani, M., Shirazi, A.A.B: A new approach to content-based file type detection. In: IEEE Symposium on Computers and Communications, 2008, pp. 1103–1108, July 2008
9. Cao, D., Luo, J., Yin, M., Yang, H.: Feature selection based file type identification algorithm. In: 2010 IEEE International Conference on Intelligent Computing and Intelligent Systems, vol. 3, pp. 58–62 (2010)

10. Ahmed, I., Lhee, K., Shin, H., Hong, M.: Content-based File-type identification using cosine similarity and a divide-and-conquer approach. IETE Tech. Rev. **27**(6), 465 (2010)
11. Ahmed, I., Lhee, K., Shin, H., Hong, M.: Fast content-based file-type identification. In: 7th Annual IFIP WG 11.9 International Conference on Digital Forensics, pp. 65–75 (2011)
12. Amirani, M.C., Toorani, M., Mihandoost, S.: Feature-based type identification of file fragments. Secur. Commun. Netw. **6**(1), 115–128 (2013)
13. Evensen, J.D., Lindahl, S., Goodwin, M.: File-type detection using naïve Bayes and n-gram analysis. In: Norwegian Information Security Conference, NISK, vol. 7, no. 1. Fredrikstad (2014)
14. Vafaie, H., De Jong, K.: Genetic algorithms as a tool for feature selection in machine learning. In: International Conference on Tools with AI, pp. 200–203 (1992)
15. Qian, J., Zhuo, L., Zheng, J., Wang, F., Li, X., Ai, B.: A genetic algorithm based wrapper feature selection method for classification of hyper spectral data using support vector machine. Geogr. Res. **27**(3), 493–501 (2008)
16. Jourdan, L., Dhaenens, C., Talbi, E.: A genetic algorithm for feature selection in data-mining for genetics. In: Proceedings of the 4th Metaheuristics International Conference (2001)
17. Hall, M.: Correlation-based feature selection for machine learning. The University of Waicato (1999)
18. Harris, R.: Using artificial neural networks for forensic file type identification. Master's Thesis, Purdue Univ. (2007)
19. Kohavi, R.: A study of cross-validation and bootstrap for accuracy estimation and model selection. Int. J. Conf. Artif. Intell. **14**(12), 1137–1143 (1995)
20. Fei-Fei, L., Fergus, R., Perona, P.: Learning generative visual models from few training examples: an incremental Bayesian approach tested on 101 object categories. Comput. Vis. Image Underst. **106**(1), 59–70 (2007)
21. CoolUtils, Powerful image converter yet easy-to-use (2017)
22. T.E.I of Crete, E-Thesis (2015). http://nefeli.lib.teicrete.gr/search/. Accessed 26 Oct 2015
23. The MathWorks Inc., MATLAB. The MathWorks Inc., Natick, Massachusetts (2016)
24. Hall, M., Frank, E., Holmes, G., Pfahringer, B., Reutemann, P., Witten, I.H.: The WEKA data mining software. ACM SIGKDD Explor. Newsl. **11**(1), 10 (2009)
25. Goldberg, D.E.: Genetic Algorithms in Search, Optimization and Machine Learning. Oct 1989
26. Salzberg, S.L.: In: Quinlan, J.R. (ed) C4.5: Programs for Machine Learning. Morgan Kaufmann Publishers, Inc. (1993); Mach. Learn. **16**(3), 235–240 (1994)
27. Chang, C.-C., Lin, C.-J.: LIBSVM: a library for support vector machines. ACM Trans. Intell. Syst. Technol. **2**(3), 27:1–27:27 (2011)
28. NCH Software, Convert Between All Popular Image Formats with Pixillion (2017)

Konstantinos Karampidis M.Sc. was born in Athens Greece in 1971. He received his diploma in Electrical Engineering in 1994 from the Technological Educational Institute of Crete and his Master's degree from the same institute. He is currently working on his Ph.D. in image steganalysis, at the Department of Information & Communication Systems Engineering at Aegean University. His current research interests include Digital Forensics, Computational Intelligence, Deep Learning. He has authored 2 Journal and 2 International Conference publications.

Ioannis Deligiannis M.Sc. was born in Athens Greece in 1988. He received his B.Sc. and Master's degree from the Technological Educational Institute of Crete and currently is enrolled as a Ph.D. student at University of Patras. His current research interests are in the fields of Intelligent and Embedded Systems and is currently participating in a European research project respectively. Finally, he has authored 1 book chapter, 1 Journal and 12 International Conference publications.

Giorgos Papadourakis Ph.D. was born in Patras Greece in 1959. He received his B.Sc. degree from the Michigan Technological University in 1978, his Master's degree from the University of

Cincinnati in 1981, and his Ph.D. Degree from the University of Florida in 1986, all in Electrical Engineering. Since 2002 is a professor at the Department of Informatics Engineering of the Technological Educational Institute of Crete. His current research interests include Intelligent System applications, firmware and software development, telemedicine, Engineering Education, Open Distance Learning and Assessment of Technology to Health Care Systems. He has authored (editor) of 17 books, over 28 Journal and 148 International Conference publications. He has coordinated or participated in over 98 National, European and International projects.

Part V
Theoretical Advances and Tools for Data Analytics

Chapter 13
Deep Learning Analytics

Nikolaos Passalis and Anastasios Tefas

Abstract The recent breakthroughs in Deep Learning have provided powerful data analytics tools for a wide range of domains ranging from advertising and analyzing users' behavior to load and financial forecasting. Depending on the nature of the available data and the task at hand Deep Learning Analytics techniques can be divided into two broad categories: (a) unsupervised learning techniques and (b) supervised learning techniques. In this chapter we provide an extensive overview over both categories. Unsupervised learning methods, such as Autoencoders, are able to discover and extract the information from the data without using any ground truth information and/or supervision from domain experts. Thus, unsupervised techniques can be especially useful for data exploration tasks, especially when combined with advanced visualization techniques. On the other hand, supervised learning techniques are used when ground truth information is available and we want to build classification and/or forecasting models. Several deep learning models are examined ranging from simple Multilayer Perceptrons (MLPs) to Convolutional Neural Networks (CNNs) and Recurrent Neural Networks (RNNs). However training deep learning models is not always a straightforward task requiring both a solid theoretical background as well as intuition and experience. To this end, we also present recent techniques that allow for efficiently training deep learning models, such as batch normalization, residual connections, advanced optimization techniques and activation functions, as well as a number of useful practical suggestions. Finally, we present an overview of the available open source deep learning frameworks that can be used to implement deep learning analytics techniques and accelerate the training process using Graphics Processing Units (GPUs).

N. Passalis (✉) · A. Tefas
Aristotle University of Thessaloniki, Thessaloniki, Greece
e-mail: passalis@csd.auth.gr

A. Tefas
e-mail: tefas@aiia.csd.auth.gr

G. A. Tsihrintzis et al. (eds.), *Machine Learning Paradigms*, Intelligent Systems Reference Library 149, https://doi.org/10.1007/978-3-319-94030-4_13

13.1 Introduction

The recent breakthroughs in Deep Learning have provided powerful data analytics tools for a wide range of domains ranging from advertising and analyzing users' behavior [19], to load and financial forecasting [68, 79, 80]. Developing more accurate models using deep learning techniques provides several benefits to domain experts and users. For example, better understanding the users' behavior allows for deploying more targeted advertising campaigns or developing products that better fit the needs of the customers. Financial forecasting models can provide useful indicators about the state of financial markets protecting the investors from unwanted losses, while accurate load prediction allows for more precise resource planning lowering the cost of acquiring the relevant goods/services, e.g., electricity.

In this chapter the most important deep learning tools are presented. However, developing and deploying a deep learning model is not always a straightforward task. To this end, recent techniques that allow for efficiently training deep learning models as well as a number of useful practical suggestions are provided through this chapter to equip the reader with the necessary tools for overcoming many difficulties that may arise. The rest of the chapter is structured as follows. First, the used notation is briefly introduced and the necessary preliminaries are provided in Sect. 13.2.

Next, in Sect. 13.3, the *unsupervised* deep learning models are introduced. Unsupervised learning methods are able to discover and extract the information from the data without using any ground truth information and/or supervision from domain experts. Thus, these techniques can be especially useful for *data exploration tasks* [16]. An unsupervised deep learning model, the *autoencoder* [30], is presented in detail, as well as several variants of it, such as denoising autoencoders [85], sparse autoencoders [31] and contractive autoencoders [69]. Also, it is demonstrated how to combine the aforementioned methods with advanced visualization techniques, such as the t-SNE algorithm [47].

Section 13.4 provides an overview of *supervised* deep learning models. Supervised learning techniques are used when ground truth information is available and we want to build classification and/or forecasting models. First, the Multilayer Perceptrons (MLPs) [25], that were among the first practical neural networks that were used to handle pattern recognition tasks with satisfactory results, are presented. However, when well-defined spatial or temporal relationships exist within the input data, more advanced models can be used. Convolutional Neural Networks (CNNs) are neural architectures that are able to take into account the spatial arrangement of the input features [43], while Recurrent Neural Networks (RNNs) are capable of exhibiting dynamic temporal behavior when processing sequences of arbitrary length, such as texts or timeseries, allowing them to model the long-term relationships between distant words or points [10, 21]. Both the CNNs and the RNNs are also presented in Sect. 13.4. Apart from unsupervised and supevised deep learning techniques, deep reinforcement learning methods have also achieved state-of-the-art results in various control tasks and games [52]. However, the study of reinforcement learning techniques is out of the scope of this chapter that focuses on deep learning analytics.

Developing, training and deploying deep learning models is a challenging and demanding task. To alleviate this issue, a number of open source deep learning frameworks have been developed. These frameworks are capable of greatly simplifying the process of developing and deploying a deep learning model. Also, most of them provide transparent support for accelerating the training process using Graphics Processing Units, significantly reducing the training time. Section 13.5 provides an overview of some deep learning frameworks, along with a short example of how to use one of them to develop a deep learning model. Finally, Sect. 13.6 concludes this chapter.

13.2 Preliminaries and Notation

The necessary preliminaries are provided and the used notation is introduced in this Section. Let $\mathscr{X}$ be a collection of objects, e.g., images, text documents, videos, etc. These objects may be *unstructured* and they may need several preprocessing steps before they can be used for any learning task. Therefore, some *features* must be identified and extracted from each object. This process is called *feature extraction* [24]. Feature extraction may involve very simple techniques, such as directly using the raw pixel values of an image, or more advanced ones, such as using scale-invariant features [45], or constructing a dictionary of the words that appear in a collection of documents and expressing each document using this dictionary [13]. After the feature extraction process, each object is represented by a vector $\mathbf{x} \in \mathbb{R}^n$, where n is the dimensionality of the extracted vectors. Most deep learning techniques require these vectors to have constant dimensionality, i.e., n is not allowed to vary between different objects. However, some recent techniques, e.g., CNNs with global pooling [61], and RNNs [21], can also *directly* handle features with variable dimensionality (e.g., images with different sizes or texts with different lengths). At this point, it should be stressed that the fundamental idea behind deep learning is to extract a very basic set of features and then let the model learn *automatically* learn more complex features instead of using hand-engineered features, e.g., [5, 45]. Indeed, it has been demonstrated that deep learning leads to great improvements over using handcrafted features [41]. It is also useful to define the *data matrix* $\mathbf{X} \in \mathbb{R}^{N \times n}$ that contains N data samples (each of dimensionality n). For unsupervised learning tasks only the data matrix is available, while for supervised learning tasks the data matrix is accompanied by a label vector $\mathbf{t} \in \mathbb{R}^N$ that contains the labels that denote the category of each sample or continuous values that describe the attribute that we want to predict. Finally, the mathematical notation used in the rest of this chapter is summarized in Table 13.1.

Table 13.1 Mathematical notation used in this chapter

Notation	Meaning
$\mathbb{R}/\mathbb{R}^n/\mathbb{R}^{n\times m}$	the set of real numbers/real n-dimensional vectors/real $n \times m$ matrices
$\mathscr{X}$	a set of objects (appropriately defined in the context)
x	a scalar number
$\mathbf{x}$ or $\mathbf{X}$	a vector/matrix/tensor (appropriately defined in the context)
$[\mathbf{x}]_i$	the i-th element of the vector $\mathbf{x}$
$[\mathbf{X}]_{ij}$	the element in the i-th row and j-th column of the matrix $\mathbf{X}$
$\mathbf{x}_i$	the i-th vector of a collection of N vectors (appropriately defined in the context)
$\mathbf{x}^{[t]}$ or $\mathbf{X}^{[t]}$	the vector/matrix used at the t-th timestep of a process
$[\mathbf{a}, \mathbf{b}]$	the concatenation (stacking) of two vectors $\mathbf{a}$ and $\mathbf{b}$
$\lvert\lvert\mathbf{x}\rvert\rvert_1/\lvert\lvert\mathbf{x}\rvert\rvert_2$	the l^1/l^2 norm of vector $\mathbf{x}$
$\lvert\lvert\mathbf{X}\rvert\rvert_F$	the Frobenius norm of matrix $\mathbf{X}$
$\mathbf{X}^T$	the transpose of matrix $\mathbf{X}$
$\mathbf{a} \odot \mathbf{b}$	the elementwise (Hadamard) product between the vectors $\mathbf{a}$ and $\mathbf{b}$
$\frac{\partial \mathscr{L}}{\partial \mathbf{x}}$	the partial derivative of function $\mathscr{L}$ with respect to vector $\mathbf{x}$
$\log(x)$	the natural logarithm of x
$tanh(x)$	the hyperbolic tangent of x
$max(a, b)$	the maximum between two numbers a and b
$exp(x)$	the exponential function with base e
$E[x]$	the expected value of a set of observations represented by the variable x (appropriately defined in the context)
$Var[x]$	the variance of a set of observations represented by the variable x (appropriately defined in the context)
$f(\mathbf{x})$	a function $f(\cdot)$ that is applied *elementwise* on the vector $\mathbf{x}$ (only when $f(a)$ is a scalar unary function and $a \in \mathbb{R}$)

13.3 Unsupervised Learning

In many cases, we do not know yet what we are looking for in the data, but we want to extract all the *valuable* information they contain for further processing and analysis, such as visualizing the data in a low-dimensional space [47], discovering interesting patterns using cluster analysis [4], or detecting outliers [3]. This learning scenario is called *unsupervised learning* [8], since we are interested in uncovering the latent structures of the data without using any *supervision* from domain experts or ground truth information, e.g., labels. Unsupervised learning is becoming increasingly important in the Big Data era allowing us to exploit the vast amount of the available unlabeled data.

Autoencoders [30], along with Restricted Boltzmann Machines [71], are among the major deep learning tools used for unsupervised learning tasks. Both of them are

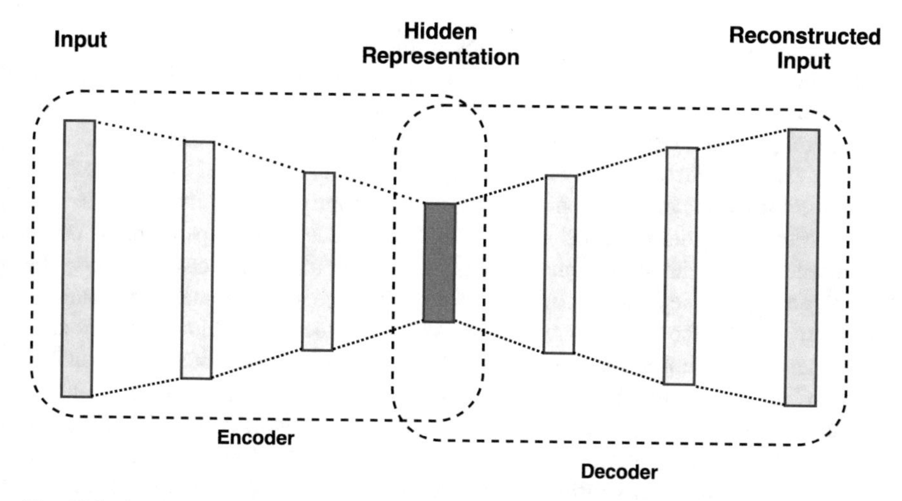

Fig. 13.1 A deep autoencoder: the encoder is used to map the input into a lower dimensional representation (possibly through multiple intermediate mappings), while the decoder reconstructs the original input using the reverse procedure

capable of reducing the dimensionality of the data, similar to classical dimensionality reduction techniques [83], while maintaining most of the useful information contained in the data. The main focus of this section is on deep autoencoding techniques, since they are the most frequently used unsupervised deep learning technique in the last few years. The rest of this section is structured as follows. First, the basic structure of an autoencoder is described in detail. Then, it is demonstrated how autoencoders can be used as the building block to form powerful deep neural networks capable of extracting useful higher level information from the data. Also, we demonstrate how to use advanced visualization techniques, such as the t-distributed stochastic neighbor embedding (t-SNE) algorithm [47], or the Similarity Embedding Framwork (SEF) [59], with the representation learned with deep auto-encoders to perform data exploration tasks. Finally, some useful variants of autoencoders that are capable of extracting representations with specific properties, e.g., sparse representations, are presented.

13.3.1 Deep Autoencoders

Autoencoders are *neural networks* trained to *reconstruct* their input through (a typically) lower dimensional *intermediate* representation. An autoencoder is composed of two parts: (a) an *encoder* that maps the original input into the intermediate representation (also called *hidden representation*) and (b) a decoder that reconstructs the original input from the intermediate representation. If the autoencoder manages to reconstruct its input using the intermediate representation, then that means that the intermediate representation maintains all the information contained in the original image. The aforementioned process is illustrated in Fig. 13.1. Also, note that the encoder and the decoder might use several internal intermediate representation before outputting the final one giving rise to *deep* (or *stacked*) autoencoders.

Typically, we *force* the network to strip out the noise from the data and extract only the important information by using a significantly smaller intermediate representation. To understand this consider an autoencoder that maps its input into a representation of the same dimensionality as the input. In this case, learning the identity mapping, i.e., simply copying the input into the intermediate representation, leads to perfect reconstruction of the input of the autoencoder. However, this kind of autoencoder is not useful, since it is not able either to reduce the dimensionality of the data or to extract the important information from them. On the other hand, when the intermediate representation is smaller, the network is forced to *focus* on the important features of the data in order to be able to efficiently compress the representation. However, it worths mentioning that it was experimentally verified that even when the dimensionality of the intermediate representation is equal (or even larger) than the dimensionality of the input, the autoencoders are still able, under certain conditions, to extract a useful representation [6].

13.3.1.1 Autoencoders

An autoencoder is formally defined as follows. Given an n-dimensional input vector $\mathbf{x} \in \mathbb{R}^n$ the autoencoder employs the encoder $f(\mathbf{x}) = \mathbf{h} \in \mathbb{R}^m$ to extract the intermediate representation $\mathbf{h}$. As it was already mentioned, typically the dimensionality of the intermediate representation is significantly smaller, i.e., $m \ll n$. Then the original input is reconstructed using the decoder $g(\cdot)$, i.e., $\tilde{\mathbf{x}} = g(\mathbf{h}) = g(f(\mathbf{x})) \in \mathbb{R}^n$. The reconstruction error is typically measured using the squared error function:

$$\mathscr{L}_{ae} = \sum_{j=0}^{n} ([\mathbf{x}]_j - [\tilde{\mathbf{x}}]_j)^2 = ||\mathbf{x} - \tilde{\mathbf{x}}||_2^2, \tag{13.1}$$

where the notation $[\mathbf{x}]_j$ is used to refer to the j-th element of the vector $\mathbf{x}$ and $||\mathbf{x}||_2$ denotes the l^2 norm of the vector $\mathbf{x}$. Note that a set of data $\mathscr{X} = \{\mathbf{x}_1, \mathbf{x}_2, ..., \mathbf{x}_N\}$, where N denotes the number of data points, is usually used to train an autoencoder. Therefore, it makes more sense to evaluate the quality of the autoencoder by averaging the reconstruction error over the training set. That is, Eq. (13.1) is extended as follows:

$$\mathscr{L}_{ae}^{(batch)} = \frac{1}{N} \sum_{i=1}^{N} \sum_{j=1}^{n} ([\mathbf{x}_i]_j - [\tilde{\mathbf{x}_i}]_j)^2 = \frac{1}{N} \sum_{i=1}^{N} ||\mathbf{x}_i - \tilde{\mathbf{x}_i}||_2^2, \tag{13.2}$$

When the input features of the autoencoder can be interpreted as probabilities, i.e., they are bounded into the [0, ..., 1] interval, other loss functions can be also used to

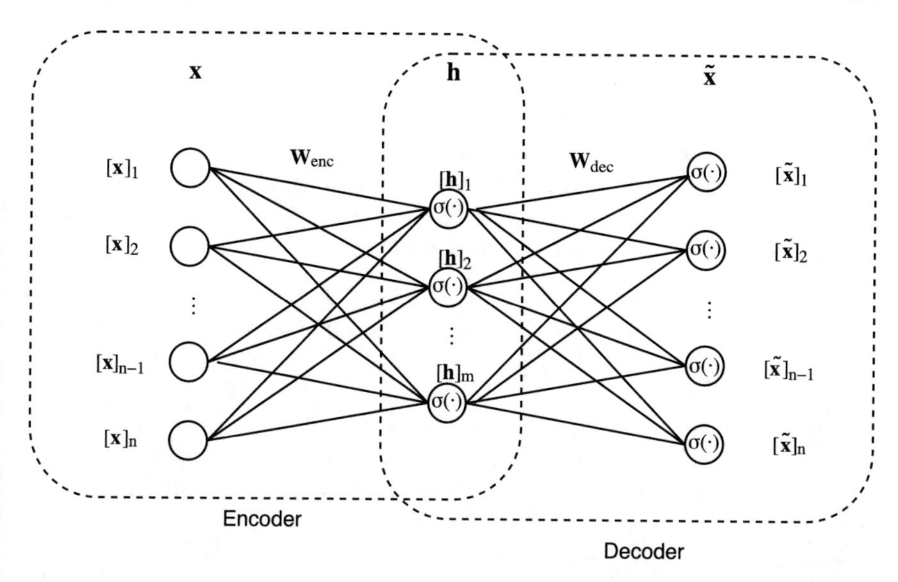

Fig. 13.2 The basic structure of an autoencoder. First, the input **x** is transformed into a lower dimensional representation **h** through a (non-linear) fully connected layer (encoder). Then, the decoder is used to reconstruct the original input using only the compressed low dimensional representation. The bias terms used in the hidden and the output layers are omitted for presentation purposes

measure the reconstruction error. One of them is the binary cross-entropy loss:

$$\mathcal{L}_{ae-ent} = -\sum_{i=0}^{n} [\mathbf{x}]_j \log([\tilde{\mathbf{x}}]_j) + (1 - [\mathbf{x}]_j)\log(1 - [\tilde{\mathbf{x}}]_j). \tag{13.3}$$

Both the encoder and the decoder are usually defined as simple fully connected neural layers [25]. The detailed architecture of an autoencoder is shown in Fig. 13.2. The encoder consists of m neurons, while the decoder of n neurons. Therefore, the encoder is defined as:

$$f(\mathbf{x}) = \sigma(\mathbf{W}_{enc}\mathbf{x} + \mathbf{b}_{enc}), \tag{13.4}$$

where $\mathbf{W}_{enc} \in \mathbb{R}^{m\times n}$ are the weights of the fully connected layer used in the encoder and $\mathbf{b}_{enc} \in \mathbb{R}^m$ contains the corresponding bias terms for each neuron. Note that the output of each neuron is passed through the activation function $\sigma(\cdot) \in \mathbb{R}$. Several choices exist for the activation function $\sigma(\cdot)$. Among the most widely used activation functions are the sigmoid function $\sigma_{sigm}(u) = \frac{1}{1+e^{-x}}$ [25], the hyperbolic tangent function $\sigma_{tanh}(u) = tanh(u)$ [25], and the (more modern) rectifier linear unit activation (also called *ReLU*) $\sigma_{ReLU}(u) = max(0, u)$ [23]. The decoder is similarly defined as:

$$f(\mathbf{x}) = \sigma(\mathbf{W}_{dec}\mathbf{h} + \mathbf{b}_{dec}), \tag{13.5}$$

where $\mathbf{W}_{dec} \in \mathbb{R}^{n \times m}$ are the weights of the fully connected layer used in the encoder and $\mathbf{b}_{dec} \in \mathbb{R}^{n}$ contains the corresponding bias terms for each neuron. Typically, the weights of the encoder and the decoder are tied together, i.e., $\mathbf{W} = \mathbf{W}_{enc} = \mathbf{W}_{dec}^{T}$. Note that the data must be appropriately scaled to match the output range of the activation function used in the decoder. For example, if the sigmoid activation function is used, then the data $\mathbf{x}$ must lie in the interval [0...1], since the network is not capable of producing values outside this range.

Using a non-linear activation function significantly increases the expressive power of the autoencoder allowing for modeling more complex phenomena (note that even the simple *XOR* gate cannot be implemented in an neural network without using non-linear activation functions [25]). However, using activation functions that get easily saturated, such as the sigmoid activation, can sometimes have a detrimental effect on training the network due to a phenomenon known as *vanishing gradients* (more details regarding these phenomena are given later in this section).

We have defined the structure of an autoencoder, but we have not provided a way to actually train the autoencoder to fit the avaiable data and learn a useful intermediate representation. Similar to most neural network architectures the well known back-propagation algorithm can be used to this end [25]. *Back-propagation* provides an efficient to way to update the weights of a neural network using the *gradient descent* optimization algorithm. For the autoencoder described above the updates for each iteration t of gradient descent are calculated as:

$$\mathbf{W}^{[t+1]} = \mathbf{W}^{[t]} - \eta \frac{\partial \mathcal{L}_{ae}^{(batch)}}{\partial \mathbf{W}^{[t]}}, \tag{13.6}$$

$$\mathbf{b}_{enc}^{[t+1]} = \mathbf{b}_{enc}^{[t]} - \eta \frac{\partial \mathcal{L}_{ae}^{(batch)}}{\partial \mathbf{b}_{enc}^{[t]}}, \tag{13.7}$$

and

$$\mathbf{b}_{dec}^{[t+1]} = \mathbf{b}_{dec}^{[t]} - \eta \frac{\partial \mathcal{L}_{ae}^{(batch)}}{\partial \mathbf{b}_{dec}^{[t]}}. \tag{13.8}$$

where η is the learning rate of the algorithm. The learning rate is typically set to a small positive value (e.g., $\eta = 0.001$) that allows for smooth convergence of the optimization process.

Even though plain gradient descent can be used to learn the parameters of a network, sometimes it can be difficult to select the appropriate learning rate for optimizing deep neural networks. This becomes even worse when the magnitude of the gradients is different for different parameters of the network, requiring using separate learning rates for each set of parameters. These problems are addressed by more advanced gradient descent-based optimization techniques that are able to *adapt* to the behavior of each parameter, such as the Adagrad [18], the Adadelta [92], and more recently, the Adam [39].

Even though some of these techniques also perform a kind of *learning rate annealing*, i.e., they effectively reduce the learning rate as the training process progresses [39], it is often useful to use an explicit learning rate annealing schedule [72]. Temporal averaging techniques can also improve the convergence of the training process [39, 60].

To accelerate the convergence of the optimization process usually mini-batch gradient descent is used, i.e., small batches of the data are used to calculate the loss function defined in Eq. (13.2) instead of the whole dataset. After performing a sufficient number of iterations, the weights of the autoencoder converge. Note that the general problem of minimizing the loss function (13.2) is non-convex and the gradient descent algorithm usually ends up to a local minimum. However, as it was throughly demonstrated in the neural network literature [12], these local minima are usually good enough for most practical problems.

When a linear activation function is used, i.e., $\sigma(u) = u$, and the squared error is used as the loss function (as defined in Eq. (13.2)), then the autoencoder actually performs Principal Compoment Analysis (PCA) [2]. However, when non-linear activation functions are used, the expressive power of the autoencoder increases significantly and different solutions from the plain PCA are obtained. Furthermore, autoencoders are capable of learning representations oriented to the task at hand, such as sparse representations and representations more robust to noise, as demonstrated in Sect. 13.3.2.

The well known MNIST dataset [44], that contains 70,000 images of ten handwritten digits (0–9) is used to demonstrate the ability of autoencoders to learn compact representations from which we can reconstruct the original images. The MNIST dataset is split into 60,000 train images (used to train the used models) and 10,000 test images (used to evaluate the trained models). The experimental results are shown in Fig. 13.3. As the dimensionality of the hidden representation increases, more information is retained, leading to better reconstruction of the input images. However, even when the original image (that contains $28 \times 28 = 784$ pixels) is compressed into a vector of just 20 values, the autoencoder is able to reconstruct a quite good approximation of each digit.

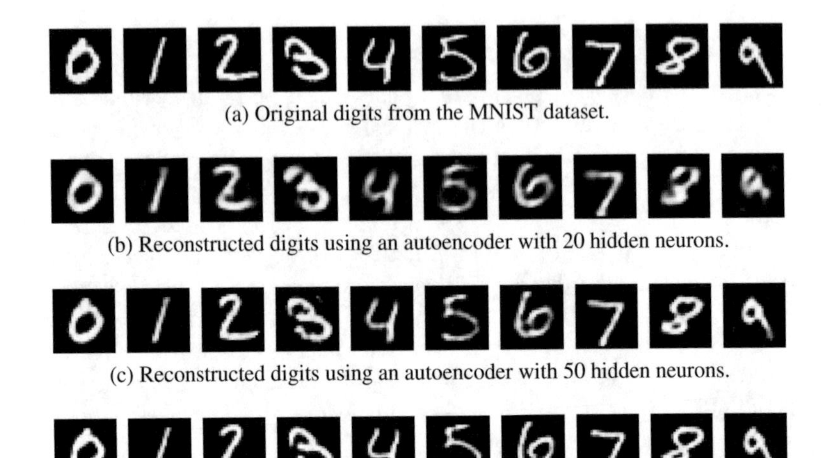

(a) Original digits from the MNIST dataset.

(b) Reconstructed digits using an autoencoder with 20 hidden neurons.

(c) Reconstructed digits using an autoencoder with 50 hidden neurons.

(d) Reconstructed digits using an autoencoder with 100 hidden neurons.

Fig. 13.3 Comparing the effect of the size of the learned representation of an autoencoder on the reconstruction accuracy

Achieving the lowest possible reconstruction error does not always guarantee that we have learned the best representation. Usually increasing the size of the intermediate representation lowers the reconstruction error. But this comes with the risk of either learning a mapping close to the identity mapping or simply *memorizing* the input samples without extracting a useful representation. This phenomemon is known in the machine learning literature as *overfitting*. To avoid overfitting the models, a separate validation set (that is not contained in the train set) must be used to evaluate the ability of the model to *generalize* on new unknown samples.

13.3.1.2 Stacked Autoencoders

Multiple autoencoders can be stacked to form *Deep Autoencoders* (also called Stacked Autoencoders in the literature [85]), as shown in Fig. 13.1. Deep autoencoders can learn increasingly complex feature representations, effectively extracting the high level information from the data without any supervision from domain experts/ground truth information. First, the encoding function of the autoencoders is applied in series and then the original representation is reconstructed by applying the corresponding decoders in reverse order. There are several proposed ways to train deep autoencoders. In the early years of Deep Learning, a layer-wise pretraining procedure was followed [7]. That is, each autoencoder was first independently trained (using the representation extracted from the previous encoder) and then the whole architecture was finetuned using the regular back-propagation algorithm (as described above).

But why we couldn't just directly apply the back-propagation algorithm? The answer lies on the *vanishing gradients* phenomenon that greatly affected the early deep learning models [22, 54]. Even though this phenomenon might has different causes for different models, it is mainly caused when highly non-linear activation functions are used without properly initializing the model. This led to saturating the activation functions, that in turn led to very small gradients (the derivative of a saturated activation function has very small magnitude). This problem only got worse as the depth of the network increased, making impossible to train the first layers of the network. Using the proposed layer-wise pre-training approach allowed for overcoming this problem, since it was ensured that most neurons will avoid operating in their saturation regions. However, later it was found that using appropriate initialization schemes, such as the Glorot initilization [22], and activation functions, such as the ReLU activation [23], allow for overcoming the aforementioned problem without using this time consuming layer-wise pre-training. Therefore, this kind of pre-training is usually no longer used in modern deep learning architectures.

Using deep autoencoders with a linear activation function does not increase the expressive power of the model. Indeed, it is easy to show that such deep network is equivalent to a single autoencoder where its weights are appropriately set (simply by multiplying the weight matrices of the used autoencoders).

13.3.1.3 Visualizing the Data

Deep autoencoders can be directly used to visualize the data in two or three dimensions simply by setting the appropriate size for the hidden representation, i.e., $m = 2$ or $m = 3$. However, in such low dimensional spaces several special phenomena occur, such as the well-known *crowding problem* [47], which refers to the inability of low dimensional spaces to accurately model the distances of a high-dimensional space. To understand why this happens, consider a 10-dimensional space, where it is possible to place 11 points in such way that are all equidistant to each other. However, there is no way to achieve this in a 2- or 3-dimensional space. Therefore, any low-dimensional mapping of these 11 points will fail to accurately model their relationships in the high dimensional space.

To reduce the effect of the aforementioned limitations, advanced visualization techniques, such as the t-distributed stochastic neighbor embedding (t-SNE) algorithm [47], or the Similarity Embedding Framework (SEF) [59], were proposed. However, even these techniques usually perform better when the dimensionality of the input data is low. Therefore, deep autoencoders can be combined with the t-SNE algorithm or the SEF to provide more accurate and meaningful visualizations of the data that can be used for data exploration/interactive data analysis/etc. To do so, first a deep autoencoder is used to reduce the dimensionality of the data by a significant factor (e.g., the 784-dimensional MNIST images can be reduced to just 50 dimensions) and then the t-SNE algorithm or the SEF is used for visualizing the resulting vectors in a very low-dimensional space. That way, better visualizations can be obtained than directly applying autoencoders to visualize the data. Note that the SEF can be also used to visualize *out-of-sample* data, i.e., data that were not contained in the original training set, allowing for interactive visualization of larger datasets.

13.3.2 Autoencoder Variants

According to the task at hand, the autoencoders can be appropriately adapted to ensure that useful intermediate representations will be learned. In this Subsection three variants of the classical autoencoder, the denoising autoencoder, the sparse autoencoder and the contractive autoencoder, are presented.

13.3.2.1 Denoising Autoencoders

To make an autoencoder more robust to noise we can stochastically corrupt its input during the training process while requesting to reconstruct the original uncorrupted input. The most straightforward way to do so is to randomly zero some of the input features for each sample [85]. The percentage of the corrupted input features usually ranges from 10 to 50%. Forcing the network to *guess* the missing values of the input makes the autoencoder more robust to noise, while allowing for performing denoising and preventing overfitting the autoencoder even when large intermediate representations are used. Note that this process is actually very similar to the dropout technique [76], that also randomly zeros some of the input features. Also, note that other types of noise, such as simple Gaussian noise can be also applied to the input images. The ability of denoising autoencoders to recover the original image, even though a significant amount of Gaussian noise has been introduced, is demonstrated in Fig. 13.4.

13.3.2.2 Sparse Autoencoders

The reasoning behind sparse autoencoders is to associate only some of the intermediate neurons with specific regions of the input space. This way interesting patterns in the data can be discovered without overfitting the autoencoder even when the intermediate representation is larger that the actual input. Ideally, a sparse autoencoder might be able to assocciate specific neurons with the actual classes of the ground truth information (when they exist). In other cases, it might be able to uncover features of the input data that were previously unknown even to the domain experts.

(a) Stochastically corrupted digits from the MNIST dataset using Gaussian noise.

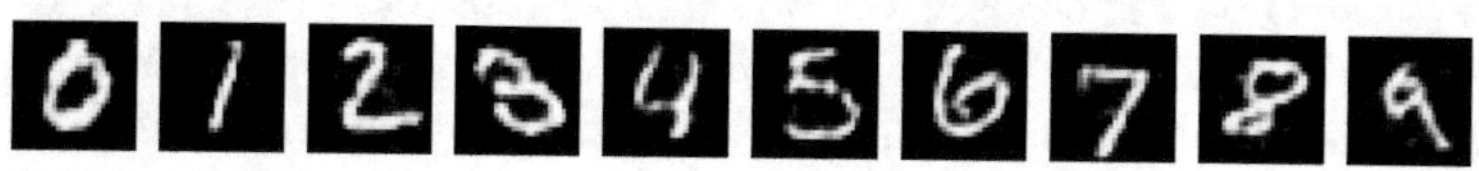

(b) Reconstructed digits from the noisy input using an denoising autoencoder with 200 hidden neurons.

Fig. 13.4 Demonstrating the ability of denoising autoencoders to recover the original image even when a significant amount of noise is introduced

There are several ways to enforce sparsity constraints on the learned representation, e.g., by using k-sparse autoencoder [49], or by adding a sparsity term in the loss function that penalizes non-sparse intermediate representations [31]. The latter approach is described in this Subsection. More specifically, the activations of the intermediate representation are penalized using the Kullback-Leibler divergence function [31]. To this end, the mean activation of the intermediate representation is measured as:

$$\hat{\rho} = \frac{1}{N} \sum_{i=1}^{N} f(\mathbf{x}_i). \tag{13.9}$$

Then, we demand that the average activation of each neuron is close to a predifined sparsity parameter ρ. Typically, the sparsity parameter is set to a small positive number near 0, e.g., $\rho = 0.05$. This effectively forces most neurons to be deactivated, i.e., their output to be near zero, most of the time, while only allowing a few (different) neurons to fire for each sample. This sparsity constraint is enforced by minimizing the Kullback-Leibler divergence between two Bernoulli random variables with means ρ and $[\hat{\rho}]_j$ (for each encoder neuron j). Therefore, the sparsity constraint is defined as:

$$\mathscr{L}_{sparsity} = \sum_{j=1}^{m} \rho \log(\frac{\rho}{[\hat{\rho}]_j}) + (1 - \rho) \log(\frac{(1 - \rho)}{(1 - [\hat{\rho}]_j)}), \tag{13.10}$$

while the loss function used for the optimization is now defined as the weighted sum between the reconstruction loss and the sparsity penalty:

$$\mathscr{L}_{sae}^{(batch)} = \mathscr{L}_{ae}^{(batch)} + \alpha_{sparse} \mathscr{L}_{sparsity}, \tag{13.11}$$

where the parameter α_{sparse} controls the importance of the sparsity constrains.

13.3.2.3 Contractive Autoencoders

Another way to make autoencoders more robust to noise is to ensure that the intermediate representation will be insensitive to small perturbations of the input. Similarly to the sparse autoencoder, this can be achieved by adding an appropriate term to the used loss function that penalizes autoencoders that are sensitive to small perturbations of their input. A way to measure the sensitivity of the learned representation is to use the Frobenius norm of the Jacobian $J_f(\mathbf{x})$ of the encoder [69]:

$$\mathscr{L}_{con} = \sum_{i=0}^{N} ||J_f(\mathbf{x}_i)||_F^2 = \sum_{i=0}^{N} \sum_{j=0}^{n} \sum_{l=0}^{m} \left(\frac{\partial [f(\mathbf{x}_i)]_l}{\partial [\mathbf{x}_i]_j} \right)^2. \tag{13.12}$$

Minimizing Eq. (13.12) ensures that the learned representation will be robust to small changes of the input. Again, the total loss function can be defined as the weighted sum of the reconstruction loss and the contractive penalty:

$$\mathscr{L}_{cae}^{(batch)} = \mathscr{L}_{ae}^{(batch)} + \alpha_{con} \mathscr{L}_{con}, \tag{13.13}$$

where the parameter α_{con} again controls the contractive penalty.

There are also several other interesting autoencoder variants, such as the variational autoencoders [40], that are also able to *generate* samples of the underlying data distribution, and discriminative autoencoders [53, 70], that are able to introduce supervised information into the learned representation.

Apart from autoencoders, *clustering* techniques can be also used to extract useful information from the data without supervision [38, 57, 64, 90]. Clustering is the task of grouping a set of objects into groups (clusters), where each object is as similar as possible to the objects of its cluster and as dissimilar as possible to the objects of the other clusters. Similarly to visualization techniques, clustering can be used with the representation extracted from deep autoencoders, since most clustering algorithms tend to work better with low-dimensional data. Note that a few deep learning-based extensions of clustering techniques also exist [29, 42].

13.4 Supervised Learning

In contrast to unsupervised learning, supervised learning techniques are used when ground truth information, e.g., label annotations for the data, is available. Supervised

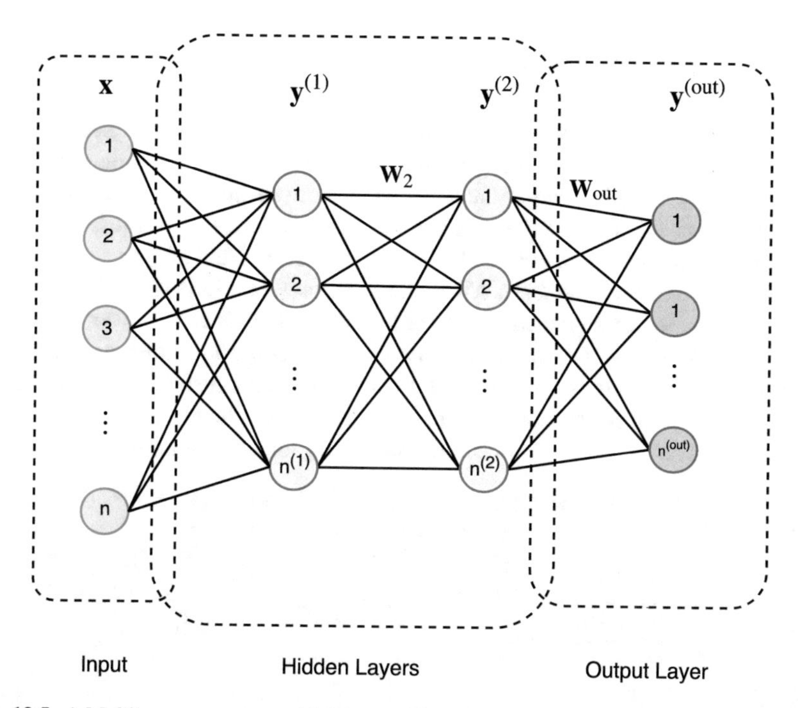

Fig. 13.5 A Multilayer perceptron with 3 layers. First, the input $\mathbf{x}$ is transformed into a sequence of lower dimensional representations $\mathbf{y}^{(1)}$ and $\mathbf{y}^{(2)}$ through two (non-linear) fully connected layers. Then, a fully connected layer is used to obtain the final output of the network. When the MLP is used for classification, each output neuron corresponds to a category (class) that the network recognizes. The input sample $\mathbf{x}$ belongs to the category that corresponds to the output neuron with the largest activation. For regression tasks the output neurons directly regress the values of the predicted attributes. The bias terms have been omitted for presentation purposes

learning can either refer to *classification* problems, i.e., predicting the category of a sample, or to *regression* problems, i.e., predicting a (continuous) value for a numerical attribute. For example, recognizing which animal is depicted in an image is a typical classification problem, while predicting the future load (as a numerical value) of an energy distribution system based on the current conditions is a typical regression problem. Both tasks can be handled with great success by deep neural models, such as Convolutional Neural Networks [41], and Recurrent Neural Networks [37], that are able to provide powerful classification and forecasting tools. Note that selecting the appropriate model depends mainly on the task at hand as well as on the nature of the data.

The rest of this section is structured as follows. First, a simple deep learning model, the Multilayer Perceptron (MLPs), is introduced. Then, the more powerful Convolutional Neural Networks (CNNs) and the Recurrent Neural Networks (RNNs) are described and several properties of the models are discussed. Finally, through this section, we provide guidelines and tips on selecting the most appropriate model for the task at hand.

13.4.1 Multilayer Perceptrons

Multilayer Perceptrons (MLPs) were among the first practical neural networks that were used to handle pattern recognition tasks with satisfactory results [25]. Even though they are nowadays largely superseded by more advanced neural architectures, they are still used either as a part of more complex models, e.g., as the last layers of a Convolutional Neural Network or a Recurrent Neural Network, or as standalone models for less demanding tasks. MLPs are composed of a series of fully connected layers, similar to the autoencoders described in Sect. 13.3.1. However, instead of reconstructing their input, the output layer is responsible for either predicting the class of the input vector or the values for some numerical attributes. Therefore, the loss functions used for training the network must appropriately measure how well the predictions of the network match the supplied ground truth information.

When well-defined temporal or spatial relationships exist between the values contained in the input feature vector, then the CNN and/or the RNN models are usually better candidates since they are able to take into account such relationships. Typical examples of data that exhibit such kind of relationships are images, text and time-series.

The structure of an MLP with 3 layers is shown in Fig. 13.5. The input vector $\mathbf{x} \in \mathbb{R}^n$ is first transformed to the intermediate representations $\mathbf{y}^{(1)} \in \mathbb{R}^{n^{(1)}}$ and $\mathbf{y}^{(2)} \in \mathbb{R}^{n^{(2)}}$ before producing the final output of the network $\mathbf{y}^{(out)} \in \mathbb{R}^{n^{(out)}}$. More formally, the output of each layer is calculated as:

$$\mathbf{y}^{(1)} = \sigma^{(1)}(\mathbf{W}_1\mathbf{x} + \mathbf{b}_1), \tag{13.14}$$

$$\mathbf{y}^{(2)} = \sigma^{(2)}(\mathbf{W}_2\mathbf{y}^{(1)} + \mathbf{b}_2), \tag{13.15}$$

and

$$\mathbf{y}^{(out)} = \sigma^{(out)}(\mathbf{W}_{out}\mathbf{y}^{(2)} + \mathbf{b}_{out}), \tag{13.16}$$

where $\mathbf{W}_1 \in \mathbb{R}^{n^{(1)} \times n}$, $\mathbf{W}_2 \in \mathbb{R}^{n^{(2)} \times n^{(1)}}$ and $\mathbf{W}_{out} \in \mathbb{R}^{n^{(out)} \times n^{(2)}}$ are the weights of each of the layers and $\mathbf{b}_1 \in \mathbb{R}^{n^{(1)}}$, $\mathbf{b}_2 \in \mathbb{R}^{n^{(2)}}$ and $\mathbf{b}_{out} \in \mathbb{R}^{n^{(out)}}$ are the corresponding bias terms. Note that different activation functions can be used for each of the layers (denoted by $\sigma^{(1)}(\cdot)$, $\sigma^{(2)}(\cdot)$ and $\sigma^{(out)}(\cdot)$ respectively). MLPs with more or less layers can be similarly defined. However, in contrast to Convolutional Neural Networks that can have more than 1000 layers [32], MLPs are usually limited to 3–4 layers, since increasing the number of layers usually does not improve the accuracy of the network, while increasing the risk of overfitting.

As it was already mentioned in Sect. 13.3, the ReLU activation function [23], along with more sophisticated initilization techniques [22, 51], made it possible for training deep neural networks. The vanishing gradients problem was tackled by the piece-wise linear behavior of the ReLU activation, while still allowing the network to model non-linear phenomena. Recall that the ReLU function is defined as:

$$\sigma_{ReLU}(u) = max(0, u) = \begin{cases} u & \text{if } u > 0 \\ 0 & \text{otherwise.} \end{cases} \tag{13.17}$$

Note that when a neuron outputs a negative value, i.e., the output of the ReLU is 0, it receives no gradients and it becomes essentially trapped in its state.

To overcome this issue, several variants of the ReLU function have been proposed. The *Leaky ReLU* overcomes this issue by providing a small gradient when the neuron is not activated [46]. The Leaky ReLU is thus defined as follows:

$$\sigma_{LReLU}(u) = \begin{cases} u & \text{if } u > 0 \\ \alpha_{ReLU} u & \text{otherwise,} \end{cases} \tag{13.18}$$

where α_{ReLU} is a small positive number, i.e., $\alpha_{ReLU} = 0.01$. The behavior of the ReLU is compared to the Leaky ReLU in the following plot:

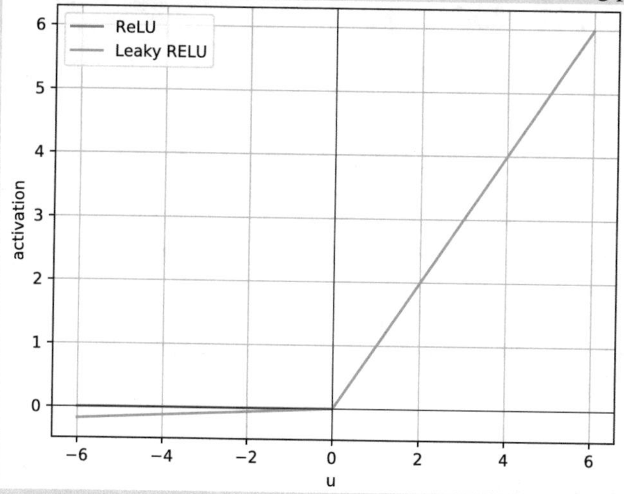

Note that for positive input both functions have the same behavior, but the Leaky ReLU provides a small activation for negative values allowing for training the corresponding neurons. *Parametric ReLU* further improves the behavior of the Leaky ReLU function by allowing for learning the slope parameter α_{ReLU} [27].

When MLPs are used for classification tasks, each of the output neurons corresponds to each of the categories that the network has to predict. That is, an input sample $\mathbf{x}$ belongs to the category that is associated with the output neuron with the largest activation. To obtain a probability distribution over the predicted categories and easily transform arbitrary large or small values that the network might produce into a specific range, the softmax activation is usually used for the output layer:

$$\sigma^{(out)}_{softmax}([\mathbf{u}^{(out)}]_j) = \frac{exp([\mathbf{u}^{(out)}]_j)}{\sum_{i=1}^{n^{(out)}} exp([\mathbf{u}^{(out)}]_i)}, \tag{13.19}$$

where $\mathbf{u}^{(out)} = \mathbf{W}_{out}\mathbf{y}^{(2)} + \mathbf{b}_{out}$ is the pre-activation response of the output neurons.

Before being able to actually produce useful results, the MLPs must be trained. Let $\mathcal{X} = \{\mathbf{x}_1, \mathbf{x}_2, ..., \mathbf{x}_N\}$, be a train set composed of N training samples. Each sample is represented as an n-dimensional feature vector that describes the properties of the corresponding sample. These vectors might contain essentially anything that can be useful for predicting the attributes we are interested in. Examples of such features include the raw values of an image [43], measurements of various sensors [82], hand-crafted features [63], and others.

Normalizing the input features is crucial to avoid saturating the activation functions and ensuring that every feature is equally important. Usually the features are normalized either to a specific interval, i.e., to [0, ..., 1], (this kind of normalization is called *min-max* scaling) or to have zero mean and unit variance (called *standardization* or *z-score* scaling).

In supervised learning each training feature vector $\mathbf{x}_i$ is also accompanied by a label $\mathbf{t}_i$ that describes the attribute that we want to predict. For regression tasks, the label $\mathbf{t}_i \in \mathbb{R}$ is just a real number that contains the target value. For classifications tasks several ways exist to encode the information of different categories. The most frequently used approach is to use vectors that contain as many values as the different categories that exist in our data, i.e., $\mathbf{t}_i \in \mathbb{R}^{n^c}$, where n^c is the number of categories [25]. Then, each dimension of this vector is associated with a specific class (one-hot encoding):

$$[\mathbf{t}_i]_j = \begin{cases} 1, & \text{if the } i\text{-th sample belongs to the } j\text{-th class} \\ 0, & \text{otherwise.} \end{cases} \tag{13.20}$$

For example, for a classification problem where the samples belong to three categories the following vectors can be used: $(1, 0, 0)$ for the first category, $(0, 1, 0)$ for the second category and $(0, 0, 1)$ for the third category. As it was already mentioned, this implies that the number of output neurons must be equal to the number of the classes, i.e., $n^{(out)} = n^{(c)}$. Also, error correcting codes can be used to make the network more robust to misclassifications [34]. Similar encodings can be used for handling *multi-*

label problems [93], i.e., problems where each sample belongs to several categories. In such cases, the number of output neurons must be appropriately adjusted.

The MLP is trained to directly predict the vectors $\mathbf{t}_i$. For both regression and classification tasks, the *mean squared error* loss can be used for training the network:

$$\mathcal{L}_{mlp} = \frac{1}{N}\sum_{i=1}^{N}\sum_{j=1}^{n^{(out)}}([\mathbf{y}_i^{(out)}]_j - [\mathbf{t}_i]_j)^2 = \frac{1}{N}\sum_{i=1}^{N}||\mathbf{y}^{(out)} - \mathbf{t}_i||_2^2. \tag{13.21}$$

However, when dealing with classification problems the mean squared error loss is usually not preferred. Instead, the *cross-entropy* loss function is used (combined with the softmax activation function to ensure that the output of the network is a proper probability distribution over the predicted categories):

$$\mathcal{L}_{mlp} = -\frac{1}{N}\sum_{i=1}^{N}\sum_{j=1}^{n^{(out)}}[\mathbf{t}_i]_j \log([\mathbf{y}_i^{(out)}]_j). \tag{13.22}$$

Note that when the softmax activation with the cross-entropy loss are used, the network actually performs a variant of non-linear multinomial logistic regression. Similarly to the autoencoders, the network can be trained using the gradient descent method:

$$\mathbf{W}^{[t+1]} = \mathbf{W}^{[t]} - \eta\frac{\partial \mathcal{L}_{mlp}}{\partial \mathbf{W}^{[t]}}, \tag{13.23}$$

where the notation $\mathbf{W}$ is used to refer to the concatenation of all the parameters of the network, i.e, $\mathbf{W} = [\mathbf{W}_1, \mathbf{W}_2, \mathbf{W}_{out}, \mathbf{b}_1, \mathbf{b}_2, \mathbf{b}_{out}]$ for the example MLP presented in this section, and η is the used learning rate.

When layers with a large number of neurons are used, then the network can *overfit* the data, i.e., almost perfectly learn the train data, while performing poorly on other unseen data. This behavior is demonstrated in the left training curve plot of Fig. 13.6, where the MNIST dataset is used to train the model. It is easy to see that after the first 10 iterations point the test loss starts increasing while the train loss still decreases and converges normally. A lot of techniques have been proposed to overcome this problem. These techniques are known in the literature as *regularization* methods.

Perhaps the simplest regularization approach is to identify the point where the network starts overfitting the data and stop the training process. This process is known as *early stopping* [91]. Also, reducing the number of neurons in the hidden layers of the network can also alleviate this problem, even though this comes at the cost of also reducing the learning capacity of the model. Furthermore, penalizing large weights, that in turn lead to large activations, reduces the risk of overfitting the data. Usually the weights are penalized by adding an appropriate term in the loss function weighted by a regularization parameter. Two common option are the l^1-normalization and the l^2-normalization. For l^1-normalization the l^1 norm of the weight matrix is used, while for the l^2-regularization the l^2 norm is used [25]. For example, for using l^2-normalization the loss function defined in Eq. (13.22) must be modified as:

$$\mathscr{L}_{mlp} = -\frac{1}{N}\sum_{i=1}^{N}\sum_{j=1}^{n^{(out)}}[\mathbf{t}_i]_j \log([\mathbf{y}_i^{(out)}]_j) + \alpha_{reg}(||\mathbf{W}_1||_2^2 + ||\mathbf{W}_2||_2^2 + ||\mathbf{W}_{out}||_2^2), \tag{13.24}$$

where the parameter α_{reg} is a small positive number, i.e., 0.001, that controls the regularization process. The l^1 normalization leads to sparse solutions, while the l^2 regularization heavily penalizes larger weights, thus leading to more dense small weights. More advanced regularization techniques include the dropout [76], and the dropconnect [87], techniques that randomly deactivate neurons and connections between neurons during the training process, as well as data augmentation techniques [9, 15]. In the right learning curve of Fig. 13.6, the ability of a regularization technique (dropout) to improve the generalization ability of the network and reduce the overfitting phenomena is clearly demonstrated. The overfitting is greatly reduced for the regularized model (the test loss does not increase to the same extent), while the network achieves an overall lower generalization error. Finally, it should be noted that the aforementioned regularization techniques can be applied to any of the models presented in this Section, e.g., they can be also used for regularizing a CNN or an RNN, further improving their accuracy.

Using very strong regularization, e.g., a large number for the regularization parameter α_{reg} or dropping a very large number of neurons during the training, could possibly lead to *underfitting*, preventing the network from learning anything useful.

MLPs can only handle fixed sized inputs, i.e., the size of the input vectors must be constant and known beforehand. However, in some cases a varying number of feature vectors might be extracted from each object (e.g., when dealing with time-series data [65], or using hand-crafted feature extractors [45]). In these cases, neural formulations of the Bag-of-Features representation can be used to compile a trainable fixed length histogram representation of the objects before feeding them to MLP [63].

Retrieval or Classification? In some cases we might be interested in *retrieving* objects similar to a reference *query* object, instead of predicting its category. In such cases, information retrieval techniques can be used [13]. When supervised information exist for some of the objects stored in the database, then the retrieval process can be fine-tuned towards the task at hand, significantly improving the retrieval precision [56, 62, 81].

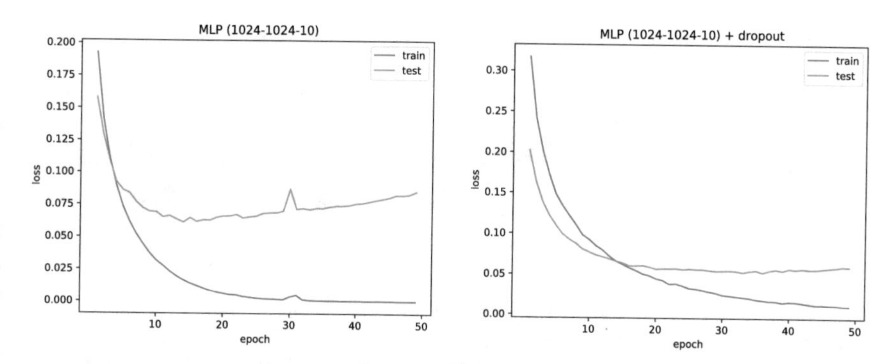

Fig. 13.6 Training a network without (left plot) and with (right plot) a regularization technique (dropout [76]). The network that does not use regularization heavily overfits the data (the test loss increases during the training), while the regularized network exhibits a significantly better behavior (the test loss remains relatively stable after the first 10 epochs)

13.4.2 Convolutional Neural Networks

Convolutional Neural Networks (CNNs) are neural architectures that are able to take into account the spatial arrangement of the input features, while implementing a technique known as *weight sharing* [28, 43, 75, 77]. Weight sharing refers to using the same weights over different regions of the input data, allowing for detecting the same features at different locations. Apart from providing *translation invariance*, weight sharing also reduces the number of parameters that we have to learn making the network more resistant to overfitting. This process was inspired by the visual cortex of many animals, where it has been observed that specialized neurons that individually respond to small regions of the visual field exist [33]. The region that each neuron covers is called *receptive field* of the neuron.

Several deep convolutional architectures with a varying number of layers have been proposed, including the GoogLeNet [77], the VGG [75] and the ResNet [28]. It worths mentioning that the ResNet architecture is capable of scaling up to more than 1000 layers by the use of *residual connections* [28]. Residual connections support the direct flow of information between non-neighboring layers effectively allowing the network to incrementally *finetune* the extracted representation and improve the generalization performance even when a large number of layers is used.

The typical architecture of a CNN that operates on 2-dimensional images is shown in Fig. 13.7. CNNs are composed of a series of convolutional and pooling layers, usually followed by an MLP classifier. Each convolutional layer contains many filters that are (implicitly) *trained* to detect a specific type of feature. Each filter is sequentially applied to all valid locations (usually where it fully overlaps with the input image) leading to a (usually smaller) *feature map* that describes the regions of the input that a specific feature has been detected. After applying several filters we end up with a volume of feature maps that are then fed to the next layer. Several convolutional layers can be stacked or pooling layers can be used in between them to reduce

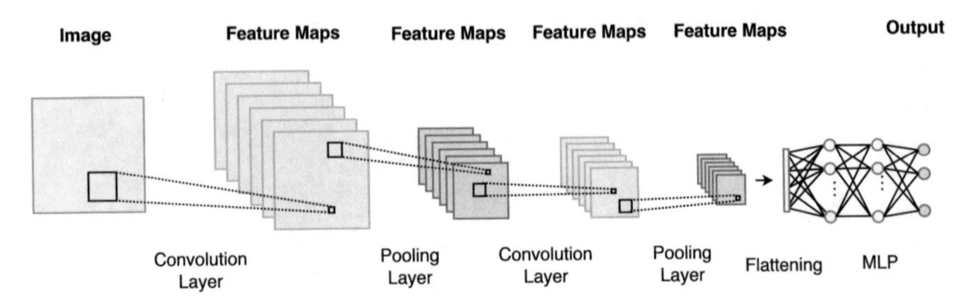

Fig. 13.7 A convolutional neural network with 2 convolutional and 2 pooling layers. The input image is fed into a convolutional layer that applies several convolutional filters leading to the first feature map that is then subsampled using a pooling layer. Then, this process is repeated before flattening the last feature maps into a vector that is fed to an MLP that performs the final classification/regression task

the size of the extracted feature maps. Pooling layers also increase the translation invariance of the neural network discarding the spatial information from the detected features. Note that the first convolutional layers detect simple features, such as plain edges, while the deeper convolutional layers have larger receptive fields and respond to more complex structures, such as object shapes. The interested reader is referred to [48], where the features detected by a CNN at various levels are visualized.

Let $\mathbf{x} \in \mathbb{R}^{n_{ch} \times k \times k}$ be a $k \times k$ part of the input image (with n_{ch} channels, e.g., $n_{ch} = 3$ for color images, $n_{ch} = 1$ for grayscale images) that is fed into a convolutional filter with size $k \times k$. Then, the response of the filter is calculated as:

$$y = \sum_{l=1}^{n_{ch}} \sigma(\sum_{i=1}^{k} \sum_{j=1}^{k} [\mathbf{x}]_{lij} [\mathbf{W}]_{lij} + b), \tag{13.25}$$

where $\mathbf{W} \in \mathbb{R}^{n_{ch} \times k \times k}$ are the weights of the filter and $b \in \mathbb{R}$ its bias. A convolutional filter is very similar to a fully connected neuron, with one important difference: a convolutional filter is repeatedly applied into the input at different neighboring locations, leading to a feature map instead of a single output value, as shown in Fig. 13.8. For example, for an image with size 4×4 and filter with size 3×3 the resulting feature map has size 2×2, since the filter is applied on 2 different positions both in the horizontal and the vertical axis. Recall that many filters are applied (each one detects a different feature type) leading to a volume of features maps. For example, for 10 filters with size 3×3 the output volume in the previous example will be $10 \times 2 \times 2$. The next convolutional layers take into account the volume of the previous feature maps, as defined in Eq. (13.25). Again, a non-linear activation function $\sigma(\cdot)$ is used to allow the neural network to exhibit non-linear behavior.

Except of the convolutional layers, pooling layers are also used in almost every CNN architecture. Among the most commonly used polling layers are the max pooling and the average pooling layers that replace a $k \times k$ region of each feature map with

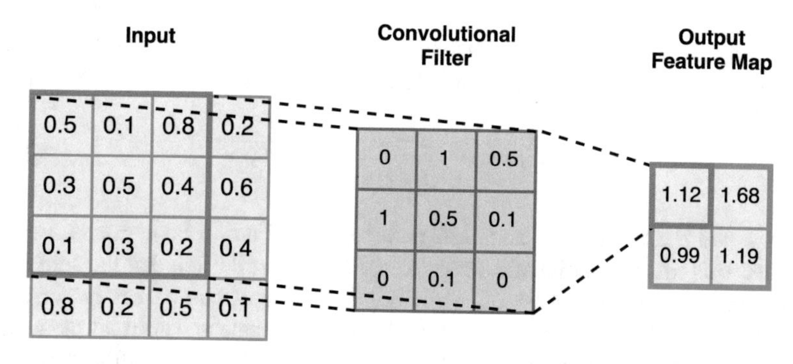

Fig. 13.8 Applying a 3×3 convolutional filter to a 4×4 input image leading to a 2×2 feature map

its max/average (respectively). The process of applying a 2×2 max polling operator on a 4×4 feature map is shown in Fig. 13.9. This allows for effectively reducing the size of the extracted feature maps by a factor of k (typically $k = 2$), as well as increasing the translation invariance of the network. Modern CNN architectures also use *global* pooling operators that completely discard any spatial information. However, global pooling also reduces the size of the subsequent MLP layers and allows the network to handle arbitrary sized images. Other advanced pooling operators include the Spatial Pyramid Pooling [26], that re-introduces spatial information into the pooled representation, the Bag-of-Features Pooling [61], and fast linear pooling techniques [58], that are capable of reducing the size of a CNN while improving its accuracy.

Another kind of layer that is also frequently used in CNNs is the *batch normalization* layer [35]. Batch normalization is a normalization technique that reduces the *internal covariate shift* that is caused during the training process, allowing for both increasing the convergence speed as well as reducing the risk of overfitting the network. Batch normalization works as follows. First, the mean $E[u]$ and the vari-

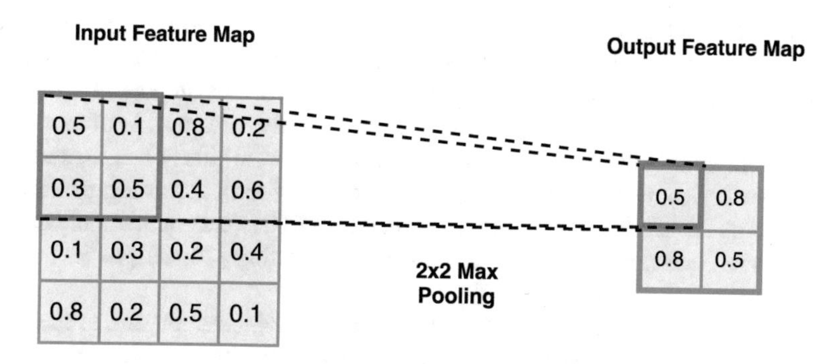

Fig. 13.9 Applying 2×2 max pooling to a 4×4 feature map leading a subsampled 2×2 feature map

ance $Var[u]$ of the activation u of a neuron (before applying the activation function) are calculated over the samples of a batch. Then, this activation is replaced by the following:

$$y_{bn} = \gamma \frac{u - E[u]}{\sqrt{(Var[u] + \epsilon}} + \beta, \tag{13.26}$$

where γ and β are the scale and shift parameters that are to be learned during the training. This transformation allows for keeping the distribution that the subsequent layers receive relatively constant, while any adjustments can be easily performed using the γ and the β parameters without having to rescale and shift all the weights of a layer. Equation (13.26) is applied separately for all the neurons/filters of a layer both during the training and the testing. Note again that batch normalization is usually applied immediately before the activation function.

Even though CNNs are mainly used for image analysis, they can be also used for any task that involves either spatial or temporal relationships between the used features. Examples of such tasks include text analysis/classification [17, 73], and timeseries analysis and forecasting [79, 94]. In these cases, 1-dimensional convolution is used (instead of 2-dimensional convolution) taking into account the temporal succession of the features, e.g., the words of a text or the points of a timeseries.

Word embeddings are powerful models that can be used with deep learning text analysis techniques, since they allow for mapping each word into a vector that captures its semantic content [66]. Word embedding models can be combined with many deep learning techniques, such as CNNs [73], RNNs [86], or even deep Neural Bag-of-Features formulations [55].

13.4.3 Recurrent Neural Networks

Multilayer Perceptrons and Convolutional Neural Networks are *feedforward* networks, since the information flows from the network's input to the network's output. In contrast, in Recurrent Neural Networks (RNNs) (part of) the output of the network is redirected to the input forming a *directed cycle*. RNNs are capable of exhibiting dynamic temporal behavior and processing sequences of arbitrary length, such as a texts or timeseries, while being able to model the long-term relationships between distant words or points. These properties rendered RNNs excellent candidates for handling such tasks, ranging from time-series forecasting [20, 80], to text understanding and generation [74, 84, 88].

Early RNNs models were notoriously difficult to train, mainly due to the problems of *vanishing* and *exploding* gradients [54]. The problem of exploding gradients describes the opposite situation of that of vanishing gradients, where the derivatives are getting continuously larger making the training procedure unstable. Intuitively, when the largest eigenvalue of the weight matrix is less than 1, then the repeated application of the weight matrix causes the gradient to vanish to zero, while when the largest eigenvalue of the weight matrix is larger than 1, then the gradients explode to infinity.

These problems were addressed by the Long Short-term Memory (LSTM) model [21], that overcomes these issues using a plain linear activation for the recurrent part of the model. Since the derivative of the linear activation is always 1 the errors are back-propagated through the previous time-steps without vanishing. That also allows the network to establish long term relationships between temporally distant features. In this section, a slightly simpler and more modern version of the LSTM model, the Gated Recurrent Unit (GRU) [10], is presented.

GRUs, like the rest of the recurrent models, receive an input vector $\mathbf{x}^{[t]} \in \mathbb{R}^n$ for each timestep and produce an output vector $\mathbf{h}^{[t]} \in \mathbb{R}^m$ that also encodes the current state of the model. One crucial component of the GRU model is the *update gate* that decides how much information will flow into the model from the previous output $\mathbf{h}^{[t-1]}$ and from the new state $\tilde{\mathbf{h}}^{[t-1]}$. The output of the update gate is calculated as:

$$\mathbf{z}^{[t]} = \sigma_{sigm}(\mathbf{W}_z[\mathbf{h}^{[t-1]}, \mathbf{x}^{[t]}]) \in \mathbb{R}^m, \tag{13.27}$$

where the notation $[\mathbf{h}^{[t-1]}, \mathbf{x}^{[t]}]$ is used to refer to the concatenation of the vectors $\mathbf{h}^{[t-1]}$ and $\mathbf{x}^{[t]}$ and $\mathbf{W}_z \in \mathbb{R}^{m\times(m+n)}$ are the weights of the update gate. The new state of the model is calculated as:

$$\tilde{\mathbf{h}}^{[t]} = \sigma_{tanh}(\mathbf{W}_h[\mathbf{r}^{[t]} \odot \mathbf{h}^{[t-1]}, \mathbf{x}^{[t]}]) \in \mathbb{R}^m, \tag{13.28}$$

where $\mathbf{W}_h \in \mathbb{R}^{m\times(m+n)}$ are the weights for calculating the new state, the operator $\odot$ refers to the element-wise product between two vectors (Hadamard product) and $\mathbf{r}^{[t]} \in \mathbb{R}^m$ is the output of the gate that controls how much information of the current state will be used to compute the new state. The output of this gate is calculated as:

$$\mathbf{r}^{[t]} = \sigma_{sigm}(\mathbf{W}_r[\mathbf{h}^{[t-1]}, \mathbf{x}^{[t]}]) \in \mathbb{R}^m, \tag{13.29}$$

where $\mathbf{W}_r \in \mathbb{R}^{m\times(m+n)}$ are the weights of the gate. Finally, the output of the GRU is computed as the weighted sum between the old and the new state according to the output of the update gate $\mathbf{z}^{[t]}$:

$$\mathbf{h}^{[t]} = (1 - \mathbf{z}^{[t]}) \odot \mathbf{h}^{[t-1]} + \mathbf{z}^{[t]} \odot \tilde{\mathbf{h}}^{[t]}. \tag{13.30}$$

The bias terms have been omitted from the above equations for presentation purposes.

GRUs can be used to produce a prediction either immediately after each timestep (by feeding the current output $\mathbf{h}^{[t]}$ to the used classifier, i.e., the MLP), as shown in Fig. 13.10, or after all the feature vectors have been fed to the model, as shown in Fig. 13.11. Depending on the task at hand, the appropriate architecture technique can be selected. All the parameters of the model are learned using a variant of the regular backpropagation algorithm, called *backpropagation through time* (BPTT) [89].

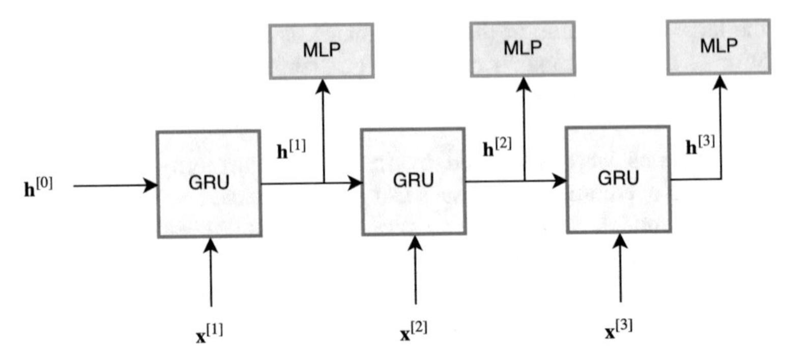

Fig. 13.10 Using a GRU to model a temporal sequence of vector $[\mathbf{x}^{[1]}, \mathbf{x}^{[2]}, \mathbf{x}^{[3]}]$ and classify each point of the sequence using an MLP. The *same* GRU and MLP are used for every timestep

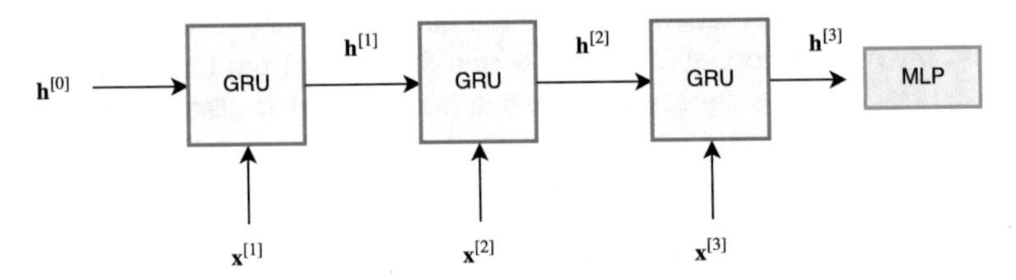

Fig. 13.11 Using a GRU to model a temporal sequence of vector $[\mathbf{x}^{[1]}, \mathbf{x}^{[2]}, \mathbf{x}^{[3]}]$ and classify the whole sequence using an MLP. The *same* GRU is used for every timestep

BPTT essential works by *unfolding* the repeated application of the weights of the network through time and then accumulating the corresponding derivatives before applying the gradient descent updates.

13.5 Deep Learning Frameworks

Implementing and deploying deep learning models can be a difficult task for a number of reasons. First, implementing the backpropagation algorithm requires analytically deriving the partial derivatives of the loss function with respect to all the parameters of the model. These derivatives must be appropriately updated when any changes are made either to the network architecture or to the loss function. As the models get deeper and more complicated this process is becoming increasingly difficult. Also, training deep models is a computationally intensive task. To this end, Graphical Processing Units (GPUs) are usually used to accelerate the training process. However, the switch between CPU and GPU is not transparent and usually requires porting and rewriting a large part of the code that implements the neural network. Finally, without a common way to store a neural network it is especially difficult to share a pretrained model, often requiring retraining the model from scratch (or writing specialized code to read the network parameters).

Several open source deep learning frameworks were developed to overcome the aforementioned difficulties. Most of these frameworks support *automatic differentiation*, i.e., only the structure of the network needs to be defined and the used framework automatically implements the backpropagation algorithm. Furthermore, these frameworks also allow for switching between the CPU and the GPU for the optimization and the deployment process transparently (in many cases this is just a matter of switching a flag in the code), and thus significantly accelerating the training process without any additional programming effort. Finally, the community usually provides many implementation of well known network architectures along with pre-trained models. This allows for easily testing, extending and deploying existing models without having to retrain them (in some cases the training process using large datasets, such as the ImageNet dataset [41], might take weeks, even when high-end GPUs are used).

Among the most well-known deep learning frameworks are Caffe [36], Tensorflow [1], CNTK [50], and Torch [14]/PyTorch [67]. Most of them provide interfaces to various languages (allowing for easily integrating the trained models into various applications). The same basic set of features is provided by all of the aforementioned frameworks, so for most deep learning tasks the choice of the framework is more a matter of personal choice. However, it worths mentioning that, at the moment, PyTorch is among the few libraries that efficiently support *dynamic computational graphs*, i.e., the structure of the network is not predefined but it can dynamically vary during the training/deployment. Note that there is also a number of other deep learning frameworks, but it is out of the scope of this chapter to provide an extensive list of them.

Higher level wrappers, such as Keras [11], also exist. These wrappers work at even higher level, hiding the complexities of the underlying deep learing framework further simplifying the process of defining and training a deep learning model. Keras provides support for three backend frameworks, Theano [78], Tensorflow, and CTNK, with almost no changes in the code. For example, defining an MLP with three layers to solve a 10-class classification problem using the Keras library requires less than 10 lines of code. An example is provided below:

```
from keras.models import Sequential
from keras.layers import Dense
from keras.optimizers import Adam

model = Sequential()
model.add(Dense(1024, activation='relu', input_shape=(784,)))
model.add(Dense(1024, activation='relu'))
model.add(Dense(10, activation='softmax'))
model.compile(loss='categorical_crossentropy',
              optimizer=Adam(lr=0.001),
              metrics=['accuracy'])
```

Then, the model can be trained using a single command (the training data are contained in the x_train variable, while the training targets in the y_train variable):

```
model.fit(x_train, y_train, batch_size=256, epochs=50)
```

while the accuracy of the model can be also easily measured accordingly:

```
print(model.evaluate(x_test,y_test))
```

The interested reader is referred to [11], for more information regarding the Keras library.

13.6 Concluding Remarks

The most important deep learning techniques, that can be used for a variety of data analytics tasks, were presented in this chapter. First, several unsupervised learning methods, that are capable of discovering and extracting useful information from the data without supervision, were presented. We also discussed how these techniques can be combined with other methods, such as visualization or clustering techniques, to allow for efficiently performing several data exploration tasks. Next, several supervised deep learning techniques, ranging from Multilayer Perceptrons (MLPs) to Convolutional Neural Networks (CNNs) and Recurrent Neural Networks (RNNs), were also presented, allowing for building powerful classification and/or forecasting models. As thoroughly discussed in this chapter, training deep learning models is not always straightforward. We presented a number of techniques and practical suggestions that can be used for overcoming several difficulties that may arise during the training of deep learning models. Finally, we briefly reviewed the available deep learning frameworks and we demonstrated how to use one of them for defining and training a simple model. That way, we provided the reader with the necessary skills and intuition to successfully use and deploy deep learning techniques.

References

1. Abadi, M., Agarwal, A., Barham, P., Brevdo, E., Chen, Z., Citro, C., Corrado, G.S., Davis, A., Dean, J., Devin, M., Ghemawat, S., Goodfellow, I., Harp, A., Irving, G., Isard, M., Jia, Y., Jozefowicz, R., Kaiser, L., Kudlur, M., Levenberg, J., Mané, D., Monga, R., Moore, S., Murray, D., Olah, C., Schuster, M., Shlens, J., Steiner, B., Sutskever, I., Talwar, K., Tucker, P., Vanhoucke, V., Vasudevan, V., Viégas, F., Vinyals, O., Warden, P., Wattenberg, M., Wicke, M., Yu, Y., Zheng, X.: TensorFlow: large-scale machine learning on heterogeneous systems (2015). www.tensorflow.org
2. Abdi, H., Williams, L.J.: Principal component analysis. Wiley Interdiscip. Rev. Comput. Stat. **2**(4), 433–459 (2010)
3. Aggarwal, C.C.: Outlier analysis. In: Data Mining, pp. 237–263 (2015)
4. Aggarwal, C.C, Reddy, C.K.: Data Clustering: Algorithms and Applications. CRC press (2013)
5. Ahonen, T., Hadid, A., Pietikainen, M.: Face description with local binary patterns: Application to face recognition. IEEE Trans. Pattern Anal. Mach. Intell. **28**(12), 2037–2041 (2006)
6. Bengio, Y., et al.: Learning deep architectures for AI. Found. Trends Mach. Learn. **2**(1), 1–127 (2009)
7. Bengio, Y., Lamblin, P., Popovici, D., Larochelle, H.: Greedy layer-wise training of deep networks. In: Proceedings of the Advances in Neural Information Processing Systems, pp. 153–160 (2007)

8. Celebi, M.E., Aydin, K.: Unsupervised Learning Algorithms. Springer (2016)
9. Chatfield, K., Simonyan, K., Vedaldi, A., Zisserman, A.: Return of the devil in the details: delving deep into convolutional nets (2014). arXiv:1405.3531
10. Cho, K., van Merriënboer, B., Bahdanau, D., Bengio, Y.: On the properties of neural machine translation: encoder–decoder approaches. Syntax Semant. Struct. Stat. Transl. p. 103 (2014)
11. Chollet, F., et al.: Keras (2015). https://github.com/fchollet/keras
12. Choromanska, A., Henaff, M., Mathieu, M., Ben Arous, G., LeCun, Y.: The loss surfaces of multilayer networks. In: Artificial Intelligence and Statistics, pp. 192–204 (2015)
13. Christopher, D.M, Prabhakar, R., Hinrich, S.: Introduction to information retrieval. In: An Introduction to Information Retrieval, vol. 151, p. 177 (2008)
14. Collobert, R., Kavukcuoglu, K., Farabet, C.: Torch7: a matlab-like environment for machine learning. In: BigLearn, NIPS Workshop (2011)
15. Cui, X., Goel, V., Kingsbury, B.: Data augmentation for deep neural network acoustic modeling. Proc. IEEE/ACM Trans. Audio Speech Lang. Process. **23**(9), 1469–1477 (2015)
16. De Oliveira, M.C.F., Levkowitz, H.: From visual data exploration to visual data mining: a survey. IEEE Trans. Vis. Comput. Gr. **9**(3), 378–394 (2003)
17. Dos Santos, C.N., Gatti, M.: Deep convolutional neural networks for sentiment analysis of short texts. In: COLING, pp. 69–78 (2014)
18. Duchi, J., Hazan, E., Singer, Y.: Adaptive subgradient methods for online learning and stochastic optimization. J. Mach. Learn. Res. 12(Jul), 2121–2159 (2011)
19. Elkahky, A.M., Song, Y., He, X.: A multi-view deep learning approach for cross domain user modeling in recommendation systems. In: Proceedings of the International Conference on World Wide Web, pp. 278–288 (2015)
20. Gers, F.A., Eck, D., Schmidhuber, J.: Applying LSTM to time series predictable through time-window approaches. In: Proceedings of the Italian Workshop on Neural Nets, pp. 193–200 (2002)
21. Gers, F.A., Schmidhuber, J., Cummins, F.: Learning to forget: continual prediction with LSTM. Neural Comput. **12**(10), 2451–2471 (2000)
22. Glorot, X., Bengio, Y.: Understanding the difficulty of training deep feedforward neural networks. In: Proceedings of the International Conference on Artificial Intelligence and Statistics, pp. 249–256 (2010)
23. Glorot, X., Bordes, A., Bengio, Y.: Deep sparse rectifier neural networks. In: Proceedings of the International Conference on Artificial Intelligence and Statistics, pp. 315–323 (2011)
24. Guyon, I., Elisseeff, A.: An introduction to feature extraction. Feature Extr. 1–25 (2006)
25. Haykin, S.S., Haykin, S.S., Haykin, S.S., Haykin, S.S.: Neural Networks and Learning Machines, vol. 3. Pearson Upper Saddle River (2009)
26. He, K., Zhang, X., Ren, S., Sun, J.: Spatial pyramid pooling in deep convolutional networks for visual recognition. In: Proceedings of the European Conference on Computer Vision, pp. 346–361 (2014)
27. He, K., Zhang, X., Ren, S., Sun, J.: Delving deep into rectifiers: surpassing human-level performance on imagenet classification. In: Proceedings of the IEEE International Conference on Computer Vision, pp. 1026–1034 (2015)
28. He, K., Zhang, X., Ren, S., Sun, J.: Deep residual learning for image recognition. In: Proceedings of the IEEE Conference on Computer Vision and Pattern Recognition, pp. 770–778 (2016)
29. Hershey, J.R., Chen, Z., Roux, J.L., Watanabe, S.: Deep clustering: discriminative embeddings for segmentation and separation. In: Proceedings of the IEEE International Conference on Acoustics, Speech and Signal Processing, pp. 31–35 (2016)
30. Hinton, G.E., Salakhutdinov, R.R.: Reducing the dimensionality of data with neural networks. Science **313**(5786), 504–507 (2006)
31. Hosseini-Asl, E., Zurada, J.M., Nasraoui, O.: Deep learning of part-based representation of data using sparse autoencoders with nonnegativity constraints. IEEE Trans. Neural Netw. Learn. Syst. **27**(12), 2486–2498 (2016)

32. Huang, G., Sun, Y., Liu, Z., Sedra, D., Weinberger, k.Q.: Deep networks with stochastic depth. In: Proceedings of the European Conference on Computer Vision, pp. 646–661 (2016)
33. Hubel, D.H., Wiesel, T.N.: Receptive fields and functional architecture of monkey striate cortex. The J. Physiol. **195**(1), 215–243 (1968)
34. Huffman, W.C., Pless, V.: Fundamentals of Error-Correcting Codes. Cambridge university press, 2010
35. Ioffe, S., Szegedy, C.: Batch normalization: adeep network training by reducing internal covariate shift. In: Proceedings of the International Conference on Machine Learning, pp. 448–456 (2015)
36. Jia, Y., Shelhamer, E., Donahue, J., Karayev, S., Long, J., Girshick, R., Guadarrama, S., Darrell, T.: Caffe: convolutional architecture for fast feature embedding (2014). arXiv:1408.5093
37. Jozefowicz, R., Zaremba, W., Sutskever, I.: An empirical exploration of recurrent network architectures. In: Proceedings of the International Conference on Machine Learning, pp. 2342–2350 (2015)
38. Kanungo, T., Mount, D.M., Netanyahu, N.S., Piatko, C.D., Silverman, R., Wu, A.Y.: An efficient k-means clustering algorithm: analysis and implementation. IEEE Trans. Pattern Anal. Mach. Intell. **24**(7), 881–892 (2002)
39. Kingma, D., Ba, J.: Adam: a method for stochastic optimization. In: Proceedings of the International Conference on Learning Representations (2015)
40. Kingma, D.P., Welling, M.: Auto-encoding variational bayes (2013). arXiv:1312.6114
41. Krizhevsky, A., Sutskever, I., Hinton, G.E.: Imagenet classification with deep convolutional neural networks. In: Proceedings of the Advances in Neural Information Processing Systems, pp. 1097–1105 (2012)
42. Law, M.T., Urtasun, R., Zemel, R.S.: Deep spectral clustering learning. In: Proceedings of the International Conference on Machine Learning, pp. 1985–1994 (2017)
43. LeCun, Y., Bottou, L., Bengio, Y., Haffner, P.: Gradient-based learning applied to document recognition. Proc. IEEE **86**(11), 2278–2324 (1998)
44. Lecun, Y., Cortes, C.: The MNIST database of handwritten digits
45. Lowe, D.G.: Object recognition from local scale-invariant features. Proceedings of the IEEE International Conference on Computer Vision **2**, 1150–1157 (1999)
46. Maas, A.L., Hannun, A.Y., Ng, A.Y.: Rectifier nonlinearities improve neural network acoustic models. In: ICML Workshop on Deep Learning for Audio, Speech and Language Processing (2013)
47. van der Maaten, L., Hinton, G.: Visualizing data using t-SNE. J. Mach. Learn. Res. 9(Nov), 2579–2605 (2008)
48. Mahendran, A., Vedaldi, A.: Visualizing deep convolutional neural networks using natural pre-images. Int. J. Comput. Vis. **120**(3), 233–255 (2016)
49. Makhzani, A., Frey, B.: K-sparse autoencoders (2013). arXiv:1312.5663
50. Microsoft. Microsoft cognitive toolkit CNTK (2015). https://github.com/Microsoft/CNTK
51. Mishkin, D., Matas, J.: All you need is a good init (2015). arXiv:1511.06422
52. Mnih, V., Kavukcuoglu, K., Silver, D., Rusu, A.A., Veness, J., Bellemare, M.G., Graves, A., Riedmiller, M., Fidjeland, A.K., Ostrovski, G., et al.: Human-level control through deep reinforcement learning. Nature **518**(7540), 529–533 (2015)
53. Nousi, P., Tefas, A.: Deep learning algorithms for discriminant autoencoding. Neurocomputing **266**, 325–335 (2017)
54. Pascanu, R., Mikolov, T., Bengio, Y.: On the difficulty of training recurrent neural networks. In: Proceedings of the International Conference on Machine Learning, pp. 1310–1318 (2013)
55. Passalis, N., Tefas, A.: Bag of embedded words learning for text retrieval. In: Proceedings of the International Conference on Pattern Recognition (ICPR), pp. 2416–2421 (2016)
56. Passalis, N., Tefas, A.: Entropy optimized feature-based bag-of-words representation for information retrieval. IEEE Trans. Knowl. Data Eng. **28**(7), 1664–1677 (2016)
57. Passalis, N., Tefas, A.: Spectral clustering using optimized bag-of-features. In: Proceedings of the 9th Hellenic Conference on Artificial Intelligence, p. 19 (2016)

58. Passalis, N., Tefas, A.: Concept detection and face pose estimation using lightweight convolutional neural networks for steering drone video shooting. In: Proceedings of the 25th European Signal Processing Conference, pp. 71–75 (2017)
59. Passalis, N., Tefas, A.: Dimensionality reduction using similarity-induced embeddings. IEEE Trans. Neural Netw. Learn. Syst. (to appear) 1–13 (2017)
60. Passalis, N., Tefas, A.: Improving face pose estimation using long-term temporal averaging for stochastic optimization. In: International Conference on Engineering Applications of Neural Networks, pp. 194–204 (2017)
61. Passalis, N., Tefas, A.: Learning bag-of-features pooling for deep convolutional neural networks. In: Proceedings of the IEEE International Conference on Computer Vision, Oct 2017
62. Passalis, N., Tefas, A.: Learning neural bag-of-features for large-scale image retrieval. IEEE Trans. Syst. Man Cybern, Syst (2017)
63. Passalis, N., Tefas, A.: Neural bag-of-features learning. Pattern Recogn. **64**, 277–294 (2017)
64. Passalis, N., Tefas, A.: Information clustering using manifold-based optimization of the bag-of-features representation. IEEE Trans. Cybern. **48**(1), 52–63 (2018)
65. Passalis, N., Tsantekidis, A., Tefas, A., Kanniainen, J., Gabbouj, M., Iosifidis, A.: Time-series classification using neural bag-of-features. In: Proceedings of the European Signal Processing Conference, pp. 301–305 (2017)
66. Pennington, J., Socher, R., Manning, C.: Glove: global vectors for word representation. In: Proceedings of the Conference on Empirical Methods in Natural Language Processing, pp. 1532–1543 (2014)
67. PyTorch. Pytorch (2017). https://github.com/pytorch/pytorch
68. Qiu, X., Zhang, L., Ren, Y., Suganthan, P.N., Amaratunga, G.: Ensemble deep learning for regression and time series forecasting. In: IEEE Symposium on Computational Intelligence in Ensemble Learning, pp. 1–6 (2014)
69. Rifai, S., Vincent, P., Muller, X., Glorot, X., Bengio, Y.: Contractive auto-encoders: explicit invariance during feature extraction. In: Proceedings of the International Conference on Machine Learning, pp. 833–840 (2011)
70. Rolfe, J.L., LeCun, Y.: Discriminative recurrent sparse auto-encoders (2013). arXiv:1301.3775
71. Salakhutdinov, R., Hinton, G.: Deep boltzmann machines. In: Proceedings of the Artificial Intelligence and Statistics, pp. 448–455 (2009)
72. Senior, A., Heigold, G., Yang, K., et al.: An empirical study of learning rates in deep neural networks for speech recognition. In: Proceedings of the IEEE International Conference on on Acoustics, Speech and Signal Processing, pp. 6724–6728 (2013)
73. Severyn, A., Moschitti, A.: Twitter sentiment analysis with deep convolutional neural networks. In: Proceedings of the International ACM SIGIR Conference on Research and Development in Information Retrieval, pp. 959–962 (2015)
74. Shen, Y., Huang, P-S., Gao, J., Chen, W.: Reasonet: learning to stop reading in machine comprehension. In: Proceedings of the 23rd ACM SIGKDD International Conference on Knowledge Discovery and Data Mining, pp. 1047–1055 (2017)
75. Simonyan, K., Zisserman, A.: Very deep convolutional networks for large-scale image recognition (2014). arXiv:1409.1556
76. Srivastava, N., Hinton, G.E., Krizhevsky, A., Sutskever, I., Salakhutdinov, R.: Dropout: a simple way to prevent neural networks from overfitting. J. Mach. Learn. Res. **15**(1), 1929–1958 (2014)
77. Szegedy, C., Liu, W., Jia, Y., Sermanet, P., Reed, S., Anguelov, D., Erhan, D., Vanhoucke, V., Rabinovich, A.: Going deeper with convolutions. In: Proceedings of the IEEE Conference on Computer Vision and Pattern Recognition, pp. 1–9 (2015)
78. Theano Development Team. Theano: A Python framework for fast computation of mathematical expressions, May 2016. arXiv:1605.02688
79. Tsantekidis, A., Passalis, N., Tefas, A., Kanniainen, J., Gabbouj, M., Iosifidis, A.: Forecasting stock prices from the limit order book using convolutional neural networks. Proc. IEEE Conf. Bus. Inf. **1**, 7–12 (2017)
80. Tsantekidis, A., Passalis, N., Tefas, A., Kanniainen, J., Gabbouj, M., Iosifidis, A.: Using deep learning to detect price change indications in financial markets. In: Proceedings of the European Signal Processing Conference, pp. 2511–2515 (2017)

81. Tzelepi, M., Tefas, A.: Deep convolutional learning for content based image retrieval. Neurocomputing **275**, 2467–2478 (2018)
82. Unal, M., Onat, M., Demetgul, M., Kucuk, H.: Fault diagnosis of rolling bearings using a genetic algorithm optimized neural network. Measurement **58**, 187–196 (2014)
83. Van Der Maaten, L., Postma, E., Van den Herik, J.: Dimensionality reduction: a comparative. J. Mach. Learn. Res. **10**, 66–71 (2009)
84. Venugopalan, S., Rohrbach, M., Donahue, J., Mooney, R., Darrell, T., Saenko, K.: Sequence to sequence-video to text. In: Proceedings of the IEEE International Conference on Computer Vision, pp. 4534–4542 (2015)
85. Vincent, P., Larochelle, H., Bengio, Y., Manzagol, P-A.: Extracting and composing robust features with denoising autoencoders. In: Proceedings of the International Conference on Machine Learning, pp. 1096–1103 (2008)
86. Vinyals, O., Toshev, A., Bengio, S., Erhan, D.: Show and tell: a neural image caption generator. In: Proceedings of the IEEE Conference on Computer Vision and Pattern Recognition, pp. 3156–3164 (2015)
87. Wan, L., Zeiler, M., Zhang, S., Cun, Y.L., Fergus, R.: Regularization of neural networks using dropconnect. In: Proceedings of the International Conference on Machine Learning, pp. 1058–1066 (2013)
88. Wang, S., Jiang, J.: Machine comprehension using match-lstm and answer pointer (2016). arXiv:1608.07905
89. Werbos, P.J.: Backpropagation through time: what it does and how to do it. Proc. IEEE **78**(10), 1550–1560 (1990)
90. Dongkuan, X., Tian, Y.: A comprehensive survey of clustering algorithms. Ann. Data Sci. **2**(2), 165–193 (2015)
91. Yao, Y., Rosasco, L., Caponnetto, A.: On early stopping in gradient descent learning. Constr. Approx. **26**(2), 289–315 (2007)
92. Zeiler, M.D.: ADADELTA: an adaptive learning rate method (2012). arXiv:1212.5701
93. Zhang, Y., Schneider, J.: Multi-label output codes using canonical correlation analysis. In: Proceedings of the International Conference on Artificial Intelligence and Statistics, pp. 873–882 (2011)
94. Zheng, Y., Liu, Q., Chen, E., Ge, Y., Zhao, J.L.: Time series classification using multi-channels deep convolutional neural networks. In: Proceedings of the International Conference on Web-Age Information Management, pp. 298–310 (2014)

Printed in the United States
By Bookmasters